MCBU

Molecular and Cell Biology Updates

Series Editors:

Prof. Dr. Angelo Azzi
Institut für Biochemie
und Molekularbiologie
Bühlstr. 28
CH - 3012 Bern
Switzerland

Prof. Dr. Lester Packer
Dept. of Molecular
and Cell Biology
251 Life Science Addition
Membrane Bioenergetics Group
Berkeley, CA 94720
USA

Bioradicals Detected
by ESR Spectroscopy

Edited by   H. Ohya-Nishiguchi
              L. Packer

Birkhäuser Verlag
Basel · Boston · Berlin

Editors

Dr. H. Ohya-Nishiguchi
Institute for Life Support Technology of
Yamagata Technopolis Foundation
683 Kurumanomae, Numagi
Yamagata, Yamagata 990
Japan

Prof. Dr. Lester Packer
Dept. of Molecular
and Cell Biology
251 Life Science Addition
Membrane Bioenergetics Group
Berkeley, CA 94720
USA

Library of Congress Cataloging-in-Publication Data

Bioradicals detected by ESR spectroscopy / edited by H. Ohya
    -Nishiguchi, L. Packer.
        p.   cm. – (Molecular and cell biology updates)
    Includes bibliographical references and index.
    ISBN 3-7643-5077-6 (hardcover : alk. paper)
    ISBN 0-8176-5077-6 (hardcover : alk. paper)
    1. Electron paramagnetic resonance spectroscopy – Congresses.
2. Spin labels – Congresses. 3. Free radicals (Chemistry) – Analysis-
-Congresses. I. Ohya-Nishiguchi, H. (Hiroaki), 1939- II. Packer,
Lester.   III. Series.
QP519.9.E433B56 1995
574.19'285 – dc20   95–39027
                                                CIP

Deutsche Bibliothek Cataloging-in-Publication Data

Bioradicals detected by ESR spectroscopy / ed. by H. Ohya-
Nishiguchi ; L. Packer. - Basel ; Boston ; Berlin : Birkhäuser, 1995
    (Molecular and cell biology updates)
    ISBN 3-7643-5077-6 (Basel...)
    ISBN 0-8176-5077-6 (Boston)
    NE: Ohya-Nishiguchi, Hiroaki [Hrsg.]

© 1995 Birkhäuser Verlag, PO Box 133, CH - 4010 Basel, Switzerland
Printed on acid-free paper produced from chlorine-free pulp.TCF ∞
Printed in Germany
ISBN 3-7643-5077-6
ISBN 0-8176-5077-6

9 8 7 6 5 4 3 2 1

Table of Contents

Preface

This book is based on two keywords: Bioradical and ESR. Bioradical is a newly coined word which encompasses paramagnetic species in biological systems, such as active oxygen radicals and transition metal ions. Research on the structure and function of bioradicals has been attracting growing attention in the field of biological science, and comprehensive investigations from many fields are helping to understand the real features of these species. ESR spectroscopy also has interdisciplinary features in that its techniques have been applied to many fields, ranging from physics to medicine. It was our hope, therefore, that this book would help to clarify many aspects of bioradicals and that significant progress would be achieved in combining basic research from many different fields. This book arises from the First International Conference on Bioradicals Detected by ESR Spectroscopy (ICBES), which was held in Yamagata, a city in the Yamagata Prefecture of Japan, in 1994. About 300 participants from 16 different countries attended this conference, and about 170 papers were presented. This book is a collection of contributions from the conference and also contains eleven chapters selected by the editorial board, based on suggestions from the members of the international editorial board of ICBES.

The Yamagata Technopolis Foundation is developing a biomedical technology for the 21st century based on life science fused with material and physical science. Based on such a technology, the Foundation plans to share its fruits all over the world. The Foundation focussed on "bioradical research" as a main research field and has been developing the field of measurement of bioradicals and its applications, such as diagnostic imaging. This research is based on new technology which originally came from the Department of Engineering of Yamagata University, and was partly supported by Science and Technology Agency of Japanese Government. To further this research the foundation opened the Institute for Life Support Technology in April of 1993. We hope that our new institute will promote local activities in the area of Yamagata through new research developments, and also that the institute would be an information center not only in Japan but throughout the world. The conference was planned as a first trial for this purpose.

I would like to express my appreciation for the great support and cooperation of many scientific associations as well as all participants in the conference. We hope and anticipate that publication of this book will contribute to the promotion of health and welfare of all people in the world. Finally, I regret to say that Professor Sohma succumbed to illness on October 4, 1994, after he served as the chairman of the programming committee, as one of the members of the international advisory board, and finally as the chairman of the editorial board selecting the chapters which are included in this book. His contribution in Chapter 4 is printed posthumously. May he rest in peace.

Hitoshi Kamada (Chairman of the Organizing Committee of ICBES)
April, 1995

Bioradicals Detected by ESR Spectroscopy
H. Ohya-Nishiguchi & L. Packer (eds)
© 1995 Birkhäuser Verlag Basel/Switzerland

Overview of bioradicals and ESR technology

H. Ohya-Nishiguchi

Institute for Life Support Technology of Yamagata Technopolis Foundation, Numagi, Yamagata, Yamagata 990, Japan

Introduction

"Bioradical" is a newly coined word which refers to paramagnetic species in biological systems, such as active oxygen radicals and transition metal ions. Research on their structure and function is attracting growing attention in the field of biological science, and comprehensive investigations from many fields are helping to understand the real features of these species. In cooperation with this trend, in April 1993 the Yamagata Technopolis Foundation founded a world-wide research

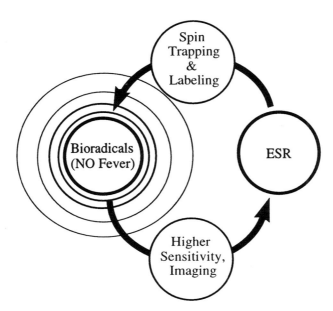

Figure 1. The correlation of bioradicals with ESR after the 1970s.

center for investigating bioradicals using ESR spectroscopy. We named this the Institute for Life Support Technology. In June 1994 we held an International Conference on Bioradicals detected by ESR Spectroscopy to celebrate the inauguration of its opening. This book is a collection of selected current topics presented at the conference. In this chapter I wish to give an introductory overview of bioradicals and ESR technology in relation to the contributions to the conference and the current activities of the institute.

In the half century since Zaboisky published his first paper in *J. Phys. USSR*, bioradicals and ESR have come to have an intimate relationship. For about 20 years, ESR research was concentrated on the direct applications of ESR techniques to biological systems *in vitro* or to model systems. Yamazaki has been a pioneer in this field (see *Chemistry of oxygen radicals*, this vol-

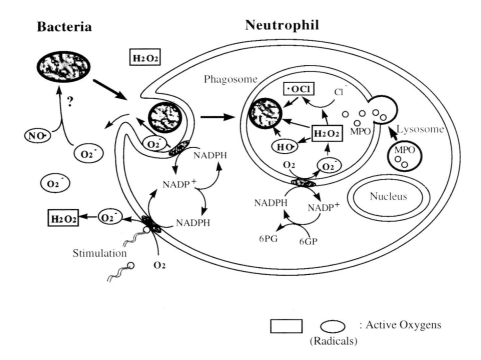

Figure 2. Main function of a neutrophil. When stimulated by bacteria, neutrophils first make many superoxide molecules (O_2^{-}). This phenomenon is called respiratory burst. The superoxides are in part converted into more potent active oxygens called hydrogen peroxides (H_2O_2) and molecular oxygen by dismutation reaction. The bacteria are phagocytosed by the cell. In the phagosome of the neutrophil cell, the hydrogen peroxides are further converted into much more potent hydroxide (HO·) or chloroxide (ClO·) radicals with the help of myeroperoxidase (MPO), and attack the engulfed bacteria. This series of radicals formed is called active oxygen or reactive oxygen species (ROS), which are the representatives of bioradicals that protect the nucleus of the cell. Here we must take into account the participation of nitrogen oxide.

ume). In addition, Hyde developed many kinds of ESR instruments at Varian Associates during these years. In the 1970s spin trap and spin label methods were developed and applied to many biological systems, as shown in Figure 1. Their easy handling and consistent performance continues to accelerate their application to living systems. Janzen and Konaka and Marsh are the pioneers in these fields. As a result of this trend, bioradical research now requires the use of machines with higher sensitivity as well as imaging techniques that allow ESR spectroscopists to interpret radical distribution in living systems better.

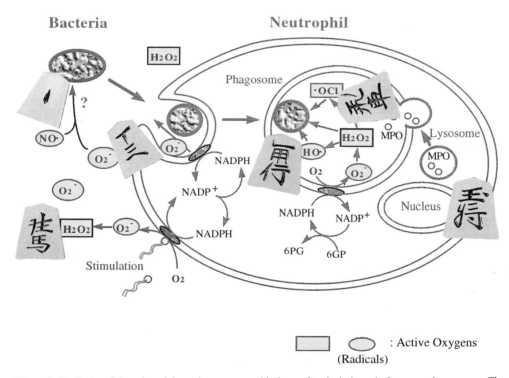

Figure 3. Similarity of the roles of the active oxygens with those of typical pieces in Japanese chess games. The superoxide has a role that can be described as that of a pawn in a chess game against bacteria. The role of hydrogen peroxide is similar to that of the knight in the same chess game. In the phagosome of the neutrophil cell, hydroxide (HO·) and chloroxide (ClO·) radicals may correspond respectively to the bishop and rook in the game. When the superoxide reacts with nitrogen oxide, what kind of new species is born and what role does it have? One of the proposed mechanisms is that the superoxide is converted into a one thousand times more powerful radical of ONOO· acting against the bacteria, just like an upside-down pawn in the figure.

Bioradicals

I will describe first bioradicals themselves as biosignals from living systems. It has been clarified in recent years that bioradicals play crucial roles in diseases, carcinogenesis, aging, and biological functions in living systems. Bioradicals are created and/or annihilated in living systems, depending on their biological functions. A typical example of the creation of active oxygen is shown in Figure 2. This figure shows the main function of a neutrophil, a kind of leukocyte in the blood of animals. The mechanism of its action is as follows: When stimulated by bacteria, neutrophils first make many superoxide molecules ($O_2^{-\cdot}$). This phenomenon is called the respiratory burst. Against bacteria, the superoxide has a role that can be described as that of a pawn in a chess game, as shown in Figure 3. The superoxides are in part converted into more potent active oxygens called hydrogen peroxides (H_2O_2). Their role is similar to that of the knight in the same chess game. Through phagocytosis the bacteria are contained in the cell, as shown in the figures. In the phagosome of the neutrophil cell, the hydrogen peroxides are further converted into much more potent hydroxide ($HO\cdot$) or chloroxide ($ClO\cdot$) radicals with the help of myeroperoxidase (MPO), and attack the ingested bacteria. These two species may correspond respectively to the bishop and rook in the game. This series of radicals formed is called active oxygens or reactive oxygen species (ROS), and they are the representatives of bioradicals which guard the nucleus of the cell as a king. Here we must take into account the participation of nitrogen oxide. When this radical reacts with the superoxide, what kind of new species is born and what kind of role does it have? One of the proposed mechanisms is that the superoxide reacts with nitrogen oxide and is converted into the one thousand times more powerful radical $ONOO\cdot$, just like an upside-down pawn in the figure.

In the presence of the series of active oxygens, metalloproteins have basic roles ranging from transportation and storage of oxygen molecules to reduction of oxygen and active oxygen species in living systems. In his chapter, Peisach will review the active site structures of metalloproteins. Yoshimura and others in the Division of Bioinorganic Chemistry of our institute are investigating nitrosyl-heme complexes and some reagents trapping nitrogen oxide by using non-heme iron complexes, which will also appear in *Metal complexes and metalloproteins*, this volume.

Protection systems

Biological systems themselves usually have built-in systems such as metalloproteins and antioxidants, which protect their cells against the active oxygens, as shown in Figure 4. If such protection is insufficient, however, the active oxygens trigger peroxidation of unsaturated fatty acids in

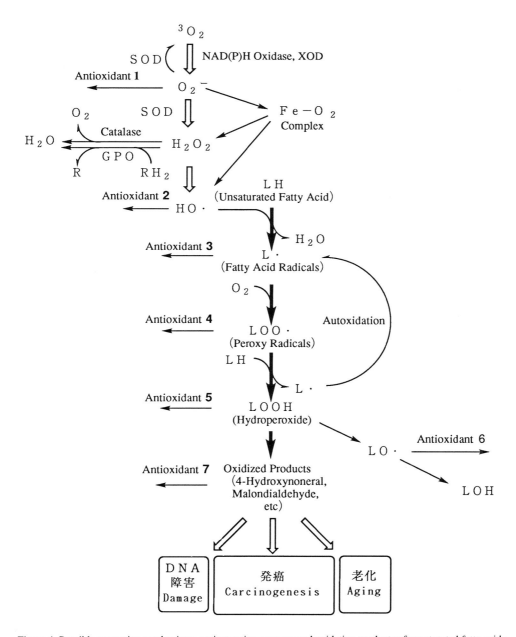

Figure 4. Possible protective mechanisms against active oxygens and oxidation products of unsaturated fatty acids in biological systems. After breaking through these defences the bioradicals induce DNA damage, carcinogenesis, or aging. Antioxidants 1 to 7 mean antioxidants neutralizing the corresponding active oxygen at each step in the figure.

Table 1. Antioxidant abilities* of typical spices surveyed by spin trapping methods

class	spices	O_2^-	HO·	t-BuOO·	$^1O_2$
leaf and flower	oregano, clove, Japanese pepper, cinnamon, ginger, sweet basil, savory, sage, thyme, *tarragon*, *mustard*, basil, rosemary, laurel, marjoram	⊕	× (○)	△	⊕
seed	anise, cardamon, caraway, cumin, coriander, celery, dill, nutmeg, fennel, white pepper	○	×	△	⊕
fruit	allspice, star anise, paprika, black pepper, red pepper, *maze*	○	× (○)	○	⊕

*Antioxidant ability: excellent ≥ ⊕ > ○ > △ > × ≥ none

the cell membrane, which may finally lead to DNA damage, carcinogenesis, or aging, through formation of oxidized products such as 4-hydroxynonenal, malondialdehyde etc. Here I have shown several kinds of antioxidants corresponding to such bioactive species. Packer, Liu and Niki will show typical examples of antioxidants and their functions *in vivo* or *in vitro*. In the Division of Medical Science Hiramatsu and her colleagues are now examining relationships between human diseases and bioradicals, and are developing antioxidants, including some used in Chinese medicine, to neutralize the bioradicals produced along each step in the figure. In the Division of Applied ESR Technology the antioxidant ability of many kinds of foods and spices are surveyed by Ogata. In Table 1 the antioxidant ability of typical spices are classified by using spin trapping techniques. Here it is noted that the antioxidant ability depends on the physical origin of the spices, leaf, flower, or seed. The spices derived from leaves have the most potent antioxidant ability, which may imply that they guard themselves with preformed antioxidant systems against the oxygen stress.

Spin trapping and spin labeling

Spin trapping and spin labeling are both indirect methods detecting unstable radicals and behavior of the surroundings, respectively. Their easy handling enables us to extend their application to biology and medicine as well as to the antioxidant ability of foods. Current uses of these methods are: 1) *in vivo* spin trapping (see, In vivo *spin trapping*, this volume), 2) selective spin trapping of active oxygen species, 3) spin probes having roles in both spin trapping and spin labeling. Fujii et al. and Lai et al. will present new *in vivo* spin trappings. In the Division of Synthetic Chemistry of

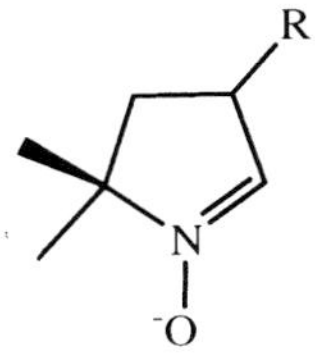

R=CH$_3$
R=CH$_2$OH(DHPO)
R=CH2-NH-nBu(BDPO)
R=C6H5(4PDMPO)
R=1-pyrydonylmethyl(DPPO)

Example:

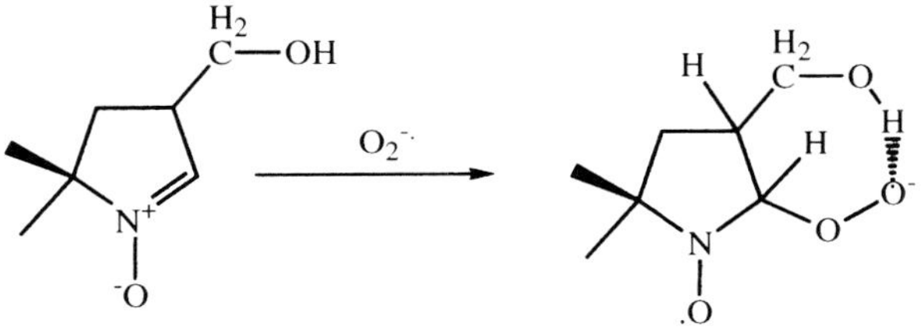

Figure 5. An example of 3-substituted DMPO stabilized by hydrogen bonding.

our institute in cooperation with some laboratories of Yamagata University, Sato et al. (1995) are making many spin traps with a selective ability specific to each active oxygen, as well as spin labels with special functions. As an example, they found that DMPO substituted with a hydroxy-methyl group at 3-position shown in Figure 5 stabilizes the superoxide adducts. This result indicates the possibility of selective spin traps in the future. Finally in this section, we should note that the combination of spin trapping and spin labeling is becoming a major interest because ESR imaging is now used in targeting a diagnosis of disease, which will be described by Swartz et al., Stösser et al. and Utsumi and Takeshita (this volume).

ESR imaging

As a next step it is a natural wish to detect bioradicals directly and to have a picture showing their distribution in the living system, in order to understand the disease state and to help in treatment of disease, as in the case of X-ray photography or magnetic resonance imaging (MRI). Unfortu-nately, the lifetimes of the bioradicals are usually very short, of the order of a microsecond or

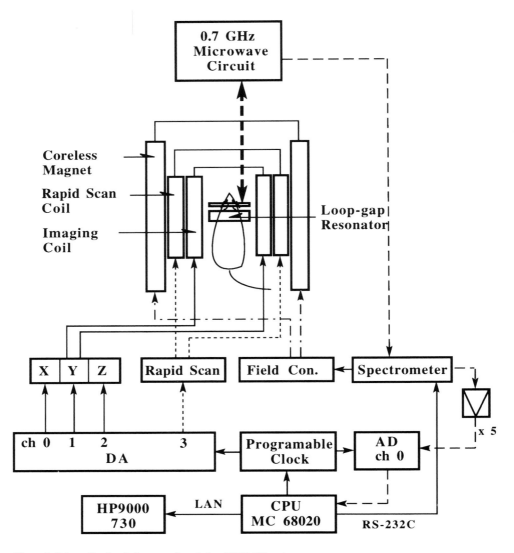

Figure 6. Schematic circuit diagram of our L-band ESR CT system.

even shorter. As a result, their concentrations are consistently too low to get clear pictures directly. Therefore, we are now using new techniques to detect them both directly and indirectly.

ESR imaging is a state-of-the-art technology closely related to MRI, measuring and analyzing *in situ* bioradicals formed in living bodies. Sotgiu et al. and Symons will present 300 MHz ESR imaging systems in their chapter. Sohma develops an open space ESR measurement at K-band in his chapter, which will be a milestone towards an alternative ESR imaging technique. We have also succeeded in observing the distribution of several spin labels in the brains and lungs of small animals by using newly developed 700 MHz ESR imaging systems. For *in vivo* ESR or ESR imaging we have found L-band microwave to be more favorable than X-band, because, as was already known, the sample space at L-band is very large and the energy change due to dielectric loss is very low. Based on these considerations, we began to construct L-band ESR imaging systems several years ago. The 0.7 GHz version was especially effective in obtaining an image with high sensitivity. Figure 6 shows our instrument. The constructed ESR imaging system consists of a signal generator, a microwave circuit, a loop-gap resonator, a coreless magnet with a field gradient coil and rapid scanning coil, and a spectrometer, followed by a computer system for data processing. The computer system was used for the ESR data processing to obtain many images. After integrating each ESR spectrum, we carried out some statistical treatment for a better image. The most reasonable picture was arrived at through a deconvolution of each spectrum, followed by imaging. The upper left figure in Figure 7 shows a three dimensional ESR imaging picture of a rat head after injection of C-PROXYL into the peritoneal cavity of the rat, and the right figure is a cutting image at the plane shown in the left figure. The position of the brain is shown by shadow imaging. It is noted here that the label-rich area is outside of the brain. The lower two are from the rat using the 16-DS radical, a spin-labeled fatty acid. In this case, the label-rich area is just inside the brain position. It is concluded from these findings that the distribution of label reagents depends on the chemical property of labels and only 16-DS distributes in the brain, since it is able to pass through the blood brain barrier. Furthermore, it is interesting to note where in the brain the label-rich position is. Such observation enables us to investigate *in vivo* the functional mechanisms of the radicals, whose formation is correlated with pathological changes. We are now constructing a large-scale ESR imaging system with 3D- and 4D-display for human bodies.

New ESR observation methods

It should be noted here that such pictures as above are made only by spin label reagents, and direct bioradical observation has not yet been successful. So we must consider the sensitivity of ESR measurements. We still need one or two orders of magnitude higher sensitivity to detect bioradicals directly. The Division of ESR Technology in our institute is developing some new observation techniques in addition to the developments of the imaging systems. We call these hybrid ESR observation methods. In Figure 8 several ESR effects of a magnetic moment, $\delta\mu$, in-

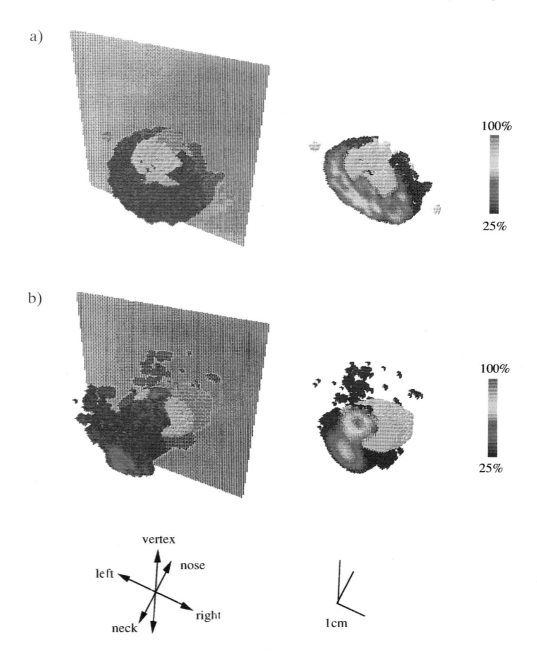

Figure 7. Typical examples of imaging figures of a rat head after injection of (a) C-PROXYL and (b) 16-DS.

duced by microwave radiation are summarized. In the figure, (1) and (2) correspond respectively to microwave absorption, and induction, usually used in our cwESR measurements and pulse ESR measurements; (3) is a heat production (Δ) due to ESR phenomenon; (4) is a magnetic flux change ($\delta\phi$), due to the change of the static μ_z component; (5) is a Raman scattering after light irradiation in the presence of the moments in the light path; (6) is a conductivity change due to the resonance of magnetic moments in conductive materials.

ESR absorbs the irradiated microwave energy, which is usually detected by a microwave detector as a change in microwave power proportional to μ_{xy} in conventional ESR spectrometers, as shown in Figure 8. This method is the most popular, used since the pioneer work by Zaboisky. ESR also induces a microwave proportional to μ_{xy}, which is detected by a probing coil or the resonator itself as free induction decay or spin echo after microwave pulse(s). In the following sections I will describe some hybrid ESR methods in details.

PADMR

The ESR phenomenon always produces heat as shown in the figure, which can be detected as a sound by using a microphone in a closed sample container, when the incident microwave is modulated by amplitude modulation through a PIN diode. Such photo-acoustic detection of magnetic resonance (PADMR), developed by Oikawa et al. (1989) in our laboratory, shows a sensitivity of about 10^{16} spins/gauss at X-band region. The detection of heat is a special technique but is expected to be simple and effective in the range of microwave frequency higher than one hundred GHz. It should be noted here that the ESR signal detected is an absorption signal different from the derivative curve usually detected by the conventional ESR method.

SQUID ESR and AFM-ESR

Figure 8 shows schematically a method detecting magnetic flux change due to resonance by a superconductive split ring. SQUID-ESR (Chamberlin et al., 1979) is now attracting more attention as a sensitive method detecting magnetic flux change, because the SQUID sensor is becoming increasingly sensitive. The change of static magnetic moment along the z-axis on resonance, $\delta\mu_z$, induces the change in the magnetic flux at the position of the sensor, which is detected as an ESR signal through a pick-up coil. An alternative method detecting $\delta\mu_z$ that appeared recently is magnetic force detection of magnetic resonance at the position of a cantilever holding a sample, which was developed by Rugar et al. (1992).

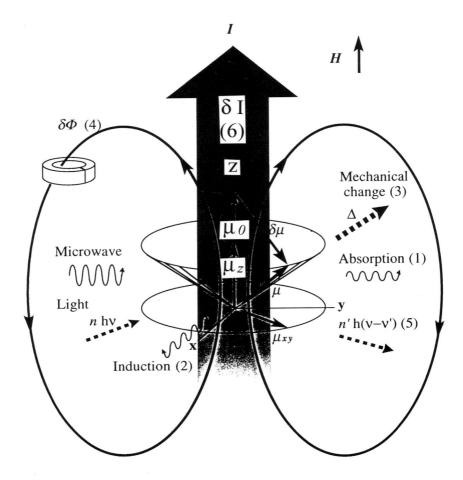

Figure 8. Several ESR effects of a magnetic moment induced by microwave radiation in the resonance condition.

Raman Heterodyne ESR

Raman Heterodyne ESR is a technique that was developed in Australia (Manson et al., 1992). This method measures ESR Raman scattering induced by ESR, which can be detected as a quantum beat by a photodetector, followed by the usual ESR detection system. A coherent light is introduced to the sample and transmitted light to a photo diode. The heterodyne beat detected is introduced to the usual ESR spectrometer system. Another approach to quantum beat observation

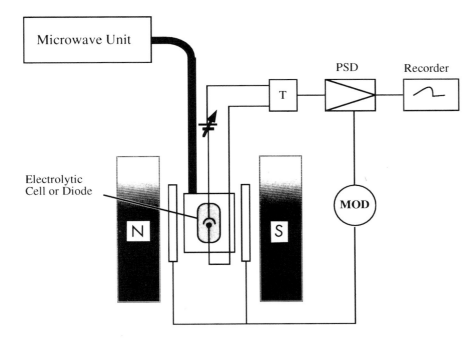

Figure 9. A schematic circuit diagram of ED-ESR.

in the microwave region, multiquantum ESR, has been carried out by Hyde et al., which will be described in *Multiquantum ESR*, this volume.

ED-ESR and STM-ESR

Electrical detection of ESR signals consists of two methods: one is called ED-ESR (Rong et al., 1992), which is schematically shown in Figure 9, and the other is STM-ESR. The ED-ESR is an up-to-date method detecting paramagnetic species in electrolyte solutions and deep traps in semiconductors, the sensitivity being of the order of 10^3 to 10^4. In this case ESR is observed as an electric current change modulated by a low-frequency field modulation. We are now developing this method to apply to biological systems such as microbes and cells. Another approach is STM-ESR, developed by Manassen et al., which detects free radicals one by one on a conductive surface by using a conventional STM technique, applying a static magnetic field (see *New ESR technologies*, this volume).

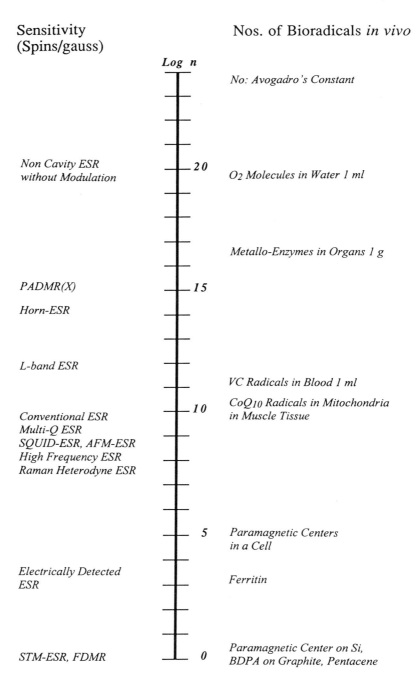

Figure 10. Sensitivity of ESR observation methods and numbers of paramagnetic species in biological systems.

The observation techniques of (1) to (6), except for (4) in Figure 8, are currently being applied at our institute. In Figure 10 are summarized the sensitivities of the ESR observation methods described above in conjunction with a rough estimate of the numbers of bioradicals *in vivo*. Here, non-cavity ESR without modulation is a primitive ESR observation method showing the upper limit of the sensitivity of the methods so far developed. On the other hand, the STM ESR and FDMR (Fluorescence-Detected Magnetic Resonance: Koeler et al., 1993) reach the lower limit of sensitivity of these methods. Each method has respective area of application. One can select the most suitable for the target of research. Combination of these techniques with the ESR imaging techniques described above is probably most suitable for biological systems in the future. For example, a most interesting one for us is the paramagnetic centers in a biological cell. The ED-ESR may be most suitable for investigating such systems.

References

Chamberlin, R.V., Moberly, L.A. and Symko, O.G. (1979) High sensitivity magnetic resonance by SQUID detection. *J. Low Temperature Physics* 35: 337.

Koeler, J.A., Disselhorst, J.A.J.M., Donkers, M.C.J.M., Groenen, E.J.J., Schmidt, J. and Moerner, W.E. (1993) Magnetic resonance of a single molecular spin. *Nature* 363: 242.

Manson, N.B., Fisk, P.T.H. and He, X.-F. (1992) Application of the Raman Heterodyne technique for the detection of EPR and ENDOR. *Appl. Magn. Reson.* 3: 999.

Oikawa, K., Ogata, T., Ono, M. and Kamada, H. (1989) Development of photoacoustically detected magnetic resonance (PADMR) system. *Abstracts of 2nd Japan-China Bilateral ESR Symposium*: 118–119.

Rong, F.C., Gerardi, G.J., Buchwald, W.R., Poindexter, E.H., Umlor, M.T., Keeble, D.J. and Warren, W.L. (1992) Electrically detected magnetic resonance of a transition metal related recombination center in Si *p-n* Diodes. *Appl. Phys. Lett.* 60: 610.

Rugar, D., Yannoni, C.S. and Sidles, J.A. (1992) Mechanical Detection of Magnetic Resonance. *Nature* 360: 563.

Sato, R., Ito, K., Takeishi, M., Niwa, R., Konaka, R. and Kamada, H. (1995) Synthesis and evaluation of novel spin traps. *Magnetic Resonance in Medcine* 6: 132.

Bioradicals Detected by ESR Spectroscopy
H. Ohya-Nishiguchi & L. Packer (eds)
© 1995 Birkhäuser Verlag Basel/Switzerland

Chemistry of oxygen radicals

I. Yamazaki*

National Center for the Design of Molecular Function, Utah State University, Logan, Utah 84322-4630, USA
*Present address: 3-6, North 27, West 11, Sapporo 001, Japan

Introduction

At the first International Symposium on Oxidases and Related Redox Systems (1964, Amherst), George (1965) discussed fundamental aspects of physico-chemical properties of oxygen under the title "The Fitness of Oxygen". His lecture was presented upon the 50th anniversary of a remarkable book, *The Fitness of the Environment*, written by Henderson (1913). Based on thermodynamic data of hydrogen-oxygen and oxygen-oxygen bonds, George calculated one-electron reduction potentials for the four-step reduction of oxygen to water *via* oxygen free radical species (Tab. 1). George concluded that the physical and thermodynamic properties of oxygen are suitable as a terminal oxidant in the energy metabolism of life.

At the same symposium, Yamazaki et al. (1965) reported the mechanism for peroxidase-oxidase reactions in which one-electron reduced species of O_2 (perhydroxyl or superoxide radical) are involved as reaction intermediates. However, in those days, free oxygen radicals were considered not to occur in biochemical and biological systems.

Chemistry and biochemistry of superoxide radical

Peroxidase-oxidase reaction

Peroxidases are well-investigated enzymes which catalyze H_2O_2-dependent oxidation of various electron donors. It has been known that peroxidases catalyze the oxygen-consuming oxidation of certain electron donors, such as dihydrooxyfumarate (DHF) and triose reductone. The reduction of methylene blue (MB) was also found to occur during the oxidation of triose reductone catalyzed by a peroxidase (Yamazaki et al., 1957). To explain the reduction of nonspecific electron-acceptors, such as O_2, MB and cytochrome c, Yamazaki (1958) proposed a free radical mechanism for peroxidase reactions,

$$\text{peroxidase} + H_2O_2 \longrightarrow \text{compound I} \qquad [1]$$
$$\text{compound I} + AH_2 \longrightarrow \text{compound II} + AH\cdot \qquad [2]$$
$$\text{compound II} + AH_2 \longrightarrow \text{peroxidase} + AH\cdot \qquad [3]$$

and concluded that certain free radicals (A'H·) are able to reduce MB and O_2.

$$A'H\cdot + MB \longrightarrow A' + MBH\cdot \qquad [4]$$
$$A'H\cdot + O_2 \longrightarrow A' + O_2^{-\cdot} + H^+ \qquad [5]$$

Yamazaki and Piette (1963) obtained an approximate value of 1:2 for the ratio of $[H_2O_2]_{consumed}$ to $[\text{compound III}]_{formed}$ in the following consecutive reactions,

$$2 \text{ DHF} + H_2O_2 \xrightarrow{\text{peroxidase}} 2 \text{ DHF}\cdot + 2 H_2O \qquad [6]$$
$$\text{DHF}\cdot + O_2 \longrightarrow \text{DKS} + O_2^{-\cdot} + H^+ \qquad [7]$$
$$\text{peroxidase} + O_2^{-\cdot} + H^+ \longrightarrow \text{compound III} \qquad [8]$$

where DKS is diketosuccinate and compound III is an inactive form of peroxidase which is formed from the reaction of compound II with H_2O_2 (Nakajima and Yamazaki, 1987). In the case of horseradish peroxidase (HRP), compound III decomposes to the native enzyme in a few minutes at room temperature. The stability of compound III depends on the type of peroxidase and increases in the cases of lactoperoxidase (Nakamura and Yamazaki, 1969) and diacetyldeuteroheme-substituted HRP (Makino et al., 1976). For peroxidase-catalyzed O_2^- consuming oxidation of DHF, we proposed a chain reaction,

$$\text{DHF} + O_2^{-\cdot} + H^+ \longrightarrow H_2O_2 + \text{DHF}\cdot \qquad [9]$$

Table 1. One-electron reduction potentials (V) for 4-step reduction of O_2 at pH 7

	George (1965)	Naqui et al. (1986)
O_2/O_2^-	−0.45	−0.33
O_2^-/H_2O_2	0.98	0.95
$H_2O_2/HO\cdot$	0.38	0.38
$HO\cdot/H_2O$	2.33	2.33

Reaction 9 is markedly accelerated in the presence of a catalytic amount of Mn^{2+} (Yamazaki and Piette, 1963).

$$O_2^{-\cdot} + Mn^{2+} + H^+ \longrightarrow H_2O_2 + Mn^{3+} \qquad [10]$$
$$Mn^{3+} + DHF \longrightarrow DHF\cdot + Mn^{2+} + H^+ \qquad [11]$$

Xanthine oxidase

The mechanism by which O_2 mediates the transfer of electrons from xanthine oxidase to cytochrome c has been the subject of considerable discussion. Fridovich and Handler (1961) suggested the formation of superoxide during the reduction of O_2 by xanthine oxidase, and later McCord and Fridovich (1968) concluded that the superoxide radical exists free in solutions and is the actual reductant for cytochrome c. Nakamura and Yamazaki (1969) reported that the one-electron flux (x) from xanthine oxidase to O_2 depends on the O_2 concentration, being 0.3 ($\kappa = 0.6$) in an air-equilibrated solution and 0.7 ($\kappa = 1.4$) in an O_2-saturated solution. In most cases, $\kappa = 2x$ (Yamazaki, 1977). The reduction of cytochrome c by the reduced xanthine oxidase (EH_2) under aerobic conditions is formulated as follows:

$$EH_2 + O_2 \longrightarrow x\ EH\cdot + (1-x)E + x\ O_2^{-\cdot} + (1-x)\ H_2O_2 \quad [12]$$
$$cytochrome\ c^{3+} + O_2^{-\cdot} \longrightarrow cytochrome\ c^{2+} + O_2 \qquad [13]$$
$$2\ O_2^{-\cdot} + 2\ H^+ \longrightarrow O_2 + H_2O_2 \qquad [14]$$

Respiratory burst in neutrophils

Active oxygen species are produced during the respiratory burst of neutrophils. The primary product of O_2 reduction has been considered to be superoxide ions (Badwey and Karnovsky, 1980). The stoichiometric conversion of O_2 to the superoxide ion was confirmed by using diacetyldeuteroheme-substituted HRP as an $O_2^{-\cdot}$ trapper (Makino et al., 1986).

EPR measurement

The superoxide ion cannot be detected by EPR in solutions at room temperature. Using a rapid freezing technique, Nilsson et al. (1969) and Knowles et al. (1969) measured at $-170°C$ EPR spectra of superoxide ion formed in the xanthine oxidase reaction and in the O_2^- consuming oxidation of DHF catalyzed by peroxidase. Fujii and Kakinuma (1990) tried to stabilize the superoxide radical at alkaline pH by a pH jump method and applied the method to EPR measurement of the radical formed under physiological conditions.

Superoxide dismutase

The finding of an enzyme which catalyzes the dismutation of superoxide ions was a milestone in the study of oxygen radicals in biological systems. McCord and Fridovich (1969) found that preparations of myoglobin and carbonic anhydrase which inhibit cytochrome c reduction during the xanthine oxidase reaction contain a copper protein called erythrocuprein, which catalyzes dismutation of superoxide ions. The superoxide-dismutating enzyme exists ubiquitously in aerobic living cells, and there are three types of the enzyme containing copper-zinc, iron or manganese at the active site (Fridovich, 1986). The reaction mechanisms are:

$$O_2^{-\cdot} + M^n + 2 H^+ \longrightarrow H_2O_2 + M^{n+1} \qquad [15]$$
$$O_2^{-\cdot} + M^{n+1} \longrightarrow O_2 + M^n \qquad [16]$$

Kinetic nature

The proton dissociation of the superoxide radical ($HO_2\cdot = H^+ + O_2\cdot^-$) occurs at $pK_a = 4.8$ and the dismutation rate of the radical depends on pH, exhibiting a maximum value at pH 4.8 (Rabani

Table 2. Rate constants ($M^{-1}s^{-1}$) for reactions of superoxide

	Rate constant	Reference
Dismutation		
$HO_2 + HO_2$	7.6×10^5	Behar et al., 1970
$HO_2 + O_2^{-\cdot}$	8.5×10^7	Behar et al., 1970
$O_2^{-\cdot} + O_2^{-\cdot}$	less than 100	Behar et al., 1970
Oxidation by $O_2^{-\cdot}$		
Mn^{2+} (pyrophosphate)	2×10^7	Cabelli and Bielski, 1984
Fe^{2+} (formate)	1×10^7	Rush and Bielski, 1985
Reduction by $O_2^{-\cdot}$		
cytochrome c $^{3+}$	$1-1.4 \times 10^6$	Koppenol et al., 1976; Butler et al., 1975, 1982
Fe^{3+}-EDTA	2×10^6	Bull et al., 1983
	1.2×10^6	Sutton, 1985
p-Benzoquinone	1.0×10^9	Patel and Willson, 1973

and Nielsen, 1969). The apparent rate constant (k_d) at pH 7 is 4.5×10^5 $M^{-1}s^{-1}$ (Behar et al., 1970).

$$k_d; 2 \text{ superoxide radicals} \longrightarrow O_2 + H_2O_2 \qquad [17]$$

The superoxide radical acts both as a reductant and an oxidant. The rate constants for some reactions are listed in Table 2.

Redox potential

Although the redox potential for the O_2/H_2O couple is known to be 0.8 V, the oxidizing activity of O_2 is rather weak without catalysts under physiological conditions. This is ascribable to the low one-electron reduction potential of O_2. George (1965) calculated Eo' ($O_2/O_2\cdot^-$) to be –0.45 V at pH 7.0. Yamazaki and Piette (1963) predicted the value to be between 0 and –0.3. Today, the most reliable value is obtained from polarographic, equilibrium and kinetic data (Tab. 3). Quinone/semiquinone couples are commonly used for the equilibrium and kinetic measurements.

$$O_2^{-\cdot} + Q \; \underset{k_r}{\overset{k_f}{\rightleftharpoons}} \; O_2 + Q^{-\cdot} \qquad [18]$$

The equilibrium constant (K) is obtained directly by a pulse radiolysis technique (Meisel and Czapski, 1975; Ilan et al., 1976) and indirectly by measuring k_f and k_r (Wood, 1974; Sawada et al., 1975). From the known value of Eo' (Q/Q$^{-\cdot}$) the value of Eo' ($O_2/O_2^{-\cdot}$) can be calculated (Yamazaki and Ohnishi, 1966; Sawada et al., 1975),

$$\text{Eo' } (O_2/O_2^{-\cdot}) - \text{Eo' } (Q/Q^{-\cdot}) = -RT/F \ln K. \qquad [19]$$

Table 3. One-electron reduction potential of O_2

Eo' (V)	Method	Reference
–0.270	Polarography	Chevalet et al., 1972
–0.33	Kinetic	Wood, 1974
–0.33	Kinetic*	Sawada et al., 1975
–0.27	Kinetic**	Sawada et al., 1975
–0.325	Equilibrium	Meisel and Czapski, 1975
–0.33	Equilibrium	Ilan et al., 1976

* coupled with semiquinone/benzoquinone; ** coupled with cytochrome c

The values so far obtained are listed in Table 3. Since the values in Table 3 are calculated on the basis of 1 atmosphere of O_2, it is more practical to use the $Eo'(O_2/O_2\cdot^-)$ value of -0.16 V which is based on 1 M O_2.

Hydroxyl radical and Fenton reaction

Historical background

A century ago Fenton (1894) reported that H_2O_2 acts as a strong oxidant in the presence of ferrous ions. The combination of H_2O_2 and a ferrous salt is called Fenton's reagent. The nature of oxidizing intermediates produced from the Fenton reaction was controversial until very recently. Haber and Weiss (1934) proposed the formation of hydroxyl radicals in the Fenton reaction.

$$Fe^{2+} + H_2O_2 \longrightarrow Fe^{3+} + HO^- + HO\cdot \qquad [21]$$

This simple equation is used by many investigators to describe a primary event causing oxygen toxicity in biology (Lai and Piette, 1978; Floyd and Lewis, 1983). However, an alternative oxidizing species has also been proposed by chemists (Rush and Koppenol, 1986, 1987; Rahhal and Richter, 1988).

$$Fe^{2+} + H_2O_2 \longrightarrow FeO^{2+} \text{ (ferryl)} + H_2O \qquad [22]$$

Besides the typical mechanisms shown in reactions 21 and 22, slightly modified equations have been reported. The formation of $HO\cdot$ from the ferryl ion was suggested by Walling and Amarnath (1982),

$$Fe^{2+} + H_2O \longrightarrow Fe(IV)OH^{3+} \rightleftharpoons Fe^{3+} + HO\cdot \qquad [23]$$

and by Sugimoto and Sawyer (1984),

$$FeO^{2+} + H_2O \longrightarrow Fe(III)OH^{2+} + HO\cdot \qquad [24]$$

Another important question was whether or not the hydroxyl radical formed in reaction 21 is free or not. The concept of 'crypto-$HO\cdot$' or the 'caged radical' was presented (Bors et al., 1979a; Youngman, 1984; Halliwell and Gutteridge, 1986).

Stoichiometry

The formation of HO· radicals has been detected by chemical analysis of oxidation products of some molecules (Bors et al., 1978; Bors et al., 1979a; Grootvelt and Halliwell, 1986) and the spin-trapping EPR technique (Finkelstein et al., 1980; Buettner, 1987). Numerous papers have been published on spin-trapping of HO· radicals in biochemical and biological systems. Although the spin-trapping technique is the most direct method of detecting the HO· radical and provides a quantitative assay, only a few papers have described the amount of HO· radical formed in the Fenton reaction. Floyd and Lewis (1983) reported that the yield of HO· radical in the Fenton reaction is only 20%.

Using the spin-trapping EPR technique with DMPO, Yamazaki and Piette (1990) examined the stoichiometry of the Fenton reaction under various experimental conditions. In the presence of 90 μM H_2O_2, the yield of HO· radical was measured at varying concentrations of Fe^{2+} ion. As shown in Figure 1, the molar ratio of [DMPO-OH]$_{formed}$ to [Fe^{2+}]$_{added}$ greatly depends on the [Fe^{2+}] and is nearly unity at Fe^{2+} concentrations below 1 μM.

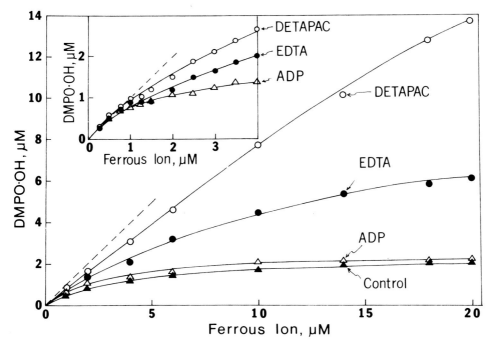

Figure 1. Stoichiometry of the Fenton reaction (Yamazaki and Piette, 1990). Broken lines show the 1:1 stoichiometry between Fe^{2+} and DMPO in reaction 21'. 90 μM H_2O_2 and 40 mM DMPO at pH 7.4.

$$Fe^{2+} + H_2O_2 + DMPO \longrightarrow Fe^{3+} + HO^- + DMPO\text{-}OH \qquad [21']$$

The molar ratio decreases with the increase in the Fe^{2+} concentration and the efficiency varies with the iron-chelators. The kind of iron chelates, however, does not significantly affect the efficiency of DMPO-OH formation at low iron concentrations.

It is now clear that the Fenton reaction as formulated in Equation [21] occurs only at very low iron concentrations. What reactions take place at higher iron concentrations? When the Fenton reaction is carried out in the presence of ethanol and DMPO, the DMPO spin adduct of ethanol radical (DMPO-Et) is observed. There is a remarkable difference in the mode of DMPO-Et formation between the iron chelates of DETAPAC and EDTA, both at 100 μM (Fig. 2). In the case of DETAPAC, the molar concentration of DMPO-Et formed is nearly identical to that of DMPO-OH lost.

$$Ethanol + HO\cdot \longrightarrow ethanol\ free\ radical + H_2O \qquad [25]$$
$$DMPO + ethanol\ free\ radical \longrightarrow DMPO\text{-}Et \qquad [26]$$

However, in the case of EDTA, the formation of DMPO-Et is much greater than the decrease in DMPO-OH. Therefore, we have concluded that ethanol is oxidized by species other than the HO·

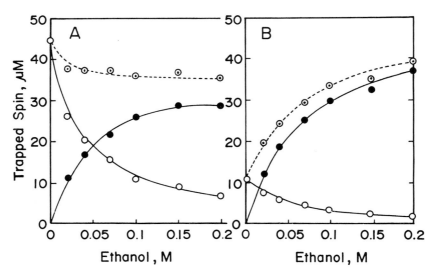

Figure 2. Effect of ethanol concentration on the DMPO-spin adducts (Yamazaki and Piette, 1990). 100 μM Fe^{2+} and 200 μM H_2O_2. ○, DMPO-OH and ●, DMPO-Et. Dotted lines show the sum of DMPO-OH and DMPO-Et. The chelator was DETAPAC in A and EDTA in B.

radical. This oxidizing species does not give rise DMPO-OH and is assumed to be an oxidized state of iron, probably the ferryl which also oxidizes ethanol,

$$\text{ethanol} + \text{Fe(IV)OH}^{3+} \longrightarrow \text{ethanol free radical} + \text{Fe}^{3+} + \text{H}_2\text{O} \qquad [27]$$

Kinetics

The significant difference in the Fenton reaction between the iron chelates of DETAPAC and EDTA can also be seen in the reaction rate. Rate constants for the reduction of H_2O_2 by different ferrous chelates are listed in Table 4. Despite the highly efficient production of HO· radical (Fig. 1), DETAPAC has often been reported as an inhibitor for the Fenton reaction. This is mostly ascribable to the slow rate of the reaction in the presence of DETAPAC. The possibility that Fe^{2+} reacts with the HO· radical or with the ferryl can be ignored under the experimental conditions where DMPO-OH formation is measured in the presence of high concentrations of DMPO or ethanol, respectively. From the slow rate constants for the reduction of DMPO-OH by Fe^{2+} (Tab. 4), it is clear that the reduction of DMPO-OH by Fe^{2+} does not disturb measurement of the yield of HO· radicals in the Fenton reaction when H_2O_2 is present in excess over Fe^{2+}.

The following kinetic experiments were performed to determine the state of the HO· radical. The question arises as to whether the HO· radical formed in the Fenton reaction is free, as formed in the photolysis of H_2O_2, or not free, regarded as "crypto-ȮH". Rate constants (k) for the reactions of the HO· radical with various hydrogen donors (RH) cannot be directly measured in the Fenton reaction. We have compared the k values for the HO· radical formed in the Fenton reaction with those for the free HO· radical formed in photolysis by measuring the k/k_{DMPO} ratios.

$$k: \qquad \text{HO·} + \text{RH} \longrightarrow \text{H}_2\text{O} + \text{R·} \qquad [28]$$
$$k_{DMPO}: \quad \text{HO·} + \text{DMPO} \longrightarrow \text{DMPO-OH} \qquad [29]$$

Table 4. Rate constants ($M^{-1}s^{-1}$) for the Fenton reaction and related reactions in the presence of iron-chelates at pH 7.4 (Yamazaki and Piette, 1990)

	EDTA	DETAPAC	ADP	Control*
Fenton reaction	1.4×10^4	4.1×10^2	8.2×10^3	2.0×10^4
Reduction of DMPO-OH by Fe^{2+}	3×10^3	1×10^3	4×10^3	2.0×10^3

* in the presence of phosphate

Table 5. The k/k_{DMPO} ratios for HO· formed in photolysis and Fenton reaction (Yamazaki and Piette, 1991)

	t-BuOH	EtOH	Benzoate
Photolysis	0.15	0.53	1.6
	0.20	0.38	1.5
Fenton Reaction			
DETAPAC	0.56*	0.59	5.3*
EDTA	0.17	0.37	3.4*
ADP	0.15	0.36	2.2
Phosphate	0.18	0.17*	1.9

* for explanation, see text.

The addition of RH to the Fenton reaction results in the decrease of DMPO-OH accumulation. We can calculate the k/k_{DMPO} ratio according to the following equation,

$$k/k_{DMPO} = [DMPO]([DMPO\text{-}OH]_o\text{-}[DMPO\text{-}OH]) / [RH][DMPO\text{-}OH] \qquad [30]$$

where [DMPO-OH] and $[DMPO\text{-}OH]_o$ are concentrations of DMPO-OH accumulated during the Fenton reactions in the presence and absence of RH, respectively (Yamazaki and Piette, 1991). The results are listed with the data obtained in photolysis systems in Table 5. If the HO· radical formed in the Fenton reaction is free, the k/k_{DMPO} ratio should be the same as that for the HO· radical formed in photolysis. The asterisked values in Table 5 are the ratios significantly deviated from those for the free HO· radical. Although the amount of experimental data is not great, we may conclude that the HO· radicals formed in the Fenton reactions are not free.

Table 6. Rate constants $(M^{-1}s^{-1})$ for the reduction of H_2O_2 by one-electron reductants

Reductant	Rate Constant	Reference
PQ^{+}·*	6.7	Levey and Ebbesen, 1983
Q^{-}·**	< 1	Sushkov et al., 1987
O_2^{-}·	3.0 ± 0.6	Koppenol et al., 1978
	$< 0.23 \pm 0.09$	Melhuish and Sutton, 1978
	0.13 ± 0.07	Weinstein and Bielski., 1979
Fe^{2+}	$10^3 - 10^4$	Table 4

* Paraquat radical; ** 9,10-anthrasemiquinone-2 sulphonate

Thermodynamic Consideration

The nature of the Fenton reaction can simply be expressed as the one-electron reduction of H_2O_2 by Fe^{2+}. Therefore, a question arises as to whether or not other one-electron reductants can replace Fe^{2+}. Many investigators have been interested in the possible one-electron reduction of H_2O_2 by free radical species, such as the superoxide ion and semiquinones. The reduction of H_2O_2 by the superoxide ion is called the Haber-Weiss reaction (Haber and Weiss, 1934).

$$O_2^{-\cdot} + H_2O_2 \longrightarrow O_2 + HO^- + HO\cdot \qquad [31]$$

After many controversial data, it is now concluded that the reduction of H_2O_2 by the superoxide ion and semiquinones is too slow to be involved in the oxygen toxicity. Several papers have reported rate constants for these reactions, which are markedly low when compared with the Fenton reaction (Tab. 6). Many other papers have also supported that H_2O_2 is not significantly reduced by $O_2^{-\cdot}$ (Bors et al., 1979b; Gibian and Ungermann, 1979) and semiquinones (Sinha et al., 1987; Kalyanaraman et al., 1991). This is surprizing because the redox potentials for $O_2/O_2^{-\cdot}$ and quinone/semiquinone couples are much lower than that for the Fe^{3+}/Fe^{2+} couple (Buettner, 1993) and semiquinones can reduce O_2 much faster than does Fe^{2+} (Tab. 7). Why does Fe^{2+} reduce H_2O_2 so easily, but semiquinones do not? It seems reasonable to assume that the one-electron reduction of H_2O_2 by semiquinones has to pass through a high energy transition state, although the $Eo'(H_2O_2/HO\cdot)$ value is relatively high (Tab. 1) and the overall cleavage reaction of the O-O bond occurs at a high electron potential.

$$H_2O_2 + Q^{-\cdot} \rightleftharpoons HO\overset{\ominus}{\cdots}OH + Q \longrightarrow HO^- + HO\cdot + Q \qquad [32]$$

Table 7. Rate constants for O_2 reduction by semiquinone and Fe^{2+}

	$M^{-1}s^{-1}$	Reference
p-Benzoquinone	4.5×10^4	Sawada et al., 1975
Duroquinone	$(2 \pm 0.5) \times 10^8$	Patel and Willson, 1973
Anthraquinone-2,6-disulphonate	5×10^8	Patel and Willson, 1973
Mitomycin	$(2.2 \pm 0.2) \times 10^8$	Butler et al., 1985
Adriamycin	$(3.0 \pm 0.2) \times 10^8$	Butler et al., 1985
Paraquat	7.7×10^8	Farrington et al., 1973
$Fe^{2+}EDTA$	6×10^2	Bull et al., 1983

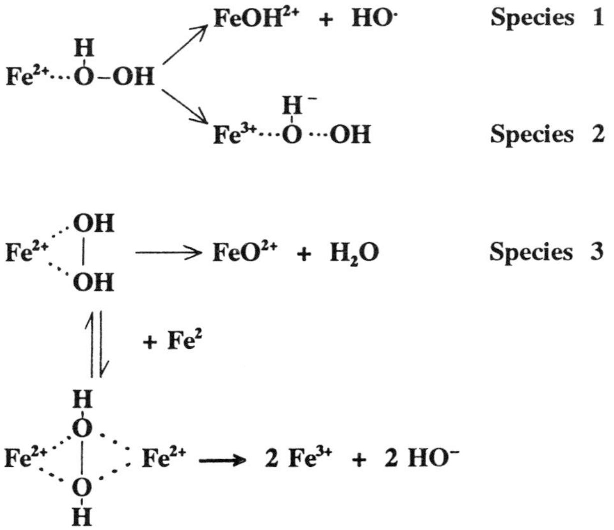

$$
\begin{array}{c}
\text{H} \\
\text{Fe}^{2+}\cdots\text{O}-\text{OH}
\end{array}
\quad
\begin{array}{l}
\nearrow \text{FeOH}^{2+} + \text{HO}^{\cdot} \qquad \text{Species} \;\; 1 \\[2em]
\searrow \;\;
\begin{array}{c}
\text{H}^{-} \\
\text{Fe}^{3+}\cdots\text{O}\cdots\text{OH}
\end{array}
\qquad \text{Species} \;\; 2
\end{array}
$$

$$
\begin{array}{c}
\cdots\text{OH} \\
\text{Fe}^{2+}\cdots\;\;| \\
\cdots\text{OH}
\end{array}
\longrightarrow \text{FeO}^{2+} + \text{H}_2\text{O}
\qquad \text{Species} \;\; 3
$$

$$\Updownarrow \; + \text{Fe}^2$$

$$
\begin{array}{c}
\text{H} \\
\text{O} \\
\text{Fe}^{2+}\cdots\;| \;\cdots\text{Fe}^{2+} \longrightarrow 2\,\text{Fe}^{3+} + 2\,\text{HO}^{-} \\
\text{O} \\
\text{H}
\end{array}
$$

Figure 3. Mechanism of the Fenton reaction. Species 1 and 2 both react with DMPO to yield DMPO-OH. The HO· radical is free in species 1, but not free in species 2. Species 3 is non-HO· oxidant, probably the ferryl.

This transition state will be stabilized when H_2O_2 is bound to the iron ion. If the HO· radical is formed through the iron-bound transition state, the radical exists as a bound form when it reacts with DMPO or RH and is supposed to have lower energy than that of the free HO radical. Our proposed scheme for the Fenton reaction is shown in Figure 3. An assumed structure for the HO· radical formed in the Fenton reaction is species 2.

Acknowledgements
The author is indebted to Dr. Shadi Farhangrazi for her careful reading of this paper. This paper is dedicated to my friend and colleague, the late Professor Lawrence H. Piette, whom I have known since 1959. He died of cancer on November 17, 1992.

References

Badwey, J.A. and Karnovsky, M.L. (1980) *Ann. Rev. Biochem.* 49: 695–726.
Behar, D., Czapski, G., Rabani, J., Dorfman, L.M. and Schwarz, H.A. (1970) *J. Phys. Chem.* 74: 3209–3213.
Bors, W., Saran, M., Lengfelder, E., Michel, C., Fuchs, C. and Frenzel, C. (1978) *Photobiochem. Photobiol.* 28: 629–638.
Bors, W., Michel, C. and Saran, M. (1979a) *Eur. J. Biochem.* 95: 621–627.
Bors, W., Michel, C. and Saran, M. (1979b) *FEBS Lett.* 107: 403–406.
Buettner, G.R. (1987) *Free Radical Biol. Med.* 3: 259–303.

Buettner, G.R. (1993) *Arch. Biochem. Biophys.* 300: 535–543.
Bull, C., McClune, G.J. and Fee, J.A. (1983) *J. Am. Chem. Soc.* 105: 5290–5300.
Butler, J., Jayson, G.G. and Swallow, A.J. (1975) *Biochim. Biophys. Acta* 408: 215–222.
Butler, J., Koppenol, W.H. and Margoliash, E. (1982) *J. Biol. Chem.* 257: 10747–10750.
Butler, J., Hoey, B.M. and Swallow, A.J., (1985) *FEBS Lett.* 182: 95–98.
Cabelli, D.E. and Bielski, B.H.J. (1984) *J. Phys. Chem.* 88: 3111–3115.
Chevalet, J., Ronelle, F., Gierst, L. and Lambert, J.P. (1972) *J. Electroanal. Chem.* 39: 201–216.
Farrington, J.A., Ebert, M., Land, E.J. and Fletcher, K. (1973) *Biochim. Biophys. Acta* 314: 372–381.
Fenton, H.J.H. (1894) *J. Chem. Soc. Trans.* 65: 899–910.
Finkelstein, E., Rosen, G.M. and Rauckman, E.J. (1980) *J. Am. Chem. Soc.* 102: 4994–4999.
Floyd, R.A. and Lewis, C.A. (1983) *Biochemistry* 22: 2645–2649.
Fridovich, I. and Handler, P. (1961) *J. Biol. Chem.* 236: 1836–1840.
Fridovich, I. (1986) *Adv. Enzymol.* 58: 61–97.
Fujii, H. and Kakinuma, K. (1990) *J. Biochem.* 108: 983–987.
George, P. (1965) *In:* T.E. King, H.S. Mason and M. Morrison (eds): *Oxidases and Related Redox Systems*, John Wiley and Sons, Inc. pp 3–36.
Gibian, M.J. and Ungermann, T. (1979) *J. Am. Chem. Soc.* 101: 1291–1293.
Grootvelt, M. and Halliwell, B. (1986) *Biochem. J.* 237: 499–504.
Haber, F. and Weiss, J.J. (1934) *Proc. R. Soc. London,* A147: 332–351.
Halliwell, B. and Gutteridge, J.M.C. (1986) *Arch. Biochem. Biophys.* 246: 501–514.
Henderson, L.J. (1913) *The Fitness of the Environment.* Macmillan, New York.
Ilan, Y.A., Czapski, G. and Meisel, D. (1976) *Biochim. Biophys. Acta* 430: 209–224.
Kalyanaraman, B., Morehouse, K.M. and Mason, R.P. (1991) *Arch. Biochem. Biophys.* 286: 164–170.
Knowles, P.F., Gibson, J.F., Pick, F.M. and Bray, R.C. (1969) *Biochem. J.* 111: 53–58.
Koppenol, W.H., van Buuren, K.J.H., Butler, J. and Braams, R. (1976) *Biochim. Biophys. Acta* 449: 157–168.
Koppenol, W.H., Butler, J. and van Leeuwen, J.W. (1978) *Photobiochem. Photobiol.* 28: 655–660.
Lai, C.S. and Piette, L.H. (1978) *Arch. Biochem. Biophys.* 190: 27–38.
Levey, G. and Ebbesen, T.W. (1983) *J. Phys. Chem.* 87: 829–832.
Makino, R., Yamada, H. and Yamazaki, I. (1976) *Arch. Biochem. Biophys.* 173: 66–70.
Makino, R., Tanaka, T., Iizuka, T., Ishimura, Y. and Kanegasaki, S. (1986) *J. Biol. Chem.* 261: 11444–11447.
McCord, J.M. and Fridovich, I. (1968) *J. Biol. Chem.* 213: 5753–5760.
McCord, J.M. and Fridovich, I. (1969) *J. Biol. Chem.* 214: 6049–6055.
Meisel, D. and Czapski, G. (1975) *J. Phys. Chem.* 78: 1503–1509.
Melhuish, W.H. and Sutton, H.C. (1978) *J.C.S. Chem. Comm.* 970–971.
Nakajima, R. and Yamazaki, I. (1987) *J. Biol. Chem.* 262: 2576–2581.
Nakamura, S. and Yamazaki, I. (1969) *Biochim. Biophys. Acta* 189: 29–37.
Naqui, A., Chance, B. and Cadenas, E. (1986) *Ann. Rev. Biochem.* 55: 137–166.
Nilsson, R., Pick, F.M. and Bray, R.C. (1969) *Biochim. Biophys. Acta* 192: 145–148.
Patel, K.B. and Willson, R. (1973) *J. Chem. Soc. Faraday Trans.* 1: 814–825.
Rabani, J. and Nielsen, S.O. (1969) *J. Phys. Chem.* 75: 3736–3744.
Rahhal, S. and Richter, H.W. (1988) *J. Am. Chem. Soc.* 110: 3126–3133.
Rush, J.D. and Bielski, B.H.J. (1985) *J. Phys. Chem.* 89: 5062–5066.
Rush, J.D. and Koppenol, W.H. (1986) *J. Biol. Chem.* 261: 6730–6733.
Rush, J.D. and Koppenol, W.H. (1987) *J. Inorg. Biochem.* 29: 199–215.
Sawada, Y., Iyanagi, T. and Yamazaki, I. (1975) *Biochemistry* 14: 3761–3764.
Sinha, B.K., Katki, A.G., Batist, G., Cowan, K.H. and Myers, C.E. (1987) *Biochemistry* 26: 3776–3781.
Sugimoto, H. and Sawyer, D.T. (1984) *J. Am. Chem. Soc.* 106: 4283–4285.
Sushkov, D.G., Gristan, N.P. and Weiner, L.M. (1987) *FEBS Lett.* 225: 139–144.
Sutton, H.C. (1985) *J. Free Radical Biol. Med.* 1: 195–202.
Walling, C. and Amarnath, K. (1982) *J. Am. Chem. Soc.* 104: 1185–1189.
Weinstein, J. and Bielski, B.H.J. (1979) *J. Am. Chem. Soc.* 101: 58–62.
Wood, P.M. (1974) *FEBS Lett.* 44: 22–24.
Yamazaki, I., Fujinaga, K. and Takehara, I. (1957) Arch. Biochem. Biophys. 72: 42–48.
Yamazaki, I. (1958) *Proceedings of the International Symposium on Enzyme Chemistry, Maruzen*, Tokyo, pp 224–229.
Yamazaki, I. and Piette, L.H. (1963) *Biochim. Biophys. Acta* 77: 47–64.
Yamazaki, I., Yokota, K. and Nakajima, R. (1965) *In:* T.E King, H.S. Mason and M. Morrison (eds): *Oxidases and Related Redox Systems*, John Wiley and Sons, Inc., pp 485–513.
Yamazaki, I. and Ohnishi, T. (1966) *Biochim. Biophys. Acta* 112: 469–481.
Yamazaki, I. (1977) *In:* W.A. Pryor (ed.): *Free Radicals in Biology.* Vol. III, Academic Press, New York, pp 183–218.
Yamazaki, I. and Piette, L.H. (1990) *J. Biol. Chem.* 265: 13589–13594.
Yamazaki, I. and Piette, L.H. (1991) *J. Am. Chem. Soc.* 113: 7588–7593.
Youngman, R.J. (1984) *Trends Biochem. Sci.* 9: 280–283.

Bioradicals Detected by ESR Spectroscopy
H. Ohya-Nishiguchi & L. Packer (eds)
© 1995 Birkhäuser Verlag Basel/Switzerland

Multiquantum ESR: Physics, technology and applications to bioradicals

J.S. Hyde, H.S. Mchaourab, R.A. Strangeway and J.R. Luglio

National Biomedical ESR Center, Biophysics Research Institute Medical College of Wisconsin, Milwaukee, Wisconsin 53226, USA

Summary. In multiquantum ESR (MQ-ESR) spectroscopy, two microwave frequencies that are derived from a common oscillator are incident on the sample. The separation of the frequencies is adjustable, but has a nominal value of 10 kHz, which is much less than the homogeneous linewidth. Successive absorption and emission of photons from both fields by the spin system leads to oscillation of the spin population at k ($\omega_1 - \omega_2$) and to the production of intermodulation sidebands at $\omega_0 \pm (k + 1/2) (\omega_1 - \omega_2)$, where ω_0 is the average frequency and k is a positive integer. These new microwave frequencies produced by the spin system are detected. Considerable technical effort has been expended to produce spectrally-pure irradiation: spurious signals are 70 dBc. Benefits of MQ-ESR include: 1) detection of pure absorption lines, rather than derivative-like shapes, 2) spectral intensities proportional to T_1, and 3) reduced linewidths. Multiquantum ENDOR and ELDOR experiments have been described. In ENDOR the effect of inducing a nuclear transition on the intensity of an intermodulation sideband is monitored. In ELDOR, the effect of pumping one transition by a pair of frequencies is monitored, by detecting sidebands produced on another transition using a weak observing microwave field. Applications of MQ-ESR that have been published to date are summarized. MQ-ESR is a new ESR methodology that has become technically feasible because of the development of loop-gap resonators as sample-containing structures.

Introduction

In this article, we consider the irradiation of a homogeneous ESR transition by two microwave sources of equal intensity. The two sources have a common time base and the separation of the two frequencies is much less than the homogeneous linewidth. In this circumstance, the spin system generates new microwave frequencies, which are called intermodulation sidebands, when the resonance condition is satisfied. Figure 1 illustrates this situation. The two sources are labeled ω_1 and ω_2 and the intermodulation sidebands are indicated by dotted lines.

The first experimental detection of intermodulation sidebands was reported from the authors' laboratory (Sczaniecki et al., 1991) and followed by a theoretical paper (Mchaourab and Hyde, 1993a). However, their existence was implicit in the theoretical literature much earlier. Anderson (1956) observed that magnetic field modulation, since it is similar to microwave frequency modulation, is analogous to irradiating the transition with a comb of microwave frequencies, i.e., modulation sidebands. His analysis of field modulation led to a double sum: a sum over intermodulation sidebands at each modulation sideband.

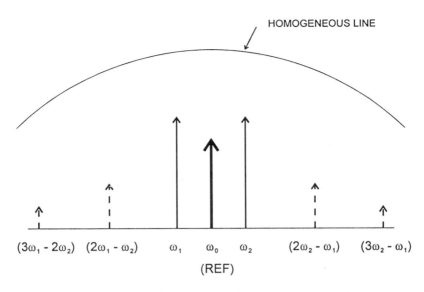

Figure 1. MQ-ESR overview. Irradiating frequencies ω_1 and ω_2, microwave mixer reference ω_0, and spin-system generated intermodulation sidebands (dotted lines) are shown superimposed on a portion of a homogeneous ESR line.

In a different context, the analysis of third-order intercepts in a maser amplifier, an explicit calculation of the intensity of intermodulation sidebands was carried out (Schultz-DuBois, 1964; Tabor et al., 1964). These authors arrived at useful closed-form expressions.

Chiarini et al. (1975) and Martinelli et al. (1977) at the University of Pisa recognized that when a spin system is irradiated by two intense microwave frequencies, the z component of magnetization oscillates at the frequency difference and at harmonics of the difference. They detected these oscillations with a pickup coil. This body of work is closely related to our own.

The principal motivation for our research in its early stages was to arrive at an experimentally robust alternative to magnetic field modulation. Our earliest papers in this field (Hyde et al., 1989; Froncisz et al., 1989; Sczaniecki et al., 1990) were practical in thrust. We sought to avoid field modulation. Additional rationales for detection of intermodulation sidebands have been developed since these first papers, and these are summarized here. But detection of pure absorption ESR spectra with good baseline and sensitivity and with avoidance of magnetic field modulation remains a major motivation for our work.

Analysis is best carried out in the average rotating frame $\omega_0 = (\omega_1 + \omega_2)/2$ in which $(\omega_2 - \omega_0)$ and $(\omega_1 - \omega_0)$ rotate clockwise and counterclockwise at the same frequency, resulting in a linear modulation, see Figure 2. If the microwave reference is at ω_0, adjustment of its phase results in a

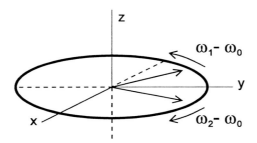

Figure 2. Average rotating frame. In this frame, ω_1 and ω_2 result in a field that is stationary along y and amplitude modulated at $(\omega_1 - \omega_2)$.

stationary vector at any desired orientation in the average rotating frame. Intermodulation sidebands occur at

$$\omega_k = \omega_0 \pm (k + 1/2) \, \Delta\omega, \qquad\qquad [1]$$

where k is an integer and $\Delta\omega = (\omega_1 - \omega_2)$. These resonances involve photons emitted and absorbed from both sources. Hence, the name of the technique: multiquantum ESR (MQ-ESR).

Detection of intermodulation sidebands is accomplished by having a third microwave frequency available that is also derived from a common time base or clock. This frequency serves as the reference at the microwave mixer of the ESR bridge. If it is placed at the average rotating frequency, both sidebands designated by a given index k are superimposed. The output of the mixer is then fed to a phase-sensitive detector (PSD) at

$$\omega_{PSD} = (k + 1/2) \, \Delta\omega. \qquad\qquad [2]$$

It is necessary to synthesize the reference frequency for the PSD, again using the same time base or clock as for $\omega_1 - \omega_2$ and $\omega_1 - \omega_0$ (or $\omega_0 - \omega_2$). Setting ω_{PSD} according to Equation [2] permits observation of any desired intermodulation sideband as determined by index k.

A key to the experimental methods of MQ-ESR is the use of radio frequency synthesizers. Although it would be possible to synthesize all *microwave* frequencies using a common time base, we have no experience, because of cost, with this approach. However, spurious products generated by the synthesis process could be harmful. We start with a single microwave oscillator, either a klystron or Gunn diode oscillator (Oles et al., 1992), and generate various new microwave frequencies where all radio frequency *differences* between the new frequencies and the original frequencies are time locked.

Theory

The general problem is to consider the interaction of the spin system with two circularly polarized electromagnetic fields. There are two ways by which the equation describing the magnetization vector can be obtained. The first is through modification of the Bloch phenomenological differential equations. The second starts with the stochastic Liouville equation of the spin-density matrix. In both cases, the interaction Hamiltonian contains two oscillating terms with different frequencies and therefore cannot be made time-independent by a simple rotation. As a result, exact analytical solutions cannot be obtained and numerical methods are required.

Using the density matrix formalism, we have derived master differential equations that describe the time evolution and/or the steady-state behavior of the density matrix under multiple frequency irradiation for two and four level systems (Mchaourab and Hyde, 1993a; Mchaourab et al., 1993a). We have taken two approaches to solve these equations. In the first, the dressed atom formalism has been used in conjunction with the semi-classical Floquet theory to transform the differential equations into an infinite system of coupled complex algebraic equations. By making reasonable assumptions concerning the maximum number of photons to be included, the system of equations can be solved numerically using standard methods. This approach has been used to investigate the dependence of MQ-ESR, MQ-ENDOR, and MQ-ELDOR signals on spectral parameters. In general, this approach permits the analysis and interpretation of the response of the spin system to any periodic perturbation, regardless of how complex the interaction Hamiltonian might be.

In the second approach, the response of the spin system is predicted by numerical integration of the differential equation. An algorithm based on the stiff Euler method was developed and used to obtain the response of the spin system to the excitation schemes used in MQ-ESR and MQ-ELDOR. The frequency response was consistent with that predicted using the dressed-atom formalism. We feel that the first approach affords significant advantages in computational efficiency. It also permits the derivation of approximate analytical expressions for the experimental observables, which is useful in the process of defining new spectroscopic displays with optimum sensitivity to specific parameters. It is clear that the theoretical analysis of multiquantum experiments will be more laborious when complex systems such as slowly-tumbling spin labels or organometallic complexes are considered. Since the first approach follows the general theory of ESR developed by Freed, it should be possible to implement, in the context of MQ-ESR, the computational strategies that were developed for the analysis of the one-quantum ESR spectra of these complex systems.

Noise and baseline stability

Non-linear processes in the microwave bridge can generate microwave frequencies that coincide with the spin-system intermodulation sidebands. AM noise on these sidebands will directly degrade the signal-to-noise ratio (SNR). In addition, if the level of the instrument-generated intermodulation sidebands becomes too high, baseline drifts due to thermal sensitivities of components of the bridge and the resonator become objectionable. The primary engineering objective in MQ-ESR is to reduce the level of these instrumental sidebands as much as possible. The various cases are summarized below.

Case 1: Instrumental intermodulation sidebands present in the incident irradiation

If they are present, they are attenuated by matching the resonator very carefully. It is better to use a low-Q resonator of the loop-gap type (Hyde and Froncisz, 1989) than a cavity resonator, since

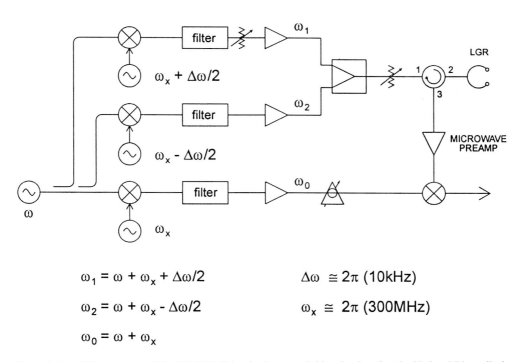

$$\omega_1 = \omega + \omega_x + \Delta\omega/2 \qquad \Delta\omega \cong 2\pi\,(10\text{kHz})$$

$$\omega_2 = \omega + \omega_x - \Delta\omega/2 \qquad \omega_x \cong 2\pi\,(300\text{MHz})$$

$$\omega_0 = \omega + \omega_x$$

Figure 3. Simplified diagram of the MQ-ESR X-band microwave bridge developed at the National Biomedical ESR Center (Medical College of Wisconsin, Milwaukee, USA).

even at perfect match the sidebands are off resonance and will be slightly reflected. In addition, because a desired value of B_1 at the sample can be achieved with a lower incident power in resonators of this type compared with cavity resonators, instrumental sidebands are further reduced in intensity.

A principal cause of intermodulation sidebands in the irradiation pattern in early MQ-ESR bridges was traced to third-order intercepts of the microwave power amplifier, which was amplifying both ω_1 and ω_2. We now form ω_1 and ω_2 and separately amplify them before combining, Figure 3. Spurious incidental intermodulation sidebands are no longer detectable in our apparatus. In principle, they can be eliminated completely.

Sinusoidal modulation of a microwave frequency is equivalent to irradiation with two microwave frequencies that have a common time base according to the trigonometric identity

$$2\cos (\Delta\omega/2)\, t \cos \omega_0\, t = \cos [\omega_0 - (\Delta\omega/2)]\, t + \cos [\omega_0 + (\Delta\omega/2)]\, t$$
$$= \cos \omega_1\, t + \cos \omega_2\, t. \qquad [3]$$

The modulation of Equation [3] is a special degenerate case, being neither pure AM nor pure FM. Not only is the amplitude modulated, but the phase is also modulated by 180°. We believe it not to be technically feasible to produce the irradiating pair by modulation according to the left side of the identity with sufficient suppression of instrumental intermodulation sidebands. One point of view is that by separate formation of the two frequencies according to the right side of the identity followed by separate power amplification and subsequent combination as illustrated in Figure 3, we have arrived at an essentially ideal sinusoidal modulation.

Case 2: Formation of instrumental intermodulation sidebands by the low-noise microwave preamplifier and the microwave mixer because of third-order intercepts in both devices (see Fig. 3)

There are two subcases: a) a small fraction of the irradiating pair (perhaps 30 or 40 dBc) coming directly from Port 1 to Port 3 of the circulator because of non-ideal circulator performance, and b) the irradiating pair reflected from the resonator, even though closely matched (again 40 or 50 dBc), and reaching Port 3. Both subcases are of fundamental concern. Although it is known that good MQ-ESR results can be obtained in many samples of practical interest, thus establishing that the noise is not unduly objectionable, we have not yet succeeded in performing a quantitative noise analysis of Case 2. Careful selection of microwave preamplifiers and microwave mixers for maximum third-order intercept is appropriate. A bucking arm around the circulator to cancel the leakage of the irradiating pair is a technical possibility.

Microwave multiquantum ESR bridge

In our previous papers, several different microwave circuits have been described in detail. Our understanding of the design requirements has continued to evolve, and the current configuration of choice is shown in Figure 3 in highly schematic form. The key ideas are two: a) Use of a synthesizer at a radio frequency ω_x. The mixers create sidebands that are ω_x apart, and by using a fairly high radio frequency, i.e., $2\pi \times 300$ MHz, design of a band pass filter that will suppress the carrier as well as all but the desired sidebands is straightforward. b) Use of a separate microwave power amplifier for each of the three desired microwave frequencies. This is a somewhat expensive but very direct solution to the problem of suppression of instrumental intermodulation sidebands in the irradiating microwaves.

An alternative to field modulation

Figure 4, from Sczaniecki et al. (1991), illustrates the microwave power saturation for individual spin-system generated intermodulation sidebands for a sample of Fremy's salt. Included is saturation data for pure DC detection (i.e., no field modulation) using a single microwave source (labeled 1-0 in the figure). It was obtained with great difficulty because of dominant source noise and baseline instability. The intensities of the intermodulation sidebands are for a single sideband, and would be two times higher if both sidebands that correspond to a given value of index k had been added as we now do routinely using the bridge design of Figure 3. Taking this factor into account, the largest signal that can be seen with ordinary single quantum ESR from Fremy's salt is about 2.5 times greater than the largest 3-quantum signal that can be seen. The theoretical analysis of Mchaourab and Hyde (1993a) provides a more detailed support for this experimental result.

For a Lorentzian line using optimum field modulation, the peak-to-peak signal intensity is the same as the peak signal intensity of the pure absorption line. Thus, MQ-ESR sacrifices a factor of 2.5 in signal intensity. However, most spectroscopists would normally use a much smaller field modulation amplitude in order to avoid lineshape distortion. In this circumstance, the MQ-ESR signal would be comparable to, or even greater than, the ordinary single quantum signal intensity. We conclude that possible loss of signal intensity in MQ-ESR is not a serious criticism of the method when used as an alternative to field modulation.

An inhomogeneous line is a summation of homogeneous lines, and, in principle, this conclusion is expected to be valid for all ESR spectra. Mchaourab et al. (1993c) studied the mixed-valence site in nitrous oxide reductase at low temperature. High quality pure absorption MQ-ESR spectra

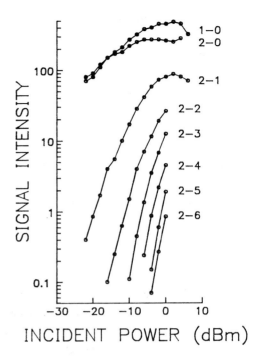

Figure 4. Peak height of intermodulation sidebands as a function of incident microwave power on Fremy's salt. The first index gives the number of irradiating frequencies, always two of equal intensity except for the uppermost trace, which is normal ESR. The second index is k.

were obtained. In this case, the overall spectral linewidth is about 400 G, and the spectroscopist would normally use a field modulation of 5 to 10 G in order to detect small inflections from hyperfine couplings. The MQ-ESR signal is much more intense than the ordinary ESR signal obtained under these conditions.

From long experience, the ESR spectroscopist often prefers to observe first harmonic, or even second harmonic, spectra, in order to enhance small inflections in the spectra. The computer-based technique of pseudomodulation (Hyde et al., 1990; Sczaniecki et al., 1991; Hyde et al., 1992) convolutes a sinusoidal modulation of spectral position with the spectrum, and is a precise mathematical equivalent of the experimental process of applying field modulation and sweeping the magnetic field. One can select the pseudomodulation amplitude, just as one selects the field modulation amplitude. The resulting spectra are identical. However, there is an advantage of pseudomodulation with respect to field modulation. Application of pseudomodulation to digitized noise will not only transform it, but will also filter the noise.

Consider the comparison of a pure absorption MQ-ESR signal and an experimental first harmonic ordinary ESR signal of the same sample using field modulation amplitude H_m. Noise levels will be identical since it is assumed that the receiver noise figures will be the same. If we now apply pseudomodulation of amplitude H_m to the MQ-ESR signal, the spectra will be found to be identical. However, the noise of the MQ-ESR signal will be reduced because of the noise filtering properties of the pseudomodulation algorithm. An additional benefit is that one can vary the pseudomodulation amplitude at will in the post-processing stage of signal analysis, which it is very tedious to do during actual signal acquisition. Detection of pure absorption with post-processing using pseudomodulation is one of the principal advantages of MQ-ESR.

There are other advantages of MQ-ESR with respect to magnetic field modulation, which are summarized here:

- No eddy current problems in resonator walls.
- No eddy current heating in cryogenic structures.
- No loss of resolution or lineshape distortion from modulation sidebands.
- Not necessary to use dewars that are transparent to field modulation.
- Can be used in high pressure or high temperature experiments where introduction of field modulation is difficult.
- Can be used for very broad ESR lines, particularly from transition metals, where sufficient field modulation amplitude with good baseline stability is technically very difficult to achieve.

Multiquantum relaxometry

Closed form solutions of the Bloch equations under CW irradiation by two closely spaced microwave frequencies can only be obtained in the limit of rather low incident power (Mchaourab and Hyde, 1993a, b). However, the differential equations are readily solved by computer under all conditions using numeric methods as described in these papers. See the *Theory* section above.

Figure 5 from Mchaourab and Hyde (1993a) shows the dependence of the multiquantum signals on T_1 and T_2 at low power. The 3-quantum signal varies approximately as $T_2^2 T_1$, compared with the usual 1-quantum dependence on T_2. A spectral display that is proportional to T_1 is one of the novel and interesting features of MQ-ESR. In addition, the dependence on T_2^2 rather than T_2 makes the display very sensitive to subtle changes in this parameter. For 5-quantum transitions, the dependence is $T_2^3 T_1^2$, enhancing these effects. The ratio of 3-quantum to 5-quantum signal intensities is $T_1 T_2$, which can also be a useful parameter (Mchaourab and Hyde, 1993b).

Signal intensities depend on $\Delta\omega$, the difference in irradiating frequencies. A method for numeric extraction of T_1 from inhomogeneously broadened lines has been described by Mchaourab et al. (1993c) and Mchaourab and Hyde (1993b). A series of 3-quantum spectra is obtained as a function of the difference in irradiating frequencies. These are called "frequency-difference swept" spectra. The signal intensity decreases as the frequency difference increases. The value of T_1 can be obtained from Equation [4],

$$T_1 = \frac{1}{\Delta\omega_{1/2}} \qquad\qquad [4]$$

where $\Delta\omega_{1/2}$ is the value of the frequency difference between ω_1 and ω_2 at which the MQ-ESR signal has decreased by 1/2. Here it is assumed that $T_2 \ll T_1$. This method was applied to spin labels and checked using saturation recovery.

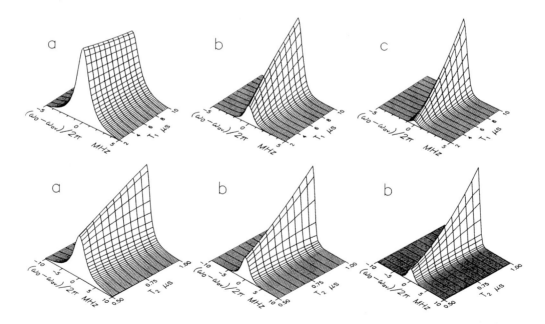

Figure 5. Dependence of the MQ-ESR signal on T_1 (upper) and T_2 (lower) for (a) 1-, (b) 3-, and (c) 5-quantum transitions.

Multiquantum oximetry and nitric oxide monitoring

An immediate consequence of multiquantum relaxometry is that the method can be used in monitoring bimolecular collisions of spin labels with fast relaxing paramagnetic species in the fluid phase. These species include molecular oxygen, nitric oxide and paramagnetic metal ions. The interaction is spin-exchange, which effectively couples the spin-label to the environment because the spin-lattice relaxation times of these species are so short. The use of site-specific mutagenesis to introduce cysteine essentially at will in proteins, with subsequent spin-labeling and measurement of bimolecular collisions with oxygen as developed by the group of W.L. Hubbell (Altenbach et al., 1990), provides a powerful motivation for this application of MQ-ESR. In our laboratory specific cysteine-labeled cecropin, a peptide involved in the immune system of insect species, was synthesized and its biophysical properties investigated using MQ-ESR oximetry (Mchaourab et al., 1993b, 1994).

It might be argued that one could always do these things by using ordinary ESR under conditions of partial saturation. However, such experiments require measurements at more than one power level, which introduces error when effects are small. In our judgment, the methodology of MQ-ESR is more reliable than CW saturation methods for oximetry. Saturation recovery is a better approach than MQ-ESR for determination of T_1. However, that equipment is not widely available. In our laboratory, we find that a *spectral* display with linear T_1-weighting is an attractive control for a saturation-recovery studies.

MQ-ESR spin-label oximetry was introduced by Mchaourab and Hyde (1993b). The first application to nitric oxide was described by Singh et al. (1994). The bimolecular collision rate of nitric oxide with the spin-label 12-SASL in phosphatidylcholine liposomes was determined in order to obtain insight into membrane partitioning of ·NO. In unpublished work from this laboratory, MQ-ESR was used to study aggregation of iron-containing complexes in solution. From the perspective of a biochemist or biophysicist, measurements of bimolecular collisions as discussed in this section are particularly attractive.

Multiquantum modulation spectroscopy including ELDOR

Hyde and Dalton (1972) introduced the method of saturation transfer spectroscopy. The field modulation frequency itself provides a time base by which quantitative statements can be made about rotational diffusion rates. By analogy, one would expect $\Delta\omega$, the separation of intermodulation sidebands in MQ-ESR, to have a potential as a "clock" in order to determine a characteristic relaxation rate. Indeed, the technique of frequency-difference swept MQ-ESR in order to determine a value for T_1 is one example. MQ-ESR can be viewed as a quasi-time domain method. By sweeping the frequency difference, kinetic information can be obtained.

The technique of multiquantum electron-electron double resonance was introduced by Mchaourab et al. (1991). From Equation [3], it is apparent that irradiation by two microwave frequencies is, except for the absence of the carrier, identical to amplitude modulation. From the work of Chiarini et al. (1975), Martinelli et al. (1977) and our own analysis, it is apparent that M_z will oscillate at $\Delta\omega$ and its harmonics. Detection of the modulation by measurement of transfer of saturation to another ESR transition is not unlike detection of oscillation of M_z by a pickup coil as employed by the Pisa group. Measurement of the saturation transfer rate can be accomplished by observing the intensities of the various harmonics and sweeping the frequency difference, $\Delta\omega$.

Multiquantum ELDOR is a more sensitive method than conventional ELDOR that is carried out using field modulation techniques. Figure 6 is a demonstration of achievable ELDOR sensitivity. The molar levels are indicated in the figure. The sample volume was about 1 μl. The high sensitivity arises primarily because both pumped and observed lines are irradiated at the line centers rather than at the peaks of the derivatives.

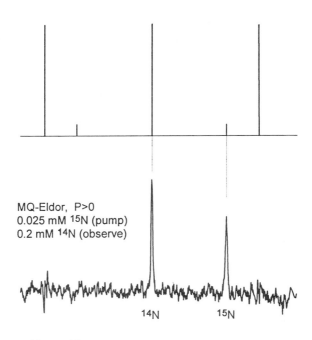

Figure 6. MQ-ELDOR of ^{14}N and ^{15}N TEMPONE, both deuterated, in water at room temperature. Pump, 2 mW; observe, 200 nW using a loop-gap resonator; $(\omega_1 - \omega_2)/2\pi = 10$ kHz; first harmonic.

Multiquantum ENDOR

The idea of multiquantum electron nuclear double resonance, MQ-ENDOR, is to investigate the effect of irradiation of a nuclear transition on electron spin-system generated intermodulation sidebands, see Figure 7. Mchaourab et al. (1993a) gives citations to the relevant literature and many theoretical and experimental details.

Tri-*t*-butyl phenoxyl at 1 mM dissolved in mineral oil served as a test sample. Two ENDOR effects were observed that were of opposite sign. At high microwave powers, generalized saturation is relieved when the nuclear resonance occurs, and the sideband intensities *increase*. This is analogous to the conventional ENDOR mechanism. At low microwave powers, the sideband intensities *decrease* when nuclear resonance is induced. This result was predicted theoretically. It appears to be a new ENDOR mechanism.

ENDOR experiments are often performed at cryogenic temperatures using high frequency field modulation. This method works well, but has always been used empirically with no attempt at theoretical analysis of the ENDOR mechanism. It seems likely that this old empirical method and the new multiquantum one are closely related, although field modulation is more like FM and MQ-ESR modulation is more like AM. Multiquantum ENDOR methodology and theory provide a fundamental opportunity to place ENDOR on a more rigorous theoretical foundation.

The experimental basis for MQ-ENDOR rests on a single model sample, and the full applicability of the method is yet to be determined.

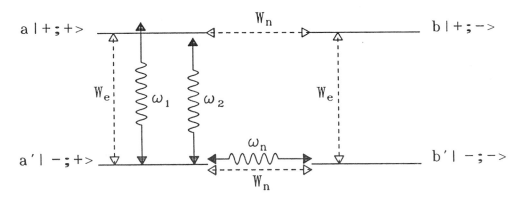

Figure 7. Energy level diagram showing a coupled $S = 1/2$, $I = 1/2$ four-level system irradiated by two microwave fields (ω_1, ω_2) and one rf field (ω_n). W_e and W_n are the electron and nuclear spin-lattice relaxation rates, respectively.

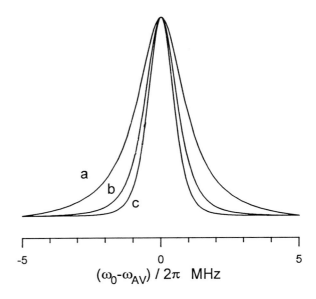

Figure 8. Theoretical field-swept lineshape of the (a) 1-, (b) 3-, and (c) 5-
quantum transitions. The width at half-
height decreases as index k increases.

Resolution enhancement

MQ-ESR homogeneous linewidths are narrower than ordinary ESR linewidths, and the width decreases as the intermodulation sideband index k increases. Theoretical lineshapes from the paper of Mchaourab and Hyde (1993a) are shown in Figure 8. They are in good agreement with experimental lineshapes. Computer-based line narrowing methods have been analyzed in some detail in Hyde et al. (1992). It is shown that there is a direct tradeoff between line narrowing and the signal-to-noise ratio. One could achieve the same linewidth starting with an ordinary ESR line and using computer-based resolution enhancement algorithms as is obtained in MQ-ESR, but there would be degradation of SNR.

If spectral resolution is important, small field modulation amplitude is used in ordinary ESR spectroscopy, further degrading the SNR. It is always preferable from the perspective of spectral resolution for the intrinsic linewidth to be as narrow as possible, and this is one of the significant benefits of MQ-ESR.

Very high resolution of 3-quantum spectra of the spin label CTPO were shown by Sczaniecki et al. (1991). This spin probe shows exceptionally well-resolved superhyperfine coupling and has been widely used for spin-label oximetry. Bimolecular collisions with O_2 cause loss of resolution. For this application, MQ-ESR is of significant benefit.

In unpublished work in cooperation with B. Kalyanaraman, well-resolved MQ-ESR spectra have been obtained from spin-trap adducts. In this application it is desirable to resolve weak couplings from protons of the adduct with the goal of providing more reliable identification.

Pure inhomogeously broadened lines will not exhibit narrower MQ linewidths. It seems certain, however, that situations will be found that are intermediate between the homogeneous and inhomogeneous limits where there will be some benefit of MQ-ESR methodology with respect to resolution.

Transition probabilities

At low powers, the signal intensity of an ordinary ESR signal varies as γB_1; a 3-quantum as $\gamma^3 B_1^3$, a 5-quantum as $\gamma^5 B_1^5$, etc. Figure 4 illustrates these dependencies experimentally. This strong dependence on the transition probability provides a basis for spectral analysis in systems exhibiting so-called forbidden transitions that in fact are only partially forbidden.

Preliminary studies in the senior author's laboratory on square-planar copper complexes have been carried out in the perpendicular part of the spectrum in order to investigate the secondary and tertiary transitions. These transitions arise from a mixing of nuclear states and occur when copper hyperfine, copper Zeeman, and copper quadrupole couplings are comparable. They involve a change of both electron and nuclear quantum numbers. The intensities are quite high, and the lines in a powder spectrum are broad. They interfere substantially with analysis in single quantum spectra, but they seem to be well suppressed in multiquantum spectra. The difference between a 3-quantum and an ordinary spectrum seems to give a spectrum of the forbidden transitions. In work in progress, we hope to analyze the difference spectra in order to obtain copper quadrupole information. These speculations are based on early experiments and illustrate the range of ideas; many further refinements no doubt will be required.

Other situations of possible interest that exhibit forbidden transitions include mixing of proton nuclear states when the proton hyperfine coupling is comparable to the proton Zeeman interaction (13 MHz at X-band), and so-called spin-flip lines that are seen in irradiated organic crystals (see Trammell et al., 1958).

Conclusions

Multiquantum spin physics seems particularly suited to ESR spectroscopy. It is not a natural approach in CW NMR because of the narrowness of the transitions. There are only a few papers in the pulse NMR literature (Zur et al., 1983; Goelman et al., 1987; Manmoto, 1990). The methods are very widely used in non-linear optics, but the relaxation times are so short that much of the interesting physics is not accessible.

Development of MQ-ESR methodology is timely. A compelling rationale is the availability of powerful computers and computer-based programs for solution of the differential equations of Bloch under substantially all conditions (e.g., MATHEMATICA, Wolfram Research). Steady improvement in microwave technology, much of which is also based on computer modeling, has made MQ-ESR technically feasible. The loop-gap resonator developed in our laboratory is also a key factor. Without this low-Q high-energy-density resonator, these experiments would be much more difficult.

One of the exciting possibilities that may be achievable using MQ techniques is the detection of time-dependent changes in relaxation parameters on a millisecond time scale. Many biological processes involve protein conformational transitions that occur in this time domain. These phenomena often involve changes in exposure of residues and local topography as well as rigid body motion of entire secondary structures. Using loop-gap resonator technology, Shin et al. (1993) showed that it is possible to detect these phenomena by observing lineshape changes in the conventional 1-quantum ESR signal of a spin-labeled side chain. In general, these structural transitions also lead to changes in the collision frequencies between spin-labeled side chains in the region of interest and paramagnetic reagents. Higher structural resolution as well as more detailed information could be obtained if it is possible directly to observe changes in T_1 or Heisenberg exchange rates. MQ-ESR and MQ-ELDOR are suited for such measurements.

The extension of 1-quantum ordinary ESR spectroscopy to multiquantum ESR spectroscopy is a natural one that seems certain to flourish as feasibility of application to an increasing range of biological, chemical, and physical problems is established.

Acknowledgement
This work was supported by grants GM27665, GM22923, and RR01008 from the National Institutes of Health.

References

Altenbach, C., Marti, T., Khorana, G. and Hubbell, W.L. (1990) *Science* 248: 1088–1092.
Anderson, W.A. (1956) *Phys. Rev.* 102: 141–167.
Chiarini, F., Martinelli, M., Pardi, L. and Santucci, S. (1975) *Phys. Rev. B* 12: 847–852.
Froncisz, W., Sczaniecki, P.B. and Hyde, J.S. (1989) *Physica Medica* 5: 163–175.
Goelman, G., Zax, D.B. and Vega, S. (1987) *J. Chem. Phys.* 87: 31–44.
Hyde, J.S. and Dalton, L. (1972) *Chem. Phys. Lett.* 16: 568–572.
Hyde, J.S. and Froncisz, W. (1989) In: G.J. Hoff (ed.): *Advanced EPR: Applications in Biology and Biochemistry*, Elsevier, Amsterdam, pp 277–306.

Hyde, J.S., Sczaniecki, P.B. and Froncisz, W. (1989) *J. Chem. Soc., Faraday Trans. 1* 85: 3901–3912.
Hyde, J.S., Pasenkiewicz-Gierula, M., Jesmanowicz, A. and Antholine, W.E. (1990) *Appl. Magn. Reson.* 1: 483–496.
Hyde, J.S., Jesmanowicz, A., Ratke, J.J. and Antholine, W.E. (1992) *J. Magn. Reson.* 96: 1–13.
Manmoto, Y. (1990) *J. Magn. Reson.* 86: 82–96.
Martinelli, M. Pardi, L., Pinzino, C. and Santucci, S., (1977) *Phys. Rev. B* 16: 164–169.
Mchaourab, H.S., Christidis, T.C., Froncisz, W., Sczaniecki, P.B. and Hyde, J.S. (1991) *J. Magn. Reson.* 92: 429–433.
Mchaourab, H.S. and Hyde, J.S. (1993a) *J. Chem. Phys.* 98: 1786–1796.
Mchaourab, H.S. and Hyde, J.S. (1993b) *J. Magn. Reson. B* 101: 178–184.
Mchaourab, H.S., Christidis, T.C. and Hyde, J.S. (1993a) *J. Chem. Phys.* 99: 4975–4985.
Mchaourab, H.S., Hyde, J.S. and Feix, J.B. (1993b) *Biochemistry* 32: 11895–11902.
Mchaourab, H.S., Pfenninger, S., Antholine, W.E., Felix, C.C., Hyde, J.S. and Kroneck, P.M.H. (1993c) *Biophys. J.* 64: 1576–1579.
Mchaourab, H.S., Hyde, J.S. and Feix, J.B. (1994) *Biochemistry* 33: 6691–6699.
Oles, T., Strangeway, R.A., Luglio, J., Froncisz, W. and Hyde, J.S. (1992) *Rev. Sci. Instrum.* 63: 4010–4011.
Schultz-DuBois, E.O. (1964) *Proc. IEEE* 52: 644–656.
Sczaniecki, P.B., Hyde, J.S. and Froncisz, W. (1990) *J. Chem. Phys.* 93: 3891–3899.
Sczaniecki, P.B., Hyde, J.S. and Froncisz, W. (1991) *J. Chem. Phys.* 94: 5907–5916.
Shin, Y.K., Levinthal, C., Levinthal, F. and Hubbell, W.L. (1993) *Science* 259: 960–963.
Singh, R.J., Hogg, N., Mchaourab, H.S. and Kalyanaraman, B. (1994) *Biochim. Biophys. Acta* 1201: 437–441.
Tabor, W.J., Chen, F.S. and Schultz-DuBois, E.O. (1964) *Proc. IEEE* 52: 656–663.
Trammell, G.T., Zeldes, H. and Livingston, R. (1958) *Phys. Rev.* 110: 630–634.
Zur, Y., Levitt, M.H. and Vega, S. (1983) *J. Chem. Phys.* 78: 5293–5310.

Bioradicals Detected by ESR Spectroscopy
H. Ohya-Nishiguchi & L. Packer (eds)
© 1995 Birkhäuser Verlag Basel/Switzerland

Scanning tunneling microscopy (STM) methods for detecting ESR of a single spin center

Y. Manassen, E. Ter-Ovanesyan and D. Shachal

Department of Chemical Physics, Weizmann Institute of Science, Rehovot 76100, Israel

The STM (Scanning Tunneling Microscope) is a new revolutionary technique which was invented approximately 15 years ago in IBM Zürich Laboratories by Binnig et al. (1982, 1983). Since this invention the STM has been found to be an extremely useful technique for many applications. The fundamental characteristic of quantum mechanical tunneling of electrons which is used by the STM is the extreme dependence of the tunneling probability and current on the tunneling barrier width. This phenomenon has been measured many times in bulk tunneling junctions. Such junctions are constructed from two conducting electrodes (c – in Fig. 1a) separated by a thin, uniform insulating layer (i – in Fig. 1a). When the thickness (d – in Fig. 1a) of the insulating layer is reduced below a certain value (≈ 100 Å) tunneling of electrons becomes possible, and

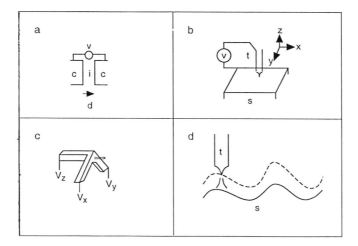

Figure 1. (a) A scheme of a bulk tunneling junction biased with a potential V. (b) A scheme of the STM. The tip position can be controlled and changed in the X, Y and Z directions. (c) A scheme of the piezoelectric tripod. (d) By maintaining constant current (and distance) and following the corrugations of the surface, the STM observes the image (the dashed line).

as a result the conductivity of the junction increases from zero to a certain value. The conductivity of the junction was found to be exponentially dependent on the thickness of the insulating layer. Decreasing this thickness by 1 Å increases the conductivity by an order of magnitude.

The STM is a special type of tunneling junction. In the STM one of the conducting electrodes is a sharp tip (t – in Fig. 1b). When this tip is placed a small distance from a conducting surface (s – in Fig. 1b), and a bias is applied between them, a tunneling current is observed (Fig. 1b). Because of the extreme dependence of the current on the distance, if the tip is sharp enough the tunneling current filament width is extremely narrow. This is so because the probability of tunneling by an atom at the edge of the tip is much greater than by neighboring atoms (due to a smaller distance from the surface). The width of the current filament is of the order of an atom (namely 2–3 Å), and it is the width of the filament, (compared with that of an electron beam in an electron microscope, for example) which gives the STM the ability to image surfaces with an atomic resolution. The tunneling is done through vacuum (under ultra-high vacuum conditions), through air or even through liquid. The tip is mounted on a piezoelectric tripod (Fig. 1c), which can control the position of the tip with an accuracy of less than 0.1 Å, by changing the voltage on the three orthogonal piezoelectric elements. The element which is perpendicular to the surface is called the Z-piezo, while those that are parallel are called the X- and Y-piezoes. While scanning the tip parallel to the surface, the voltage on the Z-piezo is controlled automatically by a feedback loop. The feedback loop changes the voltage on the Z-piezo while scanning, such that the tunneling current is kept constant. This means (due to the extreme current – distance dependence) that a constant distance is kept, and the tip follows the surface corrugations (Fig. 1d). Since the tunneling current filament is so narrow, the tip can follow corrugations of atomic size. By changing the tip surface bias voltage, it is possible to tunnel into different states of the surface, and in this way both structural and electronic information is available – with atomic resolution.

Our STM works in ultra high vacuum conditions (in a base pressure of 2×10^{-10} Torr). This pressure is required for preparing well defined clean semiconductor and metallic surfaces. The surface which has been most heavily investigated is the Si(111) surface. Bulk silicon normally has a diamond structure, where each silicon atom is bound with tetrahedral symmetry to four other atoms. When the silicon crystal is terminated, for example by a (111) plane, a very unstable structure is formed. The surface bonds are unsaturated (dangling bonds) and the surface under these conditions is unstable and very reactive. Upon heating, the surface atoms rearrange in order to reduce the number of dangling bonds. These rearrangements, called surface reconstructions, create surfaces which are relatively stable. The Si(111) surface reconstructs to a 7×7 reconstruction (7×7 means that the size of the superlattice formed by the reconstruction is 7 times larger in each direction than the original surface unit cell). The preparation of such a reconstruction requires a thermal treatment in ultra high vacuum. The silicon wafer in air is normally covered with an insulating silicon dioxide film with a thickness of several monolayers. In order to perform an atomic resolution study, this layer must be removed. Upon heating the wafer at approximately 900°C the silicon dioxide undergoes an insertion reaction with silicon to create silicon monoxide. This substance is in a gaseous state and performs a spontaneous sublimation. The high temperature also drives the remaining surface to rearrange and to create the reconstructions. A Si(111) 7 × 7 surface observed by our microscope is shown in Figure 2. It clearly shows 12 atoms in a unit cell which has 4 large holes in its corner (corner holes). This image is observed at a tip – sample bias voltage of 2V. The unit cell of the 7 × 7 surface contains 49 atoms (Takayanagi et al., 1985), so only a fraction of the atoms is seen. By changing the tip-sample bias voltage, it is possible to see other atoms which are in a different energy state (Hamers et al., 1987). The technique in which the voltage is ramped in order to find the local distribution of energy states is known as scanning tunneling spectroscopy (STS). The structure of the 7 × 7 surface is quite complicated, but since the surface is observed relatively easily, it is a test case which is necessary to show that a vacuum STM is working, before trying more difficult experiments.

Many different modifications and applications of the STM make use of the outstandingly small size of the measurement probe. Some of these modifications are directly related to the topic which is discussed here. First, several different versions of the STM use a ferromagnetic tip. In this case, the tunneling process is sensitive to the spin state of the tunneling electrons. This is because the barrier height is dependent on the exchange interactions. A first example of such an attitude was an experiment in which an antiferromagnetic Cr(100) surface was imaged by an STM with a ferromagnetic tip (Wiesendanger et al., 1990). In this surface, the terraces have alternating spin

Figure 2. An STM image of an area of 12 × 12 nm^2 of a Si(111)7 × 7 surface.

polarization. The observed height of a monoatomic step in which the higher terrace is in an up-spin state is different from that of the step in which the higher terrace is in a spin-down state. In this case, the different spin polarization of the surface is an intrinsic property of the (antiferro-magnetic) surface. A different experiment which has proved to be possible is to create a spin arti-ficial polarization of the surface. This can be done by exciting the carriers with circularly pola-rized light (Sueoka et al., 1993). The spin polarization of the carriers can be changed by varying the circular polarization of the light. An opposite experiment is when the spin polarization of the tunneling electrons from a ferromagnetic tip is detected by measuring the circularity of the light emitted by the radiative recombination of the electrons in the sample, in this case GaAs (Alvardo and Renaud, 1992).

This belongs to a different category of STM experiments which can be defined as dynamic STM experiments, where the STM is used to investigate time-dependent phenomena. This category includes many different type of studies. The following experiments are only partial, non-representative examples of STM activity in this direction. Many studies were performed in order to investigate light emission during STM operation. In one example, a study analyzed the spec-trum of light emitted under the tip while tunneling above a transition metal surface with a relati-vely large bias voltage (V ≤ 50 volts; Berndt et al., 1992). An analysis of the emitted light can provide information on the surface and the molecules adsorbed on it. These spectra can be used also as contrast for microscopy. Of course, the resolution is inferior relative to the atomic resolu-tion obtained with the STM.

Apparently similar, but in fact opposite, experiments were performed when surface bias voltages induced on an STM junction illuminated with laser radiation were measured. An example is the study of the Si(111)7 × 7 surface (Kuk et al., 1990). As a result of the illumination, a surface photovoltaic effect develops which is different for n- and p-type materials or in the neighborhood of surface defects.

The scanning noise microscope (Möller et al., 1990) can be considered as another STM measurement of dynamic phenomena. In this measurement no tip-sample bias voltage is applied. The mean square noise voltage from the junction is measured over a broad band width and is maintained constant using a feedback loop. Since the mean square noise is proportional to the gap resistance this technique also provides the means to keep a constant tip–sample distance when no tip – sample bias can be applied.

A different type of experiment uses the nonlinear characteristics of the STM to generate optical rectification and higher harmonics. A CO_2 laser was directed at the STM junction and the rectified tunnel current was used to control tip – sample spacing for image recording (Krieger et al., 1990). In another experiment a microwave voltage was applied to the tunneling gap and the higher harmonics generated were used to control the feedback loop of the STM (Kochanski,

1990). In this way, insulating surfaces can be measured. These techniques might enable optical spectroscopy to be performed on a nanometer scale.

A similar modification which is described in this work is ESR-STM (Manassen et al., 1989, 1993; Shachal and Manassen, 1991, 1992; Manassen, 1994; Moltokov, 1992; McKinnon et al., 1991; Y. Manassen, E. Ter-Ovanesyan and D. Shachal, in preparation), where this extremely small local probe is used to detect the precession frequencies of individual paramagnetic spins. In this technique, the tip scans a surface that contains isolated paramagnetic spin centers in the presence of a small external magnetic field. It is found that occasionally, when the tunneling region is close to a spin center, the time evolution of the spin center due to the magnetic field results in a periodic perturbation of the tunneling process, and a small time-dependent periodic component is found in the tunneling current. This component is in the Larmor frequency – ω_L.

In order to detect this small rf component the following modification of the STM was constructed (Fig. 3). In addition to the regular STM few major additional components are required. First of all the magnetic field is induced by two small parallel bar magnets which are mounted on the STM to induce the magnetic field H_0. The size of the magnetic field has to be measured quite carefully with a Gaussmeter exactly at the point where the tunneling is performed. The size of the field can be modified by changing the distance between the magnets. For the rf detection, it is necessary to match the output impedance of the tip to 50 Ω, which is the input impedance of the rf amplifier. The low noise rf amplifier is required for the necessary sensitivity in the spectrum analyzer. Of course, to prevent loss of an already weak rf power, a coil blocks the rf signals from entering into the DC system. The ESR-STM signals can be detected directly in the spectrum analyzer but, as will be described later, in order to improve the sensitivity of ESR-STM, phase

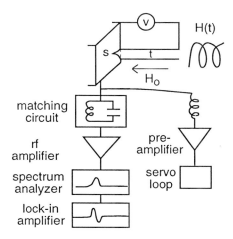

Figure 3. A scheme of the STM modified for ESR–STM experiments.

sensitive detection was also tried. A small coil was located in the neighborhood of the tip and a time-dependent magnetic field at a certain modulation frequency ω_m was applied by introducing alternating current into this coil. For detection, the signal from the spectrum analyzer was fed into a lock-in amplifier driven at the modulation frequency.

The sample which was chosen to test this technique was thermally oxidized silicon surface. The silicon surface is normally covered with a native silicon dioxide layer. This layer must be removed. After this process there are many unsaturated dangling bonds on the surface, but in order to detect ESR signals most of these bonds must be reacted. Therefore a few monolayers of silicon dioxide are grown by thermal oxidation on the clean silicon surface. Due to the lattice mismatch between silicon and silicon dioxide, one out of thousand silicon atoms remains unoxidized to create an interface spin center which is known as the P_b center. These P_b centers were heavily investigated by ESR. These studies were motivated by the role played by these centers in

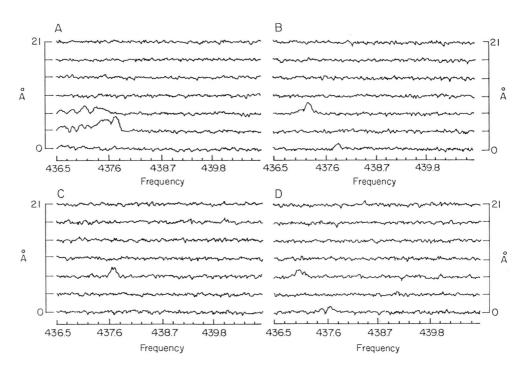

Figure 4. Four consecutive ESR-STM experiments which were performed at the same location above the surface in a magnetic field of 157 Gauss. These experiments demonstrate the reproducibility of the spatial localization of the ESR-STM signals. In addition a certain spatially dependent frequency shift was also observed. This was found to be reproducible as scanning above the same location gives the same shift in frequency on three consecutive occasions (A, B and D). (Taken from Manassen et al., 1993).

the deterioration of the performance of electronic devices. This happens because these centers are effective traps.

ESR-STM signals were observed in many magnetic fields between 139–250 Gauss (Manassen et al., 1989, 1993). So far no attempt has been made to increase the field (and the frequency) because it is expected that the problem of microwave transmission from the tip is a difficult one, and therefore at the moment experiments are done only in the rf regime. One of the most important characteristics of ESR-STM is the spatial localization of the signals. The way this localization is verified is by taking a spectrum with the analyzer at several close locations above the surface. The tip is positioned above a certain surface location, then a spectrum is taken and recorded. After that the tip is moved 3 Å forward, and another spectrum is taken. Occasionally, above a certain location, presumably the location of the spin center, a rf signal at the Larmor frequency appears. Coming back to the same location afterwards, the signal again appears at the same location. As can be seen in Figure 4, the rf signal appears when the tip is at a distance of 3 Å from the estimated location of the spin center. Not all spatially localized ESR-STM signals show the same spatial behavior. For example, in Figures 5 and 6, the signal also appears when the tunneling region was 6 Å from the estimated location of the spin center. This difference in the degree of spatial localization of the ESR-STM signals is still not understood. It might be due to a difference in the sharpness of the tip. Whether or not this is indeed the case might be answered by simultanousely performing ESR-STM and tunneling spectroscopy at the same location.

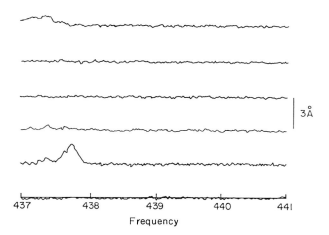

Figure 5. An ESR-STM experiment performed in a magnetic field of 157 Gauss. In this experiment a signal is observed at a relatively large distance from the estimated location of the spin center (6 Å). (Taken from Y. Manassen, E. Ter-Ovanesyan and D. Shachal, in preparation).

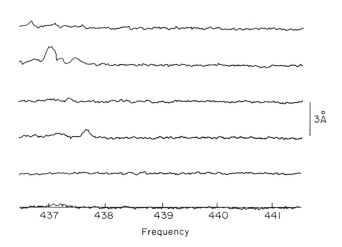

Figure 6. An ESR-STM experiment performed immediately after and at the same location as in Figure 5. This provides a second demonstration of the reproducibility of the spatial localization of the signal and of the spatially-dependent frequency shifts. (Taken from Y. Manassen, E. Ter-Ovanesyan and D. Shachal, in preparation).

An additional phenomenon which was observed in many experiments (but not in all of them) is the fact that exactly above the estimated location of the spin center, the signal disappears. This phenomenon was also noticed in the study done on two dimensional spatial localization. If there is no interest in the shape of the ESR-STM signal, but only in the location at which such a signal appears, then it is possible to look at the intensity of the rf signal at a single frequency (presumably the Larmor frequency) and to record the intensity of the signal as a function of the x and y coordinates of the surface, while the tip is rastering above the surface as in a regular STM experiment. Figure 7 shows such a two dimensional rf map. The elevated rf signal appears when the tip is above a depression in the surface, which is clearly seen in a simultanously recorded topographic STM image. This two dimensional rf map also displays an elevated signal level with a reduced level at the center, leading to a shape that looks like two lobes.

Additional important features are the line shape, the line width and the spectral diffusion of the ESR-STM signals. While a signal was always observed approximately at the Larmor frequency, it was also noticed that there are frequency fluctuations of the order of $1-2$ MHz around the Larmor frequency. Additional fluctuations were observed in the line shape and line width of the signal. Two types of central frequency fluctuations were found. While one type exhibits an entirely random behavior, we could also find spatially-dependent frequency fluctuations. These fluctuations are dependent on the relative position of the tunneling region and the spin center. Therefore, when repeating a scan over the same locations above the surface, the same fluctuations reappear. This must be associated with a certain tip – spin center interaction, as explained in greater detail later. This behavior was observed several times and examples are shown in Figures 4, 5 and 6. Increasing the magnetic field to larger values gave an increase in these fluctuations. Measurements that were done in a field of 250 G gave a frequency fluctuation of the order of 2 – 3 MHz (Manassen et al., 1993). This leads to the conclusion that these fluctuations are due to a certain tip-dependent shift in the g value of the spin center. The possible mechanism for such a shift in the g value is as a result of the spatially-dependent stress applied by the tip during the experiment.

Strain broadening is a well known phenomenon in ESR spectroscopy. Applying stress on a measured solid sample creates deviations from the ideal bond lengths and angles (strain). This affects all the ESR spectroscopic quantities which are dependent on the orbital part of the total angular momentum. In this case the components of the g tensor are affected. Strain broadening was also found in P_b centers. This case was analyzed many years ago (Watkins and Corbette, 1964). It was found that the g anisotropy in these centers is dependent on the extent of $P_{[111]}$ character. When the magnetic field is parallel to the axis of the dangling bond, $g = g_e$, where g_e is the free spin value. When the field is perpendicular to the dangling bond the g value is larger ($g = g_e + \Delta g$). In the unstrained spin center, $\Delta g = 0.01$. In the strained dangling bond $\Delta g \propto \beta^2$, where β is the $P_{[111]}$-like character in the dangling bond. For example, if the angle between the

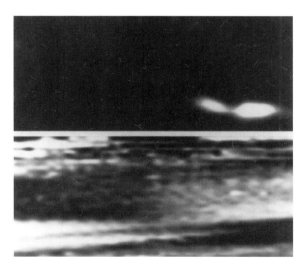

Figure 7. (Top) A two dimensional ESR-STM map observed at a field of 172 Gauss. In this image the two dimensional localization of the ESR-STM signal is demonstrated. The ESR-STM signal level at a *single* frequency was measured above an area of 2×5 nm^2. The bottom image is the regular topographic STM image recorded simultanousely with the rf image in the top. (Taken from Manassen et al., 1989).

dangling bond and the other bonds is reduced to 90° as a result of strain, it will cause an increase in Δg. Strain broadening measurements on P_b centers showed that the interaction of the spin center with the naturally strained $Si - SiO_2$ interface gives a distribution of g values (Brower, 1986). According to these measurements the strain distribution of the spin centers has a standard deviation of 0.00075 in the values of Δg. According to measurements of laser beam deflection techniques, the average stress applied to the spin center in thermally oxidized silicon (when the thickness of the film is a few monolayers) is of the order of -4.6×10^9 dyne/cm^2. These stress components are compressive and lateral (parallel to the surface) (Kobeda and Irene, 1986).

As has long been known, during measurements with the STM significant attractive forces are applied between the surface and the tip (in the normal mode of operation; when the tip is brought closer to the surface, the forces are repulsive). These forces are of the order of $0.1 - 1$ nN (Dürig et al., 1988). These forces are roughly applied to the area of a single atom. This gives stress components of the order of 10^{11} dyne/cm^2 applied to the spin center by the STM tip. These stress components are 20 times larger than those applied to the spin center by the strained interface. This might explain frequency shifts which are spatially-dependent that were observed experimentally (Figs 4, 5, and 6).

The strain induced by the tip is not the only possibility for such a shift. Since the g value is anisotropic, if for some reason the angle between the dangling bond and the spin center is changed, this also gives a frequency shift. Although the silicon surface is rather hard, it undergoes a significant deformation as a result of the extremely strong stress applied by the tip. The deformation which might be caused to a silicon surface under the same experimental conditions was found to be of the order of 1 Å (Chen and Hamers, 1991). Of course in this case if the surface close to the spin center is deformed, the direction of the dangling bond relative to the magnetic field will be changed, and the frequency of the signal will be shifted. Also in this case, the deformation and the change in frequency will be dependent on the relative distance between the spin center and the apex of the tip.

The ability to detect ESR-STM signals with phase sensitive detection is a significant step forward in developing this technique. Phase sensitive detection is commonly used in ESR spectroscopy, due to the significant improvement in sensitivity. A first attempt to use phase sensitive detection for ESR-STM was done on a sample of gold on which a BDPA free radical molecule was deposited. A signal from this sample was observed (McKinnon et al., 1991). However, the derivative signal which is expected for phase sensitive detection was not observed. A first step in applying phase sensitive detection was to observe the response of the ESR-STM signals to an external, time-dependent magnetic field. The ESR-STM signal was observed by the spectrum analyzer, when the magnetic field was modulated according to

$$H(t) = H_0 + \Delta H \cdot \cos(\omega_m t).$$

H_0 is the static magnetic field, ω_m is the modulation frequency, and ΔH is the maximum deviation from the magnetic field as a result of the modulation. As a result of the field modulation, a frequency-modulated signal is observed:

$$S(t) = S_0 \cdot \cos \{[\omega_L + \Delta\omega \cdot \cos(\omega_m t)] \, t\},$$

where $\Delta\omega$ is the maximum deviation from the central frequency as a result of the field modulation. In the case where $\Delta\omega > \omega_m$, the frequency modulated signal is known to be wide band. Figure 8 (top) shows an example of an ESR-STM signal observed when the magnetic field was modulated. Figure 8 (bottom) shows a simulated wide band frequency modulated signal which is observed from a frequency synthesizer. The modulation frequency assumed in the simulation was ω_m. The modulation intensity of the simulation was $\Delta\omega = g\beta\Delta H$. The similarity between the two signals confirms that the ESR-STM signal was indeed modulated by the magnetic field to give a wide band frequency modulated signal. The observed splitting is of the right size, and it is possible to say that the signal shown in Figure 8 (top) is a convolution of the line shape of the unmodulated signal with the modulation function shown in Figure 8 (bottom) (Y. Manassen, E. Ter-Ovanesyan and D. Shachal, in preparation).

Phase sensitive detection of ESR-STM signals was found to be possible, and because of the increase in sensitivity it is also much more effective than direct detection. Figures 9 and 10 show an example of a derivative signal observed when the signal from the spectrum analyzer was fed into a lock-in amplifier driven by the modulation frequency ω_m. In order to avoid filtering out the modulation it is important to increase the band width of the spectrum analyzer above ω_m. The sensitivity of phase sensitive detection is much greater since in all these cases, the simultanously recorded signals at the spectrum analyzer are hardly visible. In order to obtain line shape and line

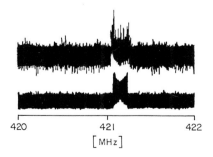

Figure 8. (Top) The response of the ESR-STM signal to a magnetic field modulation – as detected with the spectrum analyzer. The magnetic field was 151 Gauss, the modulation intensity (ΔH) was 27 mG and the modulation frequency (ω_m) was 300 Hz. (Bottom) A simulated wide band frequency modulated signal with $\Delta\omega = 75$ kHz and $\omega_m = 300$ Hz. (Taken from Y. Manassen, E. Ter-Ovanesyan and D. Shachal, in preparation).

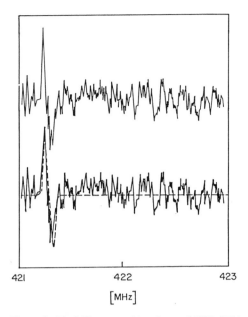

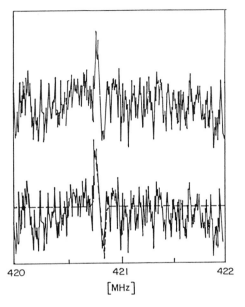

[MHz]   [MHz]

Figure 9. (Top) Phase sensitive detected ESR-STM signal with ΔH = 20 mG, H_0 = 151 G and ω_m = 300 Hz. (Bottom) The signal was successfully simulated by assuming a modulated Lorentzian (dashed curve) with $\Delta\omega$ = 56 kHz (which is the experimental modulation intensity) and T_2 = 1.25 × 10⁻⁵ seconds. (Taken from Y. Manassen, E. Ter-Ovanesyan and D. Shachal, in preparation).

Figure 10. (Top) Phase sensitive detected ESR-STM with ΔH = 23 mG, H_0 = 151 G and ω_m = 4.2 kHz. (Bottom) The signal was successfully simulated by assuming a modulated Lorentzian (dashed curve) with $\Delta\omega$ = 65 kHz (which is the experimental modulation intensity) and T_2 = 2 × 10⁻⁵. (Taken from Y. Manassen, E. Ter-Ovanesyan and D. Shachal, in preparation). This can be compared with T_2 = 10⁻⁷ observed from macroscopic ESR spectroscopy in a magnetic field of 7000 G at 100 °K (Brower, 1986).

width information, a simulation was done. In this simulation a Lorentzian line was assumed which is frequency modulated:

$$\frac{1}{\{\omega - [\omega_L + \Delta\omega \cdot \cos{(\omega_m t)}]\}^2 + (1/T_2)^2}$$

(the normalization of the Lorentzian is ignored). T_2 is the homogenous transverse relaxation time. The modulated signal was superimposed on the modulating function $\cos(\omega_m t)$. The modulation intensity is known *a priori* and the resulting line can give the approximate life time of the single spin ESR line. As can be seen from Figures 9 and 10, assuming a Lorentzian lineshape gives a reasonable agreement with the experimental result. This conclusion cannot be generalized. Many measurements, both when the signal was detected with the spectrum analyzer and when it was detected with the lock-in amplifier, gave a non-Lorentzian lineshape.

At the moment, there is no experimental evidence to indicate the mechanism of this pheno-menon. An example of a possible mechanism is magnetic interactions. A time-dependent force (Lorentz force) is applied on the tunneling electrons by a precessing spin. These interactions, however, are believed to be too weak to be responsible for this phenomenon. A reasonable con-clusion is that the magnetic field introduces a Larmor frequency component in the time evolution of the electrostatic interactions between the spin center and the tunneling electrons. A silicon atom has a significant spin-orbit coupling (19 meV). The spin-orbit coupling is responsible for the g anisotropy of the P_b center. In addition, it causes splitting of the electronic states to give the Zeeman effect. Therefore, when the electronic transition is excited into a superposition of states, the electric dipole moment of the spin center will oscillate according to

$$\mu(t) = \mu_0 + \mu_1 \cos(\omega_e) \cos(\omega_L t/2).$$

ω_e is the frequency of the electronic transition. The ω_e oscillations are modulated by the Larmor frequency. When these oscillations are close enough to the tunneling region, the barrier height oscillations will also be ω_L modulated:

$$\Phi(t) = \Phi_0 + \Phi_1 \cos(\omega_e t) \cos(\omega_L t/2).$$

Since the barrier is not rectangular, the barrier width has a similar time-dependence:

$$d(t) = d_0 + d_1 \cos(\omega_e t) \cos(\omega_L t/2).$$

In all these functions the ω_e frequency component is split into two components with a difference in frequency of ω_L. The dependence of the tunneling current on the barrier height and the barrier width is extremely nonlinear: $(J(t) \simeq e^{-\sqrt{\Phi(t) d(t)}})$, and therefore $J(\omega_L) \neq 0$. Several empirical calculations showed that can have the intensity which is indeed observed experimentally (Manassen et al., 1993).

Another possibility is that in this experiment, the Larmor frequency component is observed due to spin lattice relaxation through the Raman process. As was shown previously, because of the nonlinear nature of the current-distance and the current-barrier height dependence, if d and Φ are modulated by ω_L, $J(\omega_L) \neq 0$. However, this is precisely what happens in the Raman process of T_1 relaxation: if the propagating phonon has a frequency ω_{ph}, then the phonon scattered from the spin center will have a frequency of $\omega_{ph} - \omega_L$. In other words, in the neighborhood of the spin center, the vibrating atoms can induce a ω_L component at the Larmor frequency.

So far, no experimental evidence has indicated that either of these two mechanisms or a different one is correct. Only future experiments on different spin centers will resolve this question.

In this chapter, several different experimental studies of ESR-STM have been reviewed. Success in future studies will result in a powerful interdisciplinary technique which might provide unique information at nanoscale (or atomic) resolution.

References

Alvardo, S.F. and Renaud, P. (1992) *Phys. Rev. Lett.* 68: 247.
Berndt, R., Gimzewski, J.K. and Schlitter, R.R. (1992) *Ultramicroscopy*, 42–44: 355.
Binnig, G., Rohrer, H., Gerber, C. and Weibel, E. (1982) *Phys. Rev. Lett.* 49: 57.
Binnig, G., Rohrer, H., Gerber, C. and Weibel, E. (1983) *Phys. Rev. Lett.* 50: 120.
Brower, K.L. (1986) *Phys. Rev.* B33: 4471.
Chen, C.J. and Hamers, R.J. (1991) *J. Vac. Sci. Technol.* A9: 230.
Dürig, U., Züger, O. and W. Pohl, D. (1988) *J. Microsc.* 152: 259.
Hamers, R.J., Tromp, R.M. and Demuth, J.E. (1987) *Surf. Sci.* 181: 346.
Kobeda, E. and Irene, E.A. (1986) *J. Vac. Sci. Technol.* B4: 720.
Kochanski, G.P. (1990) *Phys. Rev. Lett.* 62: 2285.
Krieger, W., Suzuki, T., Völcker, M. and Walther, H. (1990) *Phys. Rev.* B41: 10229.
Kuk, Y., Becker, R.S., Silverman, P.J. and Kochanski, G.P. (1990) *Phys. Rev. Lett.* 65: 456.
Manassen, Y., Hamers, R.J., Demuth, J.E. and Castellano A.J.,Jr. (1989) *Phys. Rev. Lett.* 62: 2531.
Manassen, Y., Ter-Ovanesyan, E., Shachal, D. and Richter, S. (1993) *Phys. Rev.* B48: 4887.
Manassen, Y. (1994) *Adv. Mater.* 6: 401.
McKinnon, A.W., Welland, M.E., Rayment, T. and Levitt, M.H. (1991) Presented at the International Conference on Scanning Tunneling MIcroscopy, Interlaken, Switzerland.
Möller, R., Esslinger, A. and Koslowski, B. (1990) *J. Vac. Sci. Technol.* A8: 590.
Moltokov, S.N. (1992) *Surf. Sci.* 264: 235.
Shachal, D. and Manassen, Y. (1991) *Phys. Rev. B*, 44: 11528.
Shachal, D. and Manassen, Y. (1992) *Phys. Rev. B*, 46: 4785.
Sueoka, K., Mukasa, K. and Hayakawa, K. (1993) *Jpn. J. Appl. Phys.* 32: 2989.
Takayanagi, K., Tanishiro, Y., Takahashi, Y. and Takahashi, M. (1985) *Surf. Sci.* 164: 367.
Watkins, G.D. and Corbett, J.W. (1964) *Phys. Rev.* 134: A1359.
Wiesendanger, R., Güntherodt, H.-J., Güntherodt, G., Gambino, R.J. and Ruf, R. (1990) *Phys. Rev. Lett.* 65: 247.

Bioradicals Detected by ESR Spectroscopy
H. Ohya-Nishiguchi & L. Packer (eds)
© 1995 Birkhäuser Verlag Basel/Switzerland

An open space ESR spectrometer and its application to high dielectric loss samples

J. Sohma, A. Minegishi, R. Amano and H. Hara[1]

Research Institute for Integrated Sciences, Kanagawa University, Hiratsuka, Kanagawa 259–12,
[1]Microdevice Co. Komagome, Bunkyoku, Tokyo, 170, Japan

Summary. Resonance cavities are used in ESR spectrometers. A resonance cavity has two disadvantages: one is the small and closed space required for a sample, two is the high loss of sensitivity when the sample has a high dielectric loss, such as an aqueous sample. In order to improve these disadvantages, a pair of horns are used instead of a resonance cavity in ESR spectroscopy. The space between the pair of horns is open and no Q value is defined in the horns. The ESR sensitivity is decreased when the cavity is abandoned. But decrease of sensitivity is partly recovered by using a higher frequency of the microwave and also a larger volume of sample, as much as ten times the volume. The sensitivity of open space ESR spectroscopy is 10^{-6} mol/l of DPPH benzene solution. ESR was observed from solid DPPH in a pig liver of 30 g by using open space ESR spectroscopy.

Introduction

A resonance cavity is required in any ESR spectrometer used at present. The reason is to get a higher ESR sensitivity due to the high Q value of the cavity. The cavity in the ESR spectrometer has two disadvantages; one is the closed and small space needed for a sample, the other is difficulty of ESR observation of sample with a high dielectric loss, which causes a decrease of the Q value. In order to remove these disadvantages an open space ESR spectrometer was designed and constructed.

Method and instrumentation

One disadvantage of the resonance cavity is that it is closed and small. The resonance cavity was therefore abandoned in the ESR spectrometer, and an open space for an ESR sample introduced. There are two possibilities for an open space housing the ESR sample in ESR spectroscopy. One is a Fabry-Perrot resonator, the other is a couple of horns. The Fabry-Perrot resonator is very sensitive to resonance conditions. For example, the Fabry- Perrot resonance condition could be tuned for a vacant resonator without any ESR sample.

If an ESR sample is placed in the Fabry-Perrot resonator, a resonance condition is hard to find because of the dielectric constant and the geometrical shape of the sample. Therefore, it is practically impossible to tune a resonance condition after adding an ESR sample to the Fabry-Perrot resonator. In particular, it is hard to get a resonance condition in a Fabry-Perrot resonator for a sample with a high dielectric loss.

The other possibility for the open space ESR spectrometer is a couple of horns, one an emitter and the other a receiver horn. No Q value is defined for a couple of horns. The ESR sensitivity is very much decreased for an ESR spectrometer in which a couple of horns are used instead of a resonance cavity. The ESR spectrometer using horns in the X-band was reported by Ohya-Nishiguchi et al. (1986).

Decrease in ESR sensitivity in the spectrometer using a couple horns is partly recovered by using a higher frequency of the microwave, K-band (23 GHz) in our ESR spectrometer. One of these horns is an emitter the other one is a receiver. Both shape and size of the horns are shown in Figure 1. The open space between the couple of horns is the space for a sample. The open space for a sample provides us with an experimental flexibility in comparision with the closed and small space in a cavity.

A block diagram of an open space ESR spectrometer is shown in Figure 2. The detector of the spectrometer is a homodyne, and modulation is 100 KHz.

No Q value is defined in this open space ESR spectrometer, because no microwave resonance was observed in the couple of horns. Therefore no decrease in the Q value for the horns was observed even when the ESR sample has a large dielectric loss. The microwave power at the receiver horn diminishes if the ESR sample has a large dielectric loss. However, this decrease in power at the receiver horn is easily compensated for by an increase in the power of microwaves emitted from the emitter horn. Thus, the second disadvantage of the high dielectric loss sample can be overcome by using a couple of horns in an ESR spectrometer. The dielectric failure of ESR sensitivity could be compensated for by increasing the power of microwaves emitted. Instead of an ESR sample tube as used in the conventional ESR spectrometer, an ESR cell is used

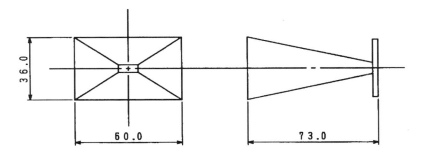

Figure 1. The shape and size of the horns used in this experiment. Numbers in the figure are in units of mm.

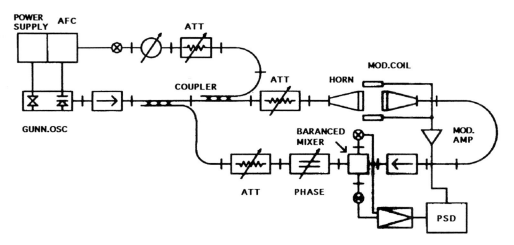

Figure 2. The block diagram of an open space ESR spectrometer at K-band.

in the open space ESR spectroscopy. The volumes of the sample cells range from 3 cm^3 to 50 cm^3. The pair of horns and a sample are set at the center of the pole pieces of the magnet.

The sensitivitity of the open space ESR spectrometer

An example of the ESR spectra observed with the open space ESR spectrometer is shown in Figure 3. The sample was DPPH benzene solution at a concentration of 5×10^{-4} mol/l. The volume of the sample in this case is 5 cm^3. The minimum concentration is 10^{-6} mol/l of DPPH

1 mT

Figure 3. An example of the ESR spectra obtained with the open space ESR spectrometer. Sample is the DPPH benzene solution at a concentration of 5×10^{-4} mol/l. Its volume is about 5 cm^3.

benzene solution. Of course, the minimum sensitivity could be decreased if the volume of the sample is increased to several tens cm^3.

Horn setting

There are two kinds of setting of horns in the open space ESR spectrometer illustrated in Figure 4. One is a parallel horn setting, the other is the cross-horn setting. In the parallel horn setting the absorption mode of the ESR resonance was observed, as shown in Figure 5a. In the

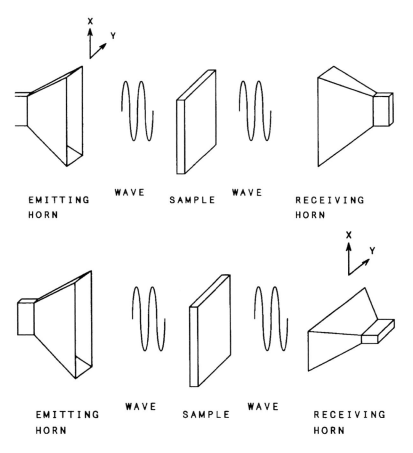

Figure 4. Two kinds of setting of horns in the open space ESR spectrometer. (Upper part) parallel horn setting, (lower part) cross-horn setting.

a) cross   b) parallel

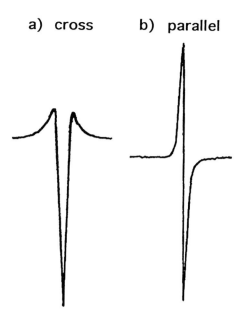

Figure 5. Detected mode of the ESR signal in (a) cross and (b) parallel setting of horns.

case of the cross-horn setting no power was detected by the receiver horn but the dispersion mode of ESR resonance was observed as shown in Figure 5b. The ESR sensitivities of the two modes are nearly the same.

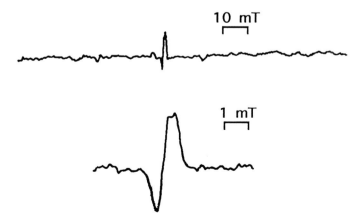

Figure 6. The ESR spectrum of a pig liver of 30 g with solid DPPH.

Application to the high dielectric loss sample

An ESR from an acidic aqueous solution of perylene, the concentration of which was 10^{-3} mol/l, was clearly observed from a sample of 3 cm^3. No Q dip was found for the sample using a conventional ESR spectrometer, and no ESR was observed in this case. The ESR spectrum from a pig liver of 30 g was detectable but uncertain. Then, solid DPPH was added to the same sample, and the ESR was clearly observed as a singlet, as shown in Figure 6.

This observation indicates the potential application of such open space ESR spectroscopy to biological systems.

References

Ohya-Nishiguchi, H., Sugito, S. and Hirota, N. (1986) *J. Magnetic Resonance* 68: 40–51.

Bioradicals Detected by ESR Spectroscopy
H. Ohya-Nishiguchi & L. Packer (eds)
© 1995 Birkhäuser Verlag Basel/Switzerland

New experimental modalities of low frequency electron paramagnetic resonance imaging

A. Sotgiu, M. Alecci, J. Brivati, G. Placidi and L. Testa

Department of Biomedical Sciences and Technologies, University of L'Aquila Via Vetoio, I-67100-L'Aquila, Italy

Introduction

In the field of electron paramagnetic resonance (EPR) at very low frequencies, the spectroscopic and imaging applications are strongly linked. A living sample is, in fact, composed of different structures and organs in which the paramagnetic probes can reach different concentrations, be exposed to different local environments, and be reduced at different rates. For this reason, the spectroscopic information from whole body measurements can be misleading because only accurate knowledge of the spatial distribution of the probe is instructive.

The present biological applications of low frequency EPR are focused on the study of the distribution and clearance of exogenous paramagnetic probes in small laboratory animals. A 2–3 mm resolution makes it possible to distinguish different organs and to investigate the *in vivo* metabolism of paramagnetic molecules.

EPR imaging has been criticized because its resolution is intrinsically lower than that of nuclear magnetic resonance (NMR) imaging. These criticisms are based on a comparison with NMR imaging in terms of the possibility of providing anatomic details. However, this comparison is inappropriate because the present purpose, and most applications of low frequency EPR imaging, are concerned with the study of paramagnetic substances in different locations and their function as localized sensors. EPR imaging is an innovative tool for pharmacokinetic studies. Topical EPR and EPR imaging are also extremely promising for *in vivo* oximetry at intravascular, interstitial and intracellular levels.

In the present article an account will be given of all the technical aspects involved in low frequency EPR imaging including: a) bridge, resonator design; b) profiling and shaping of the magnetic field and of the field gradients. Some account will then be given of the new development of low frequency pulsed techniques. Mathematical techniques play an important role in these applications and the contribution of these techniques to different applications will be described. Finally, some relevant biological applications will be presented.

Selection of the operating frequency

The selection of the frequency in a low field spectrometer is obviously dictated by the required application. Frequencies used range from 200 MHz to 2 GHz. L-band frequencies (1–2 GHz) make it possible to obtain a high sensitivity for volumes up to 20–25 cm³. Lower frequencies (200–300 MHz) are used to observe samples up to 200–250 cm³. Extremely convenient are surface loops used for topical measurements that do not require the insertion of a sample. They allow the adoption of frequencies in the range of 1–2 GHz, achieving a higher sensitivity.

In those applications in which a sample the size of a rat has to be inserted in the cavity resonator, lower frequencies between 200 and 700 MHz are required.

The present design was intended for measurements with 60 and 150 g rats. For this reason, in the continuous wave (CW) spectrometer we used a frequency of about 280 MHz and a frequency of 220 MHz for the pulsed bridge.

The sensitivity, determined in terms of minimum number of spins, is different for L-band and 280 MHz. However, the molar sensitivity is almost the same, i.e., about 1 µM. This can be explained by the following considerations. It has been shown (Alecci et al., 1989), both on a theoretical basis and in experimental models, that the electric field losses in lumped parameter resonators,

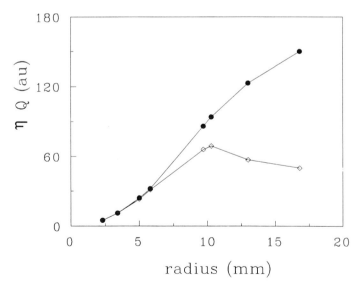

Figure 1. The sensitivity factor η Q as a function of the sample diameter for deionized water (black point) and 0.3% saline solution (white square). A re-entrant resonator 46 mm in diameter tuned at about 700 MHz was used.

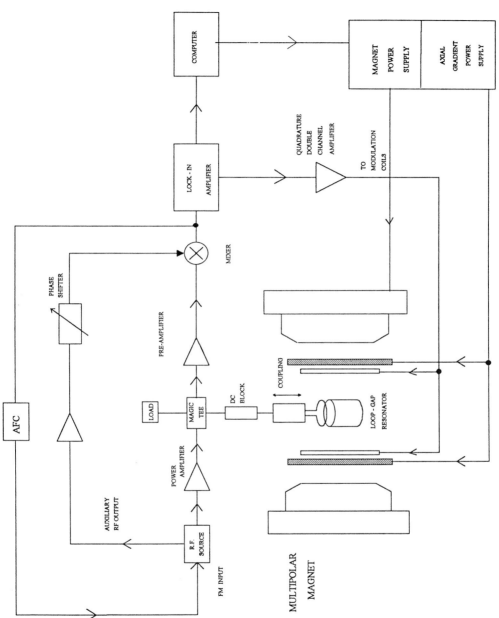

Figure 2. Low frequency CW EPR spectrometer.

loop-gap (Froncisz and Hyde, 1982; Momo et al., 1983) and re-entrant cavities (Sotgiu, 1985) can be decreased by splitting up the total capacitance. An example of this is provided by multigap resonators. The frequency of the magnetic field losses is approximately dependent on ω^2. This compensates almost exactly for the ω^2 sensitivity decrease with frequency so that, for a broad frequency range, spectrometers working at different frequencies may have the same sensitivity.

To obtain the maximum sensitivity for a given frequency, the ratio between sample and resonator dimensions must be optimized. In Figure 1 we report experimental data obtained with a re-entrant resonator at a frequency of about 700 MHz for a solution of NaCl 0.3%.

Radio frequency (RF) section

CW radio frequency bridge

To approach the saturation condition with samples ranging from 60 to 150 g, we used a power level between 25 and 200 mW. Therefore, a homodyne bridge configuration, see Figure 2, was selected.

The bridge makes it possible to operate up to 1 GHz and uses: 1) RF signal generator HP 8640B; 2) power amplifier Mini Circuit ZHL-2-8; 3) broad band hybrid junction Anzac H-1-4; 4) loop-gap resonator. A common problem in *in vivo* studies is animal movement that modifies both tuning and matching. The first problem is overcome by an appropriate automatic frequency control (AFC); the second would require some form of automatic coupling control (ACC) which has been used by some authors (Halpern et al., 1989; Brivati et al., 1991; Ishida et al., 1992). In our apparatus, by restraining the anesthetized animal, this second effect can be greatly reduced. If necessary, periodical artefacts, mainly due to breathing, can be eliminated by the use of mathematical processing.

Pulsed bridge

Transmitter
The RF source is a quartz crystal oscillator operating at 200 MHz (Fig. 3). It also provides the reference both for a SP2001 digital frequency synthesizer that, under computer control, can generate any frequency up to 200/4 MHz, and for the timing unit. For the present application, the output of the frequency synthesizer is combined in a single-side-band modulator to produce any frequency between 215 and 225 MHz.

TRANSMITTER

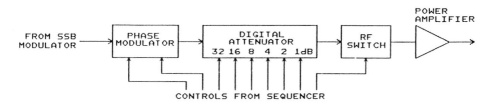

RECEIVER

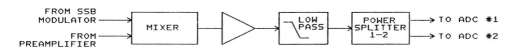

Figure 3. Radio frequency section of the pulsed bridge showing the transmitter and receiving sections.

The SP2001 maximum switching time is 20 ns. In a time of 1 μs it is then possible to scan 50 different frequency values. In particular it is possible to generate a frequency swept pulse, see Figure 4, which is known as a "chirp". This makes it possible to control the irradiated bandwidth independently of the pulse duration.

Receiver

The signal from the sample is detected in a high Q tuned coil, coaxial with the transmitter coils. During the transmitter pulse the receiver pre-amplifier is decoupled from the coil by a single pole double throw SPDT switch.

During the receiving period the frequency of the coil is swept by a high Q varicap diode to cover the same range of frequencies present in the transmitter chirp. The reference of the mixer can be either the same chirp used in transmission or a chirp displaced by a constant frequency from the transmitting chirp. For this reason, the mixer output can be either a DC signal whose amplitude at any time is proportional to the intensity of the frequency present in the receiver, or it can be centered around the frequency difference between the two chirps. In both cases there will be a direct correspondence between time and position on the sample.

An important advantage of this approach is that the observation bandwidth of the system can be drastically reduced. A block scheme of the main elements of the RF section is shown in Figure 3; a complete description will be given elsewhere (L. Testa and A. Sotgiu, in preparation).

Timing unit

The timing unit of Figure 5 generates the sequences for the frequency synthesizer, for switching and changing the amplitude of the RF, for the analog to digital converters (ADC), and for the summing unit. The frequency content of the RF pulse can be controlled by changing the pulse duration, the time profile of the pulse and its frequency.

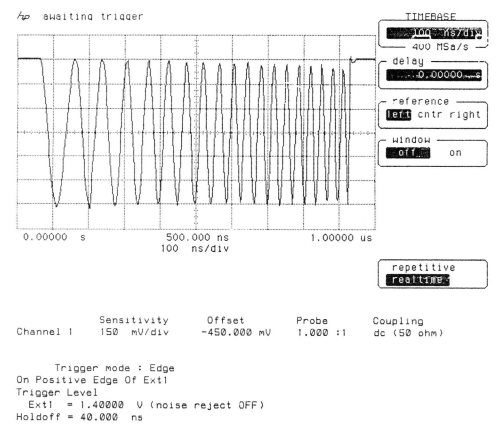

Figure 4. Radio frequency chirp.

This unit has been designed as a logic sequencer operating at 33 MHz, with 16 channels with a depth of 16 kWords. It includes all the logic circuits necessary to implement signal digital conversion and averaging. This avoids the time required to transmit the signal to the host computer.

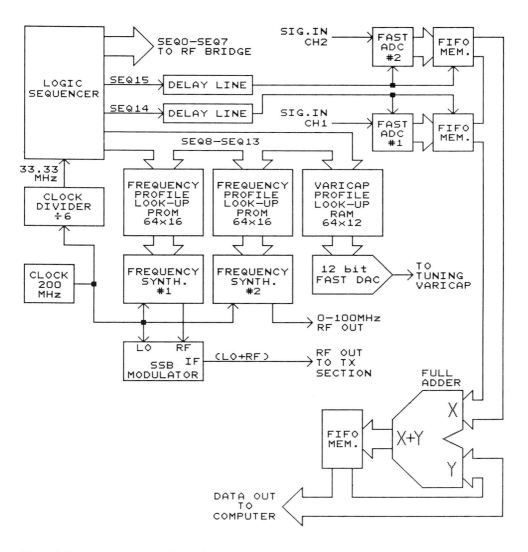

Figure 5. Sequence generator. This unit includes a fast adder that allows accumulation of the data without the introduction of delays due to data transfer to the computer.

The system clock is derived from the same quartz crystal oscillator used to generate the RF. This frequency is divided by 6, providing a clock of 30 ns duration. It passes to a further programmable divider where it can be divided by 1, 2, 4, 16, 64, 256, 1024, or 4096. Due to the 16 kWord of random access memory (RAM) that is used to drive the 16 lines of output, the total duration of the profile can range from 500 μs to 2 s.

Resonant cavity

Because of the cylindrical shape of the magnet, a bridged loop-gap resonator with 4 gaps was constructed. A Teflon tube with external diameter of 7 cm, 15 cm in length and 1 cm in thickness was used. A sheet of Teflon (0.2 mm in width) was inserted as dielectric between two layers of copper on the external wall of the Teflon tube. The screen was laid on the external surface of a Teflon tube of 13 cm in diameter and was made using three copper strips (4.3 cm in width). The strips were separated by a gap of 4 mm to improve the penetration of the low frequency field modulation (8 kHz) inside the cavity. This did not affect the RF currents and the Q reduction was negligible. The resonator was inductively coupled to the RF bridge by means of a loop of the same diameter as the resonator, mounted on a sliding mechanism.

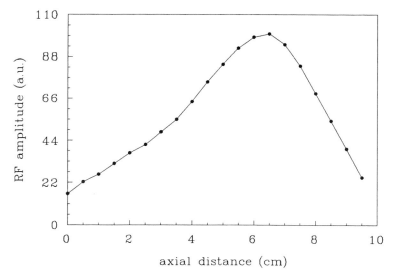

Figure 6. Longitudinal distribution of the RF field. It is determined by inserting a homogeneous sample inside the cavity and making an acquisition in the presence of a longitudinal field gradient.

The unloaded Q of the empty resonator is about 800. The cavity has been tested after the insertion of 100 cm³ of deionized water and of physiological solution. The corresponding Q values were 370 and 20 respectively. A Q of about 270 was measured for rats of about 60 g. In comparison with solutions of equivalent volume we found that the losses in living animals were equivalent to those of a 35% saline solution. The Q decreased by about 10% when a water jacket, to stabilize the temperature of the animal being studied, was inserted in the resonator.

The RF field distribution inside the resonator can be determined using image reconstruction techniques. In fact, in the presence of a field gradient along the z-axis, the spin density projection p(z) can be written as:

$$p(z) = \int_{-\infty}^{\infty} \int_{-\infty}^{\infty} n(x,y,z)r(x,y,z)dxdy \qquad [1]$$

where $n(x,y,z)$ is a sensitivity function that depends on H_1 and $r(x,y,z,)$ is the spin density distribution. By taking into account that the signal intensity is proportional to H_1^2 and that the RF field in the central region of the resonator is independent of the radius (Alecci et al., 1992), we obtain $H_1(z) = k\sqrt{p(z)}$, where k is a constant. The RF distribution of our resonator is shown in Figure 6. This profile has been used as a correction factor to avoid image distortion.

Field generation and shaping

Multipolar magnet

Low frequency imaging at a frequency well below 1 GHz requires field intensities ranging from 0.01 to 0.03 T. These field values are very low and can easily be obtained by air coils. The intensities of the field gradients, on the contrary, are quite high, in the range of 0.1 T/m. In a region of 10 cm, this can generate a field comparable to the main field.

Multipolar magnets can provide both the main field and two of the three field gradients necessary for image reconstruction (Sotgiu, 1986). The prototype designed in our laboratory is shown in Figure 7.

It consists of a cylindrical yoke and 16 poles. The yoke of the first prototype built in our laboratory was machined from a low carbon iron tube. The 16 poles have rectangular bases and are fixed longitudinally along the internal wall of the yoke. The cylindrical bore of the magnet is 27 cm.

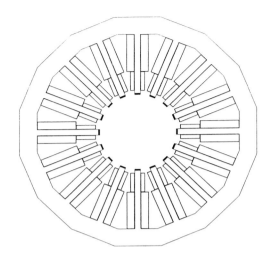

Figure 7. Section of the 16 pole multipolar magnet showing the yoke, poles and windings.

The 16 coil currents are individually controlled by computer. The field of this structure is described by the following equations:

$$B_r(r, \vartheta) = -\frac{k\mu_o}{a} \sum_{n=1}^{\infty} n\left(\frac{r}{a}\right)^{n-1}\left[C_n\cos(n\vartheta) + D_n\sin(n\vartheta)\right]$$

$$B_\vartheta(r, \vartheta) = \frac{k\mu_o}{a} \sum_{n=1}^{\infty} n\left(\frac{r}{a}\right)^{n-1}\left[C_n\sin(n\vartheta) - D_n\cos(n\vartheta)\right]$$

[2]

In these equations, a is the magnet's internal radius, r and ϑ are the polar co-ordinates in a cross section of the magnet, C_n and D_n are the coefficients of the development of the field in a series of powers, and k is a constant which depends on the iron permeability. Depending on the values of the coefficients C_n and D_n, different field configurations will be possible. A homogeneous field of intensity $\frac{k\mu_o}{a}$ A at an angle α will be obtained for:

$$C_1 = A\cos(\alpha)$$
$$D_1 = A\sin(\alpha)$$

[3]

If a field gradient along the α direction is required, then the coefficients C_2 and D_2 must also be different from zero.

The possibility of selecting the different field configurations is due to the fact that the relationship between the pole currents and the coefficients C_n and D_n can be written as:

$$I_m = \frac{\pi}{4kMN} \sum_{n=1}^{M} \frac{\gamma n}{\sin(n\gamma)} \left[C_n \cos(\frac{2\pi nm}{M}) + D_n \sin(\frac{2\pi nm}{M}) \right] \qquad [4]$$

where N is the number of turns for each pole and M is the number of poles. To take into account the nonlinearity between field and current introduced by the iron, the flux of each pole is measured by Hall sensors and the measured flux values used to correct the individual currents.

At the position of the Hall probes, where $r \sim 1$, the radial component of the field can be written as:

$$B_r(\vartheta) = \sum_{n=1}^{\infty} \left[a_n \cos(n\vartheta) + b_n \sin(n\vartheta) \right] \qquad [5]$$

where:

$$a_n = -(\mu_o/a)C_n, \qquad b_n = -(\mu_o/a)D_n \qquad [6]$$

This represents the Fourier expansion of a periodic function whose coefficients can be written as:

$$a_n = (1/\pi) \int_0^{2\pi} B_r(\vartheta)\cos(n\vartheta)d\vartheta$$

$$\qquad [7]$$

$$b_n = (1/\pi) \int_0^{2\pi} B_r(\vartheta)\sin(n\vartheta)d\vartheta$$

Taking $\vartheta = m(2\pi/M)$ and substituting the integrals by summation, we obtain:

$$C_n = -(2a/\mu_o nM) \sum_{n=1}^{M} B(2\pi m/M)\cos(2\pi nm/M)$$

$$\qquad [8]$$

$$D_n = -(2a/\mu_o nM) \sum_{n=1}^{M} B(2\pi m/M)\sin(2\pi nm/M)$$

In this way, from the measurements of the field at the probe positions, we obtain the experimental values for the coefficients C_n and D_n. If they are different from the set values, a correction will be made using the relation between current and coefficients. The block diagram of the procedure used in this correction is shown in Figure 8.

This system is obviously more complex than a traditional magnet design. However, it provides the main field and two of the three field gradients required for a three dimensional (3D) imager.

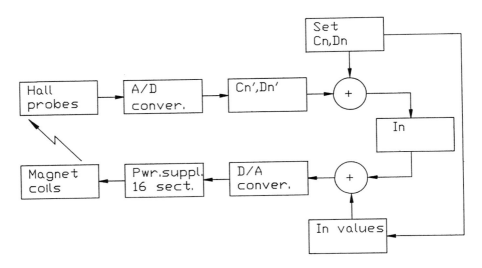

Figure 8. Automatic correction technique. This is performed by the host computer and corresponds to the evaluation of the Fourier coefficients of a periodic function defined by the positions of the Hall probes.

Moreover, the use of the Hall probes and of the mathematical techniques previously mentioned allows continuous control of the field homogeneity and gradient linearity.

Any field sweep or gradient setting is carried out under computer control by means of specifically designed cards. The host computer is a 25 MHz IBM-compatible unit which communicates

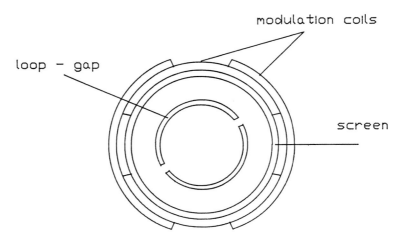

Figure 9. Loop resonator and modulation coils. The real prototype has four bridged gaps. When the currents in the two orthogonal coils used for modulation are in quadrature, the resulting field will be circularly polarized.

with the local bus of the magnet driving unit through a 12 lines optocoupled transmitter unit. The whole system operates in the following way:

- The computer evaluates the current settings for a particular configuration of the field or for a given position in the field sweep.
- These data are transmitted to the communication units and used to drive the 16-bit DAC. These analog values are used to drive the power units consisting of switching pre-regulators followed by linear stages.
- A 16-bit ADC unit reads the Hall voltage of each probe and transmits this value to the computer where these values are compared to the calibration values.
- The correction values are evaluated and sent back to the DAC.

Z-gradient coils

The third field gradient, $\partial B_r/\partial z$, is generated by the two sets of perpendicular saddle coils. By combining their effect it is possible to follow the direction of the main field. The maximum gradient intensity is about 0.025 T/m and the gradient linearity is of 4% in a cylindrical volume of 6 cm in diameter and 12 cm in length.

Modulation coils

The need to rotate the magnetic field and the gradient direction required the adoption of a circularly polarized modulation field. This was obtained by two saddle-shaped coils driven by two in-phase quadrature currents. The coils are laid on the external walls of the cavity screen (see Fig. 9) and provide a modulation field whose maximum intensity is about 1 mT.

The use of this modulation technique ensures that for any direction of the main field there is always a component of the modulation field. The rotation of the field, however, produces a phase shift that must be accounted for in each direction. Alternatively, one can process the signal by a two phase lock-in amplifier which provides the modulus of the measured signal.

Mathematical techniques

In vivo EPR spectroscopy/imaging involves numerous mathematical techniques for filtering, resolution enhancement and image reconstruction. It has been mentioned that breathing artefacts can be suppressed by filtering the spectra. Because movement due to breathing is very regular, the best way to filter this kind of noise is to perform a forward Fourier transform, suppress the unwanted frequencies from the spectrum and then apply the reverse Fourier transform.

When the EPR spectra have to be used for image reconstruction the data manipulations required are more complex and include different steps.

Spectra deconvolution

For imaging purposes, the spectra are collected by sweeping the magnetic field Ho in the presence of a field gradient G. Each spectrum b(x) represents the projection of the sample spin density f(x) along the gradient direction convoluted with the line shape h(x). In other situations, as is the case in NMR imaging, the line shape function h(x) is very narrow in comparison with the gradient intensity so that its contribution can be neglected and the reconstruction algorithms applied directly to b(x). In the case of EPR imaging, the linewidths of the studied species are not negligible and b(x) is considerably broader than f(x). In this situation, deconvolution techniques can help to increase accuracy of spin mapping.

The recorded spectrum b(x) is, in fact, the convolution of the spin projection f(x) with h(x), i.e.:

$$b(x) = \int_{-\infty}^{\infty} h(s)f(s-x)ds = f(x) \otimes h(x) \qquad [9]$$

Most of the resolution enhancement techniques (Bates and McDonnel, 1986) use the convolution theorem to obtain:

$$f(x) = FT^{-1}[B(k)/H(k)] \qquad [10]$$

where B(k) and H(k) are the Fourier transform of the measured spectra and of the lineshape function.

This conceptually simple and easy to implement technique has some negative features. In the first place, it is very sensitive to noise, giving poor results in cases of low S/N in the measured projections. In the second place, it requires a low pass filter in the frequency domain that modifies the proportions between different regions. This has only a limited effect in 1D problems but causes artefacts in 2D and 3D reconstructions.

A way to overcome the limits introduced by deconvolution in the Fourier space is provided by techniques which operate in the signal space. An example is given by spline deconvolution that we have modified and adapted to the needs of EPR imaging (Placidi et al., 1994). The technique is based on the hypothesis that in the range of definition of f(x), the function f(x-s) can be expanded in a Taylor series of powers:

$$f(x-s) = f(x) - f'(x)s + (1/2)f''(x)s^2 - (1/6)f'''(x)s^3 + \dots \qquad [11]$$

If the domain is divided into n-1 intervals, we can represent in each of them the function f(x-s) in the form:

$$f(x) = \sum_{k=1}^{n-1} S_k(x) \qquad [12]$$

where

$$S_k(x) = a_k x^3 + b_k x^2 + c_k x + d_k \qquad [13]$$

are third order polynomial in the kth interval and zero outside. In the same interval the function b(x) can be written as:

$$b(x) = \sum_{k=1}^{n-1} \overline{S}_k(x) \qquad [14]$$

where

$$\overline{S}_k(x) = A_k x^3 + B_k x^2 + C_k x + D_k \qquad [15]$$

and is zero outside the kth interval. The coefficients of the two third order developments are related by:

$$a_k = A_k/M_0$$
$$b_k = B_k/M_0 + 3a_k M_1/M_0$$

$$c_k = C_k / M_0 + 2b_k M_1/ M_0 - 3a_k M_2/ M_0$$

$$d_k = D_k / M_0 + c_k M_1 / M_0 - b_k M_2/ M_0 + a_k M_3/ M_0 \qquad [16]$$

where Mj is the jth moment of h(x), i.e.:

$$M_j = \int_{-\infty}^{\infty} t^j h(t) dt \qquad\qquad [17]$$

The deconvoluted function f(x) is then evaluated by the following procedure:

   a)   the moments M_j of the lineshape function h(x) are evaluted by numerical integration;
   b)   the function b(x) is interpolated by a cubic spline in the n–1 intervals providing the coefficients A_k, B_k, C_k, D_k for each interval;
   c)   the convoluted function f(x) is obtained by relationships [16].

The spline technique, modified for the particular requirements of EPR spectroscopy, is consistently better than Fourier deconvolution on the basis of the different parameters used to compare the two methods.

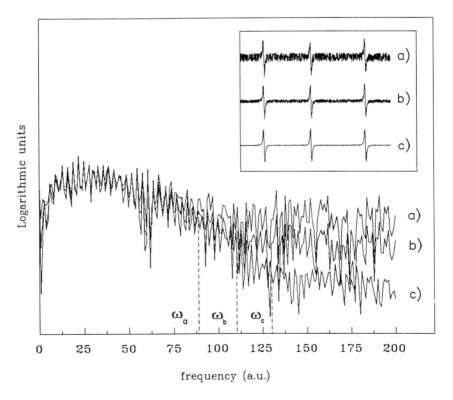

Figure 10. Fremy salt power spectrum at S/N of: (a) 10, (b) 23 and (c) 270. The intersection of the noise power spectrum and the sample power spectrum is shifted toward the low frequency at the higher noise level.

Unfortunately, the possibility of convolution to improve the resolution of the projection used in image reconstruction is limited by the noise present in the experimental data. An understanding of the limits of the spatial resolution that can be obtained may be gained by looking at the following examples.

Figures 10 and 11 show the Fourier transform of the signal obtained by two different samples at different levels of S/N. The two samples were Fremy's salt (potassium disulfonate) and PCA (2,2,5,5, tetramethyl-pyrrolidine-1-oxil-3-carboxylic acid) solutions. They are characterized by linewidths of 0.5 and 1.5 G respectively. The spectra were collected at different level of S/N ranging from 5 to 700. The power spectrum of the projection contains two distinct regions, a low frequency region where the frequency component of the sample is predominant and a high frequency region where the noise frequencies are dominant. The effect of a different noise level is to shift up and down the noise power spectrum, moving the intersections between the noise power spectrum and the signal power spectrum. In the case of a high signal to noise ratio, the sample

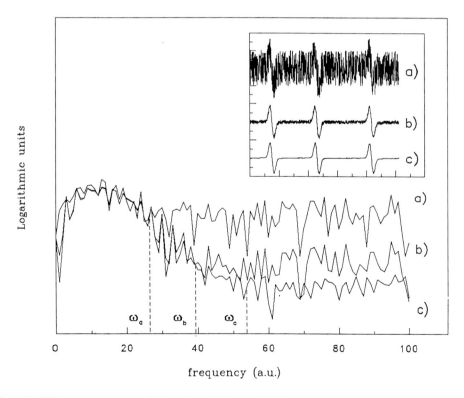

Figure 11. PCA power spectrum at S/N of: (a) 4, (b) 23 and (c) 709.

frequencies larger than the noise are more numerous, providing a higher resolution. The maximum resolution enhancement due to deconvolution is obtained in the case of narrow lines and for a high S/N. Its value can be of the order of a factor of 2-2.5 depending on the experimental conditions.

In spite of the fact that this resolution improvement is not very high, its adoption is recommended for two reasons. The first is that when a proper deconvolution technique is adopted, the resolution improvement is not at the expense of a distortion of the projections. The second is that when high values of the field gradient are used, a further increase in gradient intensity can be difficult to achieve experimentally and always has the effect of reducing the S/N. Moreover, when working with samples which have a hyperfine structure, deconvolution techniques are always required.

The use of narrow linewidth probes, instead of reducing the need for deconvolution could make its adoption even more profitable. In fact, deconvolution techniques are more effective when the kernel of the deconvolution is a small fraction of the function that must be deconvoluted.

Mathematical techniques for image reconstruction

The relaxation times T_1 and T_2 of the paramagnetic species used in EPR imaging are short in comparison with the time required to switch a field gradient on and off. Therefore, the methods developed in NMR imaging that allow selection of the observing planes cannot be used and the reconstruction techniques are almost always "reconstruction from projection".

This kind of reconstruction requires the acquisition of a regular set of projections on a polar 2D or 3D raster, the deconvolution of the measured projections with the zero gradient line shape to obtain the true projections, and the application of the reconstruction algorithm.

The algorithms normally adopted in EPR imaging are usually derived from Computer Assisted Tomography or from NMR imaging. For the purpose of the present discussion, the two that are most commonly adopted (Brooks and Di Chiro, 1976) will be presented.

Back projection and filtered back projection
In the simple 2D case, the projections of the spin density $f(x,y)$ are obtained by sweeping the main field in the presence of a field gradient in a given direction. The projections are related to the spin density by:

$$p(r, \vartheta) = \int_{r, \vartheta} f(x,y)\, ds$$

[18]

where s is the direction perpendicular to the orientation of the field gradient.

The function f(x,y) is continuous and, in principle, an infinite number of projections would be necessary for image reconstruction. However, for a limited number of projections, the approximate spin density can be obtained by the equation:

$$\hat{f}(x,y) = \sum_{j=1}^{m} p(x \cos\vartheta_j + y \sin\vartheta_j, \vartheta_j)\, \Delta\vartheta \qquad [19]$$

where ϑ_j is the jth projection angle and $\Delta\vartheta = \pi/m$ is the angular distance between projections.

The filtered back projection aims to overcome the limitation connected with this technique and in particular the presence of star artefacts (see Fig. 12).

In its practical implementation it replaces the equation for $\hat{f}(x,y)$ with the following:

$$\hat{f}(x,y) = \sum_{j=1}^{m} \tilde{p}(x \cos\vartheta_j + y \sin\vartheta_j, \vartheta_j)\, \Delta\vartheta \qquad [20]$$

where:

$$\tilde{p}(r, \vartheta) = \int_{-\infty}^{\infty} |\omega|\, P(\omega, \vartheta)\, \exp(2\pi i \omega r)\, d\omega \qquad [21]$$

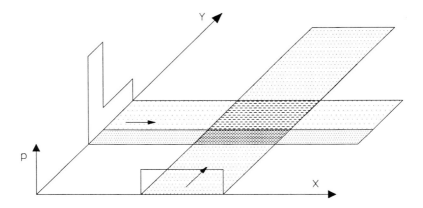

Figure 12. Reconstruction through back projection of a 2D sample. The central region represents the sample.

This guarantees that the coefficients of the image obtained by back projection are identical to the coefficients of the true image and correspond to a filtering process which increases the high frequency components.

Fourier zeugmatography

The basis of this reconstruction technique is the fact that the spin density and the Fourier transform of the projections are connected by the following relations:

$$f(x,y) = \int_{-\infty}^{\infty}\int_{-\infty}^{\infty} F(\omega_x\ \omega_y) \exp[2\pi i(\omega_x x + \omega_y y)]\ d\omega_x d\omega_y \qquad [22]$$

and

$$F(\omega,\vartheta) = \int_{-\infty}^{\infty} p(r,\vartheta) \exp[-2\pi i\omega]\ dr = P(\omega,\vartheta) \qquad [23]$$

where $F(\omega,\ \vartheta)$ is the Fourier transform of $f(x,y)$ in polar coordinates and $p(r,\vartheta)$ and $P(\omega,\vartheta)$ are the projections and their Fourier transforms.

The procedure to obtain the image from the projections then is to make the Fourier transform of the measured projections, or of those obtained after deconvolution, to interpolate in the Fourier space to obtain a Cartesian co-ordinate representation of the Fourier space, and to make a 2D inverse Fourier transform to obtain the spin density $f(x,y)$.

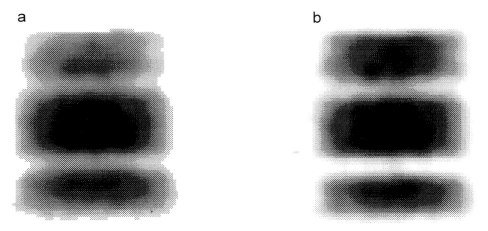

a   b

Figure 13. Comparison of reconstructions by (a) Fourier reconstruction and (b) filtered back projection. The sample consists of three slices of a gel containing a perdeuterated nitroxide. The separations between the slices are 2 and 4 mm.

Examples of applications of Filtered Back Projection and Fourier Zeugmatography are shown in Figure 13.

Selection of projections

The reconstruction techniques previously described start from a finite number of projections measured at regular angular increments. In general, to map the spin density of a n*n rectangular sample, the minimum number of distinct projections is $n_o = \pi n/4$. It can be shown, however, that the same reconstruction algorithm gives different results when applied to different sets of projections of the same object. This depends on the fact that different projections do not have the same information content (Kazantsev, 1991). An example is given in Figure 12 where the x-projection describes a uniform distribution, whereas the y-projection shows the presence of two distinct areas.

In general, if no information on the spin distribution of the sample is available, the selection of a regular set of projections is the safest choice. In other situations, however, a different choice of projections can provide a better result or can reduce the number of projections required to obtain the same result. Without going into the details of the theory, which can be found elsewhere (Placidi et al., 1994), the information content of a projection can be taken to be proportional to the function "entropy", and in the case of a projection can be written as:

UNIFORM ACQUISITION   ADAPTIVE ACQUISITION

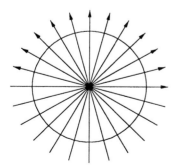
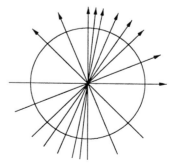

Figure 14. Regular choice of projections (left) *versus* optimal one (right).

$$E_\vartheta = \sum_{i=1}^{n} P_{i,\vartheta} \log\left(\frac{1}{P_{i,\vartheta}}\right)$$ [24]

where $P_{i,\vartheta}$ represents the ith point of the projection along the ϑ angle. If the basis of the logarithm is taken to be equal to n and no negative values are allowed, the "entropy" function will range from 0 to 1. The minimum value corresponds to a situation in which only one point is different from zero and has the value n. The maximum value corresponds to a situation in which all the n points of the projection have a value of $1/n$. The latter value relates to a minimum of information content, i.e., to a uniform spin distribution. By contrast, a low value of the entropy function indicates an uneven spin distribution. These considerations are the basis of techniques used to select the angles of the projections during the acquisition process. An example of the application of this selection process is shown in Figure 14.

These techniques can help to reduce substantially the number of projections necessary to obtain a given resolution. The application of these mathematical techniques has made it possible to obtain a resolution of about 0.2 mm with a gradient intensity of 6 G/cm, using a paramagnetic probe of about 600 mG in linewidth.

Biological applications

The present apparatus has been used for spectroscopic and imaging applications. Experiments were done on anesthetized, pressure-controlled, Wistar rats (45–65 g). The paramagnetic probe was 2,2,5,5,-tetramethyl-pyrrolidine-1-oxyl-3-carboxylic acid (PCA) injected as a bolus in the jugular vein. The final concentration in the rat was 1–3 mmol/kg.

Three kind of experiments were performed: a) study of the reduction kinetics from the whole body; b) differentiation of the kinetics from different locations by means of a longitudinal field gradient; c) study of the distribution of PCA at different times by 2-D EPR imaging.

The whole body rat PCA reduction showed a monoexponential decay. Half life depends on the PCA concentration. At 1 mmol/kg of PCA, the half-life is 13.3 ± 0.7 min. This value is of the same order as that measured at X-band in the circulating blood.

Measurements in the presence of a longitudinal field gradient showed the presence of three regions corresponding to the lower abdomen, the liver and the thorax. Corresponding PCA half lives were 19 ± 1, 17 ± 2 and 22 ± 2, respectively. The longer half-life in the thorax may be explained by reoxidation processes (Alecci et al., 1994).

Sequential 2D longitudinal maps of the probe density were obtained at the higher concentrations (3 mmol/kg) that are characterized by a longer half-life. Each reconstruction requires about

5 min and 2 or 3 *in vivo* 2D reconstructions were obtained. Also, in this case, different clearance rates may be responsible for the change in the shape of the spin distribution.

The density maps obtained by imaging techniques are strongly affected by the low S/N in the projections. This does not allow the use of high values of field gradients, imposing a serious limit on the resolution obtainable. Commercially avalaible spin probes, such as perdeuterated probes, would have allowed the achievement of both a higher sensitivity and a better resolution. However, up to now we have not used these substances because the large number of experiments involved would have made this extremely expensive.

Conclusions

Low frequency *in vivo* EPR imaging is becoming a tool of increasing importance. Although the ultimate goal is not clinical, EPR imaging is the only methodology capable of investigating *in vivo* free radical distribution. At the present stage of development, it suffers from the serious drawback of a long acquisition time. The ongoing development of pulsed techniques in this and other laboratories should reduce this time, allowing the study of new types of problems.

The problem of the linewidth of the paramagnetic probes can be partially overcome by the use of perdeuterated nitroxide probes. The newly developed narrow band single line probes, recently developed as a contrast agent for Overhauser imaging, should provide better sensitivity and resolution. EPR imaging should make possible the localization of *in vivo* processes.

References

Alecci, M., Gualtieri, G. and Sotgiu, A. (1989) Lumped parameters description of RF losses in ESR experiments on electrically conducting samples. *J. Phys. E: Sci. Instrum.* 22: 354–359.

Alecci, M., Della Penna, S., Sotgiu, A., Testa, L. and Vannucci, L. (1992) Electron Paramagnetic Resonance spectrometer for three-dimensional *in vivo* imaging at very low frequency. *Rev. Sci. Instrum.* 163: 4263–4270.

Alecci, M., Ferrari, M.,Quaresima, V., Sotgiu, A. and Ursini, C.L. (1994) Simultaneous 280 MHz EPR imaging of rat organs during nitroxide free radical clearance, *Biophys. J.* 67: 1274–1279.

Bates, R.H.T. and McDonnel, M.J. (1986) *Image Restoration and Reconstruction*, Vol. 16. Oxford Engineering Science, Oxford.

Brivati, J.A., Stevens, A.D. and Symons, M.C.R. (1991) A radiofrequency ESR spectrometer for *in vivo* imaging. *J. Magn. Reson.* 192:480–489.

Brooks, R.A. and Di Chiro, G. (1976) Principles of computer assisted tomography (CAT) in radiographic and radioisotopic imaging. *Phys. Med. Biol.* 21: 5.

Froncisz, W. and Hyde, J.S. (1982) The loop-gap resonator: a new microwave lumped circuit ESR sample structure. *J. Magn. Reson.* 147: 515–521.

Halpern, H.J., Spencer, D.P., van Polen, J., Bowman, M.K., Nelson, A.C., Dowey, E.M. and Teicher, B.A. (1989) Imaging radio frequency electron-spin resonance spectrometer with high resolution and sensitivity for *in vivo* measurements. *Rev. Sci. Instrum.* 160: 1040–1050.

Ishida, H., Matsumoto, S., Yokoyama, H., Mori, N., Kumashiro, H., Tsuchihashi, N., Ogata, T., Yamada, M., Ono, M., Kitajima, T., Kamada, H. and Yoshida, E. (1992) An ESR-CT imaging of the head of a living rat receiving an administration of a nitroxide radical. *Magn. Reson. Imag.* 110: 109–114.

Kazantsev, I.G., (1991) Information content of projections. *Inverse Problems* 7: 887–898.

Momo, F., Sotgiu, A. and Zonta, R. (1983) On the design of a slip ring resonator for ESR spectroscopy between 1 and 4 GHz. *J. Phys. E: Sci. Instrum.* 16: 43–46.

Placidi, G., Alecci, M. and Sotgiu, A. (1994) Spline-based deconvolution technique in electron paramagnetic resonance imaging. *Rev. Sci. Instrum.* 65: 58–62.

Placidi, G., Alecci, M. and Sotgiu, A. (1995) Theory of adaptive acquisition method for image reconstruction from prosections and application to EPR imaging. *J. Magn. Reson. B* 108: 50–57.

Sotgiu, A. (1985) Resonator design for *in vivo* ESR spectroscopy. *J. Magn. Reson.* 165: 206–214.

Sotgiu, A. (1986) Fields and gradients in multipolar magnets. *J. Appl. Phys.* 59: 689–693.

Whole body electron spin resonance imaging spectrometer

M.C.R. Symons

Department of Chemistry, De Montfort University, The Gateway, Leicester LE1 9BH, UK

Summary. A C.W. electron spin resonance spectrometer able to record 3D images, operating at *ca.* 250 MHz, is described. Its potential for use in whole body imaging is discussed in terms of radicals of significant interest, and of narrow-line transition metal complexes. It is concluded that, at present, it is essential to use added radicals having the narrowest possible singlet ESR spectra and that working in a pulse mode may prove to be essential for useful whole-body imaging.

Introduction

About 9 years ago, at a time when NMR imaging (MRI and "spectroscopy") was already more powerful than anyone could have foreseen, it seemed worthwhile to consider ESR imaging as being of potential use in certain specific areas. At that time I new of no activity in the field, and was extremely grateful to the Wellcome Trust for enabling us to construct a spectrometer, whilst realising that there was no guarantee of success. Some years later, when our first instrument was constructed (Brivati et al., 1991), a number of other groups had published work in this area (Eaton et al., 1991; Colacicchi et al., 1991; Fujii and Berliner, 1985; Berliner et al., 1987; Pou et al., 1991; Halpern et al., 1989). All these preliminary results were promising, but at the same time it was clear that much greater sensitivity would be required. However, it was also clear that radical processes were more important in many biological processes than had been previously realised, which made these endeavours even more worthwhile.

Since our expertise was in C.W. rather than in pulse ESR methods, we decided to construct a much larger instrument, suitable for use with humans, in the hope that this potential might one day be of use. Until recently, our imaging studies had to be confined to inanimate 'phantoms'. However, with the collaboration of Professor Graham Cherryman, we were able to contemplate extending our studies to work on suitable animals. The aim of this paper is to give an overview of some of the areas of potential interest, and a brief summary of the second instrument. This has been fully described elsewhere (J.A. Brivati, A.D. Stephens and C. Smith, personal communication).

Potential systems for imaging

There are two major classes, namely free-radicals and certain transition-metal complexes. The former can be broadly divided into those which are naturally occurring and others which are specifically added to the system.

Naturally occurring radicals

Radical intermediates are undoubtedly involved in biological processes, but in most cases their concentrations are kept low, and they never reach concentrations suitable for detection by ESR methods. Certain systems specifically contain radicals, which are occluded or encapsulated within proteins in such a way that they are immobile, and unable to react with their immediate surroundings.

The best known example is the tyrosyl radical,

$$R{-}CH_2{-}\langle\bigcirc\rangle{-}\dot{O}$$

which has a well-defined ESR spectrum. However, even in such cases, the concentrations that are involved are still too low for possible imaging experiments.

Spin-trapping

If radicals are being continuously generated, as, for example, in an immune response situation, or in the metabolism of certain ingested compounds such as tetrachloromethane (Albano et al., 1982; La Cagin et al., 1988), they normally react so rapidly that stationary concentrations are very low. However, if they can be intercepted by an added reagent that reacts to give a stable species, such as a nitroxide radical, these will accumulate, possibly to give detectable concentrations. For very reactive radicals, such as hydroxyl ($\cdot$OH), such trapping would require extremely high concentrations of the spin-traps, and hence would not be possible. However, more stable radicals may possibly be trappable. We need to discover traps that are non-toxic at relatively high concentrations, are not readily metabolised, tend to accumulate in the area of interest, and which give radicals that are retained in that area and are themselves non-toxic and not readily metabolised. Several groups are endeavouring to discover such traps, but the study is still in its infancy.

Trapping nitric oxide

One of the most stable radical intermediates known to be of great importance in many biological processes is nitric oxide (·NO). Ironically, because this is a linear π^* radical, it gives rise to an extremely broad feature in the liquid-phase that has never been detected by ESR spectroscopy (Atkins and Symons, 1962, 1967). Because of its stability, it cannot be trapped by standard traps. However, it is potentially trappable using a Diels-Alder reaction of the type shown in reaction [1] (Korth et al., 1992; Gabv et al., 1993).

$$\text{diene} \quad + \quad \cdot NO \quad \longrightarrow \quad N\dot{-}O \qquad\qquad [1]$$

Simple 1,3-dienes are not very suitable because they favour a trans-configuration rather than the cis-configuration required for [2] (Gabv and Symons, 1995). However, there are a variety of structures that seem very promising, and certain naturally occurring compounds react with high sensitivity. Hence this does seem to be a promising way forward.

Added radicals

This is a method of greater promise, and has already met with limited success (Colacicchi et al., 1993; Bourget et al., 1993; Ishida et al., 1992). Again, most nitroxide radicals are not ideal, because they are readily oxidised and reduced, though such steps may be reversible [2].

$$R_2NO^+ \quad \underset{-e^-}{\overset{+e^-}{\rightleftharpoons}} \quad R_2\dot{N}O \quad \underset{-e^-}{\overset{+e^-}{\rightleftharpoons}} \quad R_2NO^- \qquad\qquad [2]$$

The aim would be to label a drug of interest and to attempt to follow it within the body. This method has something in common with labelling with ^{13}C or ^{19}F, for NMR imaging. In some cases, the drugs themselves are readily converted by redox processes into fairly stable radicals, and it may be that modifying such drugs so as to stabilise these radical intermediates would be a useful way forward.

However, it seems that it may never be possible to prepare normal nitroxide radicals with line-widths low enough for satisfactory imaging at acceptable radical concentrations. This is, essentially, because of the stongly coupled ^{14}N nucleus and the large anisotropy of this coupling. The way forward is to use radicals that give singlets so that this mode of broadening is removed, unless sensitivity can be greatly enhanced.

Oxygen

Although dioxygen is paramagnetic, again it cannot be detected directly by ESR spectroscopy. One of its important reactions is to add to active radicals to give peroxy radicals, $R—O_2$·. Unfortunately these give rise to broad ESR features that are of no use for imaging. However, in sufficiently high concentrations, oxygen will induce line broadening in the spectra of stable radicals, and hence its local concentration established. The problem with the use of small radical probes is that many factors can alter line-widths, so that they do not specifically reflect oxygen concentration. Also, the concentration of free oxygen is always very low in tissues, and there can be steep gradients on moving away from blood capillaries. Probably the best method is to use solids that are permeable to oxygen, but not to other paramagnetic units (except, of course, ·NO) since these (carbon, and lithium phthalocyanin) can be prepared so as to give very narrow singlets that

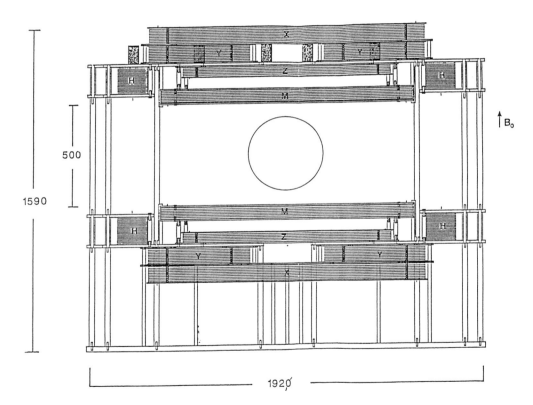

Figure 1. The magnet with modulation and gradient coils. (H) Main Helmholtz coils, shown in cross-section; (M) modulation coils; (X and Y) Anderson rectangular gradient coils (G_x, G_y); (Z) Maxwell pair (G_z). The circle indicates the diameter of the large resonator discussed in the text. (Dimensions in cm).

broaden reversibly when oxygen enters channels in the solids. It may well be that this method of oxymetry will prove to be of real use.

Transition-metal complexes

Again, despite the fact that many paramagnetic complexes occur naturally, they are of no use for imaging because relaxation effects give rise to broad, often undetectable features at ambient temperatures. Systems of possible interest are $3d^1$ complexes of Ti (III). We have found several that give single features with line widths of *ca.* 4 G in aqueous solution at ambient temperatures. Unfortunately these features do not narrow at radio-frequencies so the widths are no longer controlled by the Δg effect. We hope that by using different ligands it may be possible to reduce the line-widths still further, but at present these complexes are probably not yet suitable for image studies.

Experimental part

The whole body imaging spectrometer

This is, essentially, a scaled up version of our first spectrometer (Brivati et al., 1991), suitable, at least in size, for human imaging. Full details of the instrument and the methods developed for image collection and analysis have been given elsewhere (J.A. Brivati, A.D. Stephens and C. Smith, personal communication). Here I give a very brief overview of the spectrometer.

The magnet, field modulation and gradient coils

The magnet comprises a pair of Helmholtz coils wound with copper strip, with an internal diameter providing 0.5 m clearance for the patient (Fig. 1). The field modulation coils (1.2 m dia.), are driven by a 30 W amplifier (E.G. and G. 5209) for lock-in detection. Field modulation is at 1.13 kHz. An Acorn Archimedes 440 Rise Computer is used for direct control and data accumulation.

Field gradients are produced by water-cooled coils comprising a Maxwell pair (G_z, 1 m dia.) and two pairs of rectangular coils (1.5×0.6 m). These are powered by 3 kW supplies (Farnell AP 100.90), with maximum gradients of 40 mT m^{-1} for G_z and 15 mT m^{-1} for G_x and G_y.

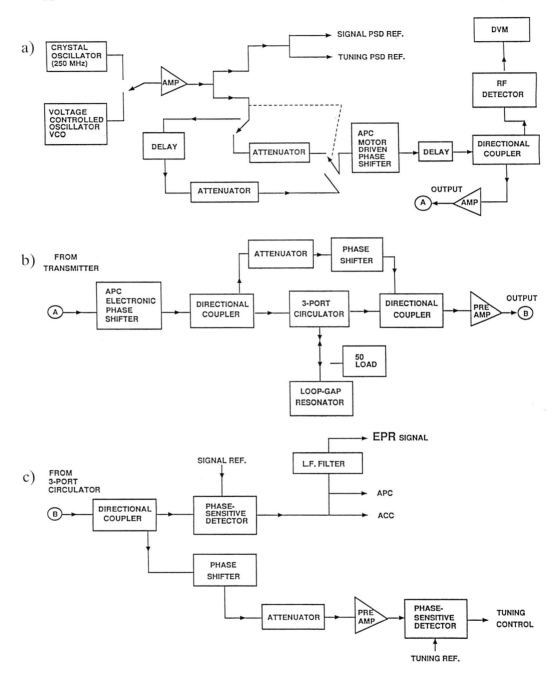

Figure 2. The RF Bridge. (a) The RF transmitter circuit, showing the alternative sources. (b) The 3-port circulator. (c) The receiver circuit.

The RF bridge

The essential features are shown in Figure 2. It is a standard homodyne reflection bridge based on the previous design (Brivati et al., 1991). The RF source is either a voltage-controlled oscillator (VCO) or a phase-locked crystal oscillator (Research Communications). After amplification, the maximum power available is *ca.* 1.5 W. Control includes automatic tuning (ATC); coupling (ACC) and phase (APC) are as previously described (Brivati et al., 1991). The receiver incorporates a gallium arsenide FET pre-amplifier (noise figure 0.6 dB, Research Communications).

Resonators

Several resonators, ranging in diameter from *ca.* 6.5 to 45 cm have been constructed, but most of the basic tests have been carried out using the small resonator (6.5 cm) which is similar to that used by Bacic et al. (1989). This comprises two copper lined concentric PTFE tubes each with a single gap, which can be twisted so as to vary θ, the angle between the two gaps. This rotation controls the resonant frequency over a fairly wide range. A copper plate positioned close to the outer gap is used for automatic tuning and a small copper plate located between the main plate and the gap is used for rapid fine tuning, controlled by a loudspeaker driven from the ATC circuit (see Fig. 3). Power is coupled inductively using a single loop (6.5 cm dia.) mounted on a sliding platform. Matching is achieved by changing the extent of overlap using a servo mechanism con-

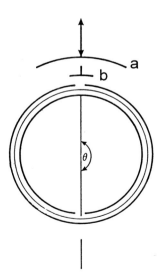

Figure 3. The 250 MHz double split ring resonator, also used as a surface probe (*ca.* 7 cm dia., 10 cm long). The two copper foil cylinders (loops) are separated by PTFE, and the E-field is largely contained in this region. θ indicates the angle between the two gaps, and (a) and (b) are the two copper plates used for major tuning (a) and fine tuning (b). Both are controlled by the ATC system.

trolled by the ACC circuit. The unloaded Q is only *ca.* 300, but this is not greatly reduced under loaded conditions.

Of the other resonators, the most interesting and potentially useful is a 45 cm three-loop two gap resonator. The main loop which can accept a whole body, has two opposed small side loops

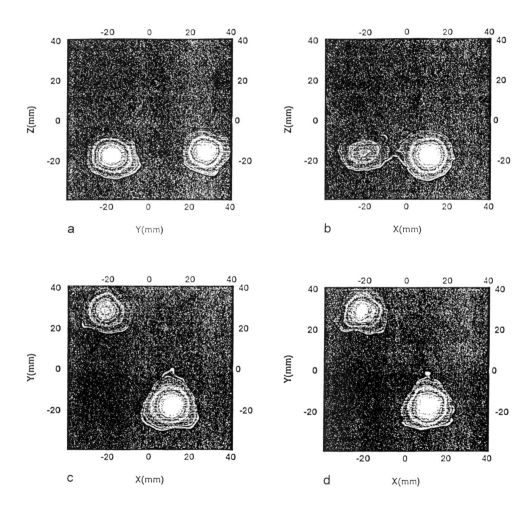

Figure 4. 3D images for two samples containing 0.8 ml of aqueous solution of 4-oxo-TMPO (0.5 mM). For representational convenience, the results show the integrated spin-densities over the x-axis (a), the y-axis (b) and the z-axis (c); (d) shows a spin-density slice at $z = -15$ mm.

(10 cm dia.), which accommodate the B_1 field. RF power is coupled *via* one of the side loops and the system is tuned using the gaps as before (Brivati et al., 1991). The best resonator is one that is only 2.5 cm long. This also has a Q of *ca*. 300, which is reduced to *ca*. 100 on insertion of a human head.

Results

Unfortunately, at the time of writing, we are still not in a position to undertake experiments on living systems. However, many experiments on solutions of nitroxides and other radicals have been undertaken, and the performance is somewhat better than that of our previous instrument (Brivati et al., 1991) with, of course, the option of using much larger samples.

We have been able to collect spectral spatial, 2D cross-sections and 3D images. These have been improved considerably using maximum entropy methods. An example, using two tubes of a dilute nitroxide, is shown in Figure 4.

Conclusions

It is clear that, although this spectrometer can accept humans, even when permission to use added radical agents is approved, we need to improve the performance considerably in order to obtain information that cannot be obtained by other methods. We hope to achieve this by using very narrow line radical spin-labels, and by moving into a pulse mode instead of the present C.W. system.

Acknowledgements
I thank the Wellcome Trust for supporting this work over the past 8 years, and the SERC for extra support. The two instruments were designed by Mr. J. Brivati who was largely concerned with their construction. Dr. A.D. Stephens was especially responsible for resonator design and construction, and now is in charge of the spectrometer. Dr. Richard Partridge and recently Dr. Colin Smith, have been responsible for all the computation aspects of image collection. Mrs. Norma Harrison carried out a wide range of tests using the smaller spectrometer.

References

Albano, E., Lott, K.A.K., Slater, T.F., Stier, A., Symons, M.C.R. and Tomasi, A. (1982) *Biochem. J.* 204: 593.
Atkins, P.W. and Symons, M.C.R. (1962) *J. Chem. Soc.* 4794.
Atkins, W. and Symons, M.C.R. (1967) *The Structure of Inorganic Radicals*. Elsevier, Amsterdam.
Bacic, G., Hilges, M.J., Magin, R.L., Walczak, T. and Swartz, H.M. (1989) *Magn. Reson. Med.* 10: 266.
Berliner, L.J., Fujii, H., Wang, X. and Lukiewicz, J. (1987) *J. Magn. Reson. in Med.* 4: 380.

Bourg, J., Krishna, M.C., Mitchell, J.B., Tschudin, R.G., Pohida, T.J., Friauf, W.S., Smith, P.D., Metcalf, J., Harrington, F. and Subramanian, S. (1993) *J. Magn. Resonance* 102: 3821.

Brivati, J.A., Stephens, A.D. and Symons, M.C.R. (1991) *J. Magn. Reson.* 92: 480.

Colacicchi, S., Ferrari, M. and Sotgiu, A. (1991) *Int. J. Biochem.* 24: 205.

Colacicchi, S., Alecci, M., Gualtieri, G., Quaresima, V., Wisinni, C.L., Ferrari, M. and Sotgiu, A. (1993) *J. Chem. Soc. Perkin Trans.* 2: 2077.

Eaton, G.R., Eaton, S.S. and Ohno, K. (1991) *EPR imaging and in-vivo EPR.* CRC Press, Boca Raton, Florida.

Fujii, H. and Berliner, L.J. (1985) *Magn. Reson. in Med.* 2: 275.

Gabv, I.M., Rai, W.S. and Symons, M.C.R. (1993) *J. Chem. Soc. Chem. Commun.* 1099.

Gabv, I.M. and Symons, M.C.R. (1995) *J. Chem. Soc. Faraday Trans.; in press.*

Halpern, H.J., Spencer, D.P., van Polen, J., Bowman, M.K., Nelson, A.C., Dowey, E.M. and Teicher, B.A. (1989) *Rev. Sci. Instrum.* 60: 1040.

Ishida, S.I., Matsumoto, S., Yokoyama, H., Mori, N., Kumashiro, H. and Tsuchihashi, N. (1992) *Magn. Reson. Imaging* 10: 109.

Korth, H.G., Ingold, K.W., Sustmann, R., de Groot, H. and Sies, H. (1992) *Angew. Chem.* 104: 915.

La Cagin, L., Connor, H.D., Mason, R.P. and Thurman, R.G. (1988) *Mol. Pharmacol.* 33: 351.

Pou, S., Peric, M., Halpern, H. and Rosen, G. (1991) *Free Rad. Res. Commun.* 12: 39.

Bioradicals Detected by ESR Spectroscopy
H. Ohya-Nishiguchi & L. Packer (eds)
© 1995 Birkhäuser Verlag Basel/Switzerland

Development of the rapid field scan L-band ESR-CT system

T. Ogata

Department of Materials Science and Engineering, Yamagata University, 4-3-16 Johnan, Yonezawa 992, Japan

Summary. The rapid field scan L-band ESR-CT system has been developed by using an air-core coil electromagnet to get three dimensional images of the distribution of free radicals in small animals. This system has an interactive data processing system for CT expression in which various CT algorithms can be compared.

Introduction

Electron spin resonance computed tomography (ESR-CT) gives us useful information on the distribution of free radicals in a sample (Eaton et al., 1991; Ohno and Tsuchihashi, 1991; Ishida et al., 1992). The development of *in vivo* ESR-CT inevitably requires a rapid field scanning in order to get a lot of the ESR data for the tomography (Demsar et al., 1988) but the rapid scanning is usually accompanied by the field delay caused by magnetic hysteresis of the iron-core and/or eddy current effects due to metal components used in the magnet. In this paper we have tried to use a newly constructed air-core electromagnet, and to make a correction for the magnetic field in the ESR spectra obtained.

In order to get ESR-CT images, we have to develop several data processing algorithms for conditioning observed data, noise rejection, estimation of projection data, reconstruction of images and so on. For this purpose, an interactive computer system is developed with on-line connection to the ESR measurement system. In an ESR-CT image reconstruction, we have to solve two inverse problems, i.e., deconvolution of observed data and reconstruction from projections. Generally, the algorithm for the inverse problem is sensitive to noise, and the data observed at the L-band ESR system is distorted by noise. Therefore, we must design a robust reconstruction technique, resistant to noise, from a few projections. In this paper, we examine and compare various methods in deconvolution and reconstruction by simulations.

Using this procedure, we show the distribution of nitroxide (aminoxyl) radicals in rats. In addition, we develop a three-dimensional expression of the ESR-CT image to observe the whole image and sectional image from any angle.

ESR-CT system

The rapid scan ESR-CT system was composed of the newly-built air-core coil electromagnet (Yonezawa Electric Wire Co., Ltd.), a field-scan coil, a field gradient coil, an L-band ESR spectrometer, and computers. The L-band ESR spectrometer, utilizing a two-gap loop-gap resonator, has been previously described (Ishida et al., 1990). The inner diameter and axial length of the resonator were 41 mm and 10 mm, respectively. The resonant frequency was about 700 MHz. The static magnetic field was generated by the main copper-wire coil. This coil was of an air-core, water cooled, two-coil Helmholtz design. The outline of this coil system is shown in Figure 1. The magnetic field could be swept by a supplementary Helmholtz coil under current regulations in one second in 15 mT scan range. We would like to call this "RAPID SCAN". The 15 mT scan range of the magnetic field is wide enough to measure the full spectra of organic free radicals under a field gradient. A three-dimensional ESR image, which was defined as an ESR-CT image

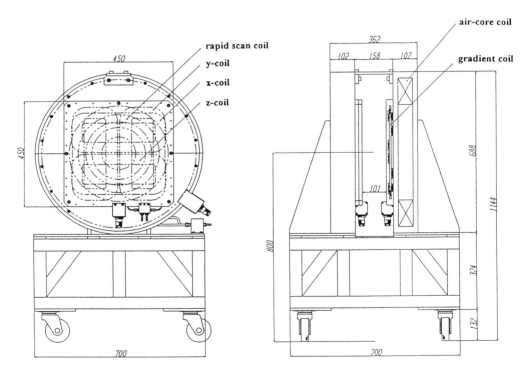

Figure 1. Outline view of air-core electromagnet system equipped with field gradient coils and rapid field scan coils. All coils are contained in aluminum casing and the total weight is 350 kg. The air-gap distance between the surface of the coil cases is 101 mm, which can be used to examine whole rats so long as a large resonator is fabricated.

in our previous paper (Ishida et al., 1992), was constructed by three-dimensional zeugmatography (Lauterbur, 1973; Lauterbur and Lay, 1980). This zeugmatography requires the linear magnetic field gradients along x-, y- and z-axes. Our gradient coils generated the linear gradient to the limit of 0.1 T/m in the range of ± 15 mm from the center. The x-, y-, and z-axes were defined as the directions along the vertical direction, on the horizontal perpendicular, or parallel to the static magnetic field direction, respectively. A pair of coils for z gradients was of an anti-Helmholtz type. The x and y gradients were generated by another pair of rectangular coils. The directions of the magnetic field gradient were varied by electronic rotations as described in our previous paper (Ishida et al., 1992). There were five units of current power supplies for the coils, controlled by the microcomputer through digital-analog (DA) converters. The microcomputer was also concerned with the data collection through an analog-digital (AD) converter. Using this system it took us about 150 s to collect 82 (9 × 9 + 1) ESR data for the reconstruction of ESR-CT images.

Correction of magnetic field strength

Rapid scan of the magnetic field is usually accompanied by the field delays, which is generally brought about by magnetic hysteresis of the core materials and/or eddy current effects in the magnetic assembling metals. Our new apparatus was composed of a simple Helmholtz coil without an iron core. This situation allowed us to evaluate the delay phenomenon while only taking eddy current losses into consideration. In this case we can obtain the real magnetic field from the delayed field by a simple mathematical treatment using a first-order low-pass filter model. The equation is as follows:

$$H = H' (t - t' (1 - \exp (- t /t'))) + H_s \qquad\qquad [1]$$

where H is magnetic field at the time t after the beginning of the field scan under the sweep rate of H', H_s the magnetic field at t = 0, and t' the delay time estimated by comparing the spectra with and without the delay.

Figures 2(a) and 2(b) show an example of the field correction for 1 mM TEMPOL aqueous solution. The fine line was obtained at a sweep rate of 10 mT/240 s which induced no delay of the magnetic field. On the other hand, the heavy line was recorded at 10 mT/1 s or 10 mT/ 2s sweep rates. In Equation [1] t' cannot be regarded as zero if the field scan is too fast. The value of t' was estimated to be 64 ms from the comparison between slow and fast scan signals at higher field. In Figures 2(a) and 2(b) the ESR signal intensity is plotted against the corrected H as a dotted line. It fitted perfectly with the fine line. This agreement indicates that the ESR spectral data could be treated according to Equation [1]. In Figure 2(c), the results obtained at a scan rate of 10 mT/2s by using an iron-core electromagnet (JEOL RE3X) are shown. As was done in the case of the air-core magnet, the magnetic field was corrected by Equation [1]. As seen in Figure 2(c), however, the field correction failed. This is due to iron losses which cannot be simulated by a simple mathematical equation such as Equation [1]. Therefore, we believe at this time that the air-core magnet was very effective in rapid scan ESR-CT imaging.

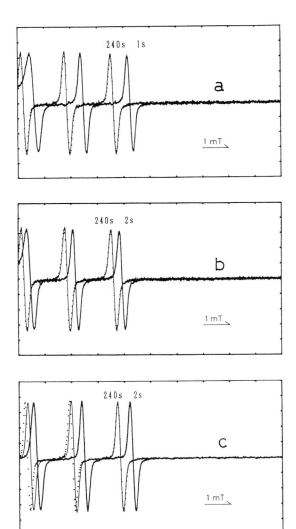

Figure 2. L-band ESR spectra of TEMPOL solution obtained at various scan rates of 10 mT/1 s (a), 10 mT/2 s (b) and 10 mT/240 s with an air-core electromagnet. The case of an iron-core electromagnet is given in (c). Small closed circles located on or near the 10 mT/240 s spectrum show the results of field correction by Equation [1].

Interactive data processing system

In our L-band ESR measurement equipment, the controls of the ESR apparatus, such as sweeping speed of the magnetic field etc. and the data acquisition were performed by a microcomputer with DOS/V operating system. We had prepared the data-handling and processing programs on this microcomputer. However, in order to improve the processing time and the tractability of the system, we introduced an Engineering Work Station (EWS) using OSF/Motif (Heller and Gilly, 1991), on HP 9000 model 715. The EWS was connected to the microcomputer by LAN.

The interactive data processing system was composed of three modules: the data base, the data processing and the display. The data base module saved and searched raw data and processed data. The data processing module had various software filters for preprocessing and deconvolution and also had algorithms for reconstruction from projections. The visualization module displayed the measured and processed data in order to let users see various situations, e.g., how experiments go and how radicals behave in a living body. These systems are connected as seen in Figure 3, and we can operate them through X Window system. Here we can easily add new programs for new algorithms on each module.

Comparison of ESR-CT algorithms

The data processing module used to obtain ESR-CT images mentioned above had three groups of programs as follows: the preprocessing tools that included the smoothing of raw data and the base-line correction from it, the deconvolution programs in which the projections from a spatial

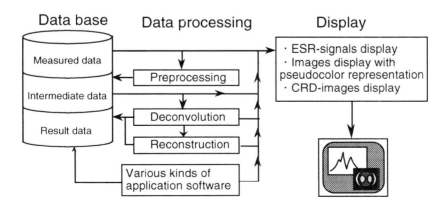

Figure 3. Data flow and software system.

distribution of a radical concentration were abstracted from the preprocessed data and the recon-struction programs to obtain a CT image of a spatial distribution of the free radical from the projection.

There are two ways of deconvolution processing, i.e., in the spatial domain and in the Fourier domain. We compared the five following methods: the low pass filter, the parametric Wiener filter, the generalized inverse filter, the least square filter with a constraint condition, and the pro-jection filter (Ogawa, 1988). The first two methods are processing in the Fourier domain. The others are in the spatial domain (Rosenfeld and Kak, 1982). First, we compared by simulations to investigate properties of each deconvolution method for noise-free data. Then, we added a random noise to the simulated data and the results were compared. A memoryless Gaussian noise was taken as a random noise and its mean was zero. The results are shown in Figure 4. The vertical and the horizontal co-ordinate are the signal-to-noise ratio (SNR) and the degree of noise (where degree is evaluated by the ratio of the standard noise against the maximum value of signal) respectively. Figure 4 shows that the projection filter, the least square filter with a constraint condition and the generalized inverse filter are comparably effective against noise.

For the reconstruction, we compared the six following methods: the filtered back projection (FBP), the modified filtered back projection (MFBP), the additive algebraic reconstruction tech-nique (ART), the simultaneous iteration reconstruction technique (SIRT), the least square method with a constraint condition (LSM), and the maximum entropy method (MEM). The first two methods are categorized as the analytical methods and others as iterative methods (Ekstrom et al., 1984). The spatial distribution of radical concentration was assumed as seen in Figure 5(a). Pro-

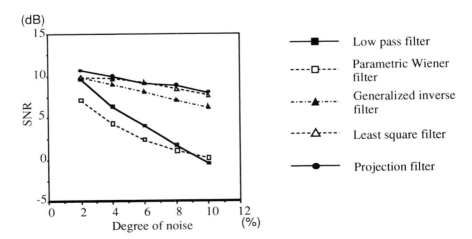

Figure 4. Performance of each filter for noisy data. SNR means the signal to noise ratio.

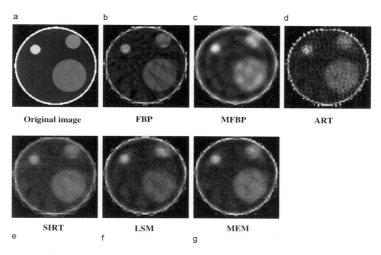

Figure 5. Comparison of the reconstruction of the original image (a) by various methods of (b) to (g).

jections were calculated analytically. For this phantom, in order to examine the properties of each method for a small number of projections we examined for 9 noise-free projections. The results are shown in Figure 5(b) to (g). Each image contained considerable artefacts, but the SIRT and the MEM could reconstruct comparably satisfactory images. The robustness for noise was then compared. The memoryless Gaussian noise was added to each simulated projection. The results are shown in Figure 6. Both the vertical and horizontal co-ordinates are similar to Figure 4. Figure 6 shows that iterative methods tend to be superior to analytical methods. The MEM and the SIRT were comparably robust for noise.

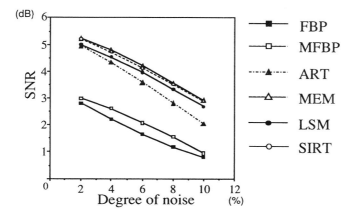

Figure 6. Performance of each reconstruction algorithm for noisy data.

Three-dimensional expression of ESR-CT

For the phantom using an agar gel of nitroxide radical and the head of a rat receiving nitroxide radicals, we made the three-dimensional expression of ESR-CT images. 4-hydroxy-2,2,6,6-tetramethyl-1-piperidinyloxy (TEMPOL), 3-carbamoyl-2,2,5,5-tetramethyl-pyrrolidine-1-oxyl (C-PROXYL), and 2-(14-carboxypropyl)-2-ethyl-4,4-dimethyl-3-oxazolidinyloxy (16DS) were used as imaging agents. Male Wister rats weighing 20–50 g were used as measuring subjects. These were anesthetized by intraperitoneal injection of pentobarbital. A phantom was prepared using an agar gel of 0.01 M TEMPOL aqueous solution. A hole was made along the central axis. Then an aqueous solution of 1 M sodium ascorbate was poured into this hole. ESR-CT images were obtained by 82 spectra at an angular increment of 20 degrees at the gradient of 0.1 T/m.

A volume element was defined to express the three-dimensional distribution of radicals. It was a small cube, whose color corresponded to the concentration of the radicals. It was arranged in a

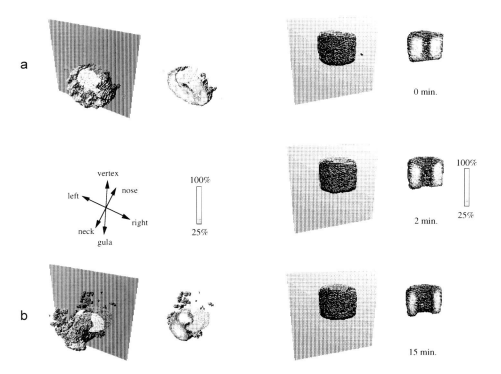

Figure 7. ESR-CT images of the rat head after injection of C-PROXYL in saline solution into the peritoneal cavity (a) and after injection of 16DS in DMF into the carotid artery (b).

Figure 8. ESR-CT images of the cylindrical phantom composed of 0.01 M TEMPOL agar gel after pouring 1 M ascorbic acid solution into a central hole.

position where the radicals distribute in three-dimensional space. From this expression of ESR-CT image, we can observe both the whole image and sectional image from any angle. The whole image with the position of the section shown by a plane is displayed on the left, and the sectional image is displayed on the right, of the figures following. As examples of the results of this method, ESR-CT images are shown in Figures 7 and 8. Figure 7(a) shows ESR-CT images of the rat head after injection of 3.3 ml of 0.2 M C-PROXYL in saline solution into the peritoneal cavity. C-PROXYL distributed just outside the brain, due to the blood brain barrier. Figure 7(b) shows ESR-CT images of the rat head after injection of 0.1 ml of 0.2 M 16DS in DMF into the carotid artery. 16DS distributed in the brain after passing through the blood brain barrier. As seen in Figure 8, 0.01 M TEMPOL in agar gel was very quickly reduced by 1 M ascorbic acid solution. The change of the hole size is due to the diffusion of ascorbate ions into TEMPOL agar gel. This three-dimensional imaging method is effective in showing where radical-rich areas are, and how radicals diffuse and react with other substances.

References

Demsar, F., Walczak, T., Morse, P. II, Bačic, G., Zolnai, Z. and Swartz, H. (1988) Detection of diffusion of oxygen by fast scan EPR imaging. *J. Magn. Reson.* 76: 224–231.

Eaton, G.R., Eaton, S.S. and Ohno, K. (1991) *EPR Imaging and in vivo EPR.* CRC Press, Boca Raton.

Ekstrom, M.P. (1984) *Digital Image Processing Techniques.* Academic Press, Orland.

Heller, D. and Gilly, D. (1991) *The Definitive Guide to the X Window System.* O'Reilly & Associates, Sebastopol.

Ishida, S., Matsumoto, S., Yokoyama, H., Mori, N., Kumashiro, H., Tsuchihashi, N., Ogata, T., Kitajima, T., Ono, M., Kamada, H. and Yoshida, E. (1990) Electron spin resonance imaging of the distribution of the nitroxide radical administered into mouse legs. *Japan J. Magn. Reson. Med.* 10: 21–27.

Ishida, S., Matsumoto, S., Yokoyama, H., Mori, N., Kumashiro, H., Tsuchihashi, N., Ogata, T., Yamada, M., Ono, M., Kitajima, T., Kamada, H. and Yoshida, E. (1992) An ESR-CT imaging of the head of living rat receiving an administration of a nitroxide radical. *Magn. Reson. Imag.* 10:109–114.

Lauterbur, P.C. (1973) Image formation by induced local ineractions: Example employing nuclear magnetic resonance. *Nature* 242: 190–191.

Lauterbur, P.C. and Lay, C.M. (1980) Zeugmatography by reconstruction from projections. *IEEE Trans. Nucl. NS* 27: 1227–1231.

Ohno, K. and Tsuchihashi, N. (1991) Advances in electron paramagnetic resonance imaging. *Trends in Physical Chemistry* 2:79–97.

Ogawa, H. (1988) Image and signal restoration (III). *J. Institute of Electronics, Information and Communication Engineers* 71: 739–748.

Rosenfeld, A. and Kak, A.C. (1982) *Digital Picture Processing,* Vol. 1. Academic Press, Orland.

Bioradicals Detected by ESR Spectroscopy
H. Ohya-Nishiguchi & L. Packer (eds)
© 1995 Birkhäuser Verlag Basel/Switzerland

Spin trapping

E.G. Janzen

*National Biomedical Center for Spin Trapping and Free Radicals, Free Radical Biology and Aging Research
Program, Oklahoma Medical Research Foundation, Oklahoma City, OK 73104, USA and
Departments of Clinical Studies and Biomedical Sciences Ontario Veterinary College MRI Facility University of
Guelph, Guelph, Ontario, N1G 2W1, Canada*

*There is a value and a danger to a label. The value is that it calls attention to something that is
very important. The danger is that it simplifies something that is actually quite complex.*

G. Beauchamp (1994) *J. NIH Res.*, Vol. 6, p. 64 in connection with discussion on the term *Pheromones*.

The spin trapping reaction

Definition

Spin trapping is defined as that chemical reaction in which a radical adds to a molecule so that the
group that was the radical (radical addend) stays with the molecule for future analysis. The
molecule which captures the radical is called the spin trap. The addition product is called the spin
adduct.

In its simplest form the reaction is a case of A + B gives C:

$$R\cdot + ST \longrightarrow R—SA\cdot \qquad\qquad [1]$$
$$\text{(spin trap)} \qquad \text{(spin adduct)}$$

Other possibilities will be discussed later.

The requirements for a "good" spin trapping reaction are as follows:

Rate

A good spin trapping reaction must start with a fast rate of radical addition to the spin trap. In
principle there are three ways to influence the rate of the spin trapping reaction:

Choice of reactive radical

Free radicals vary greatly in their reactivity. In biological systems the hydroxyl radical is probably the most reactive radical encountered (in chemical systems perhaps the fluorine atom, dioxygen radical cation or others could be as reactive as the hydroxyl radical).

On the other end of the scale there are many neutral radicals and radical ions in biological systems which are not reactive in this sense; i.e., these radicals do not add to molecules which serve as spin traps. Examples are phenoxyl radicals (e.g., tocopheroxyl) or ascorbyl anion radicals. Therefore, one must know something about the intrinsic reactivity of a radical when considering the possibility of capturing that radical by spin trapping. Not all reactive radicals add to spin traps with the same rate.

In general, sterically unhindered radicals based on the elements of the first horizontal row in the periodic table are highly reactive radicals or atoms:

$H_3C\cdot$ (methyl), $H_2N\cdot$ (aminyl), $HO\cdot$ (hydroxyl), $F\cdot$ (fluorine atom)

However, the reactivity of each radical type in this series is strongly influenced by two factors: resonance stabilization and steric hindrance. Aryl groups reduce the reactivity of the radical if unpaired electron delocalization is possible by resonance. Steric crowding around the spin center strongly decreases the reactivity of the radical. Frequently both factors combine to make addition of a free radical sluggish. Thus for example the triphenylmethyl radical cannot be spin trapped and is easily detected in the presence of a spin trap (Janzen and Blackburn, 1969). Tocopheroxyl can also frequently be detected along with spin adducts in extracts of rat liver after free radical-producing metabolism (Poyer and Janzen, unpublished work):

Here both steric hindrance and resonance stabilization combine to prevent spin trapping.

All other cases should be viewed with these features in mind. For example, one would expect the relative rate of addition of alkyl radicals to go as follows:

$H_3C\cdot > CH_3\,CH_2\cdot > (CH_3)_2CH\cdot > (CH_3)_3\,C\cdot$

This seems to be the trend (Janzen and Haire, 1990). The spin trapping of *tert*-butyl radicals or 2-cyanopropyl radicals is very slow (Janzen et al., 1990a). The necessary rate studies on allyl and *bis*-allyl spin trapping have not been done, but again one would expect spin trapping to be slow.

$$CH_2=CH-\dot{C}H_2 > CH_2=CH-\dot{C}H-CH=CH_2$$

The latter would be relevant to the question of spin trapping radicals derived from poly-unsaturated fatty acid chains in lipids. Both resonance and steric hindrance in a secondary radical site would greatly influence the rate of radical addition to a spin trap:

$$-CH=CH-\dot{C}H-CH=CH-$$

$$-\dot{C}H-CH=CH-CH=CH- \qquad -CH=CH-CH=CH-\dot{C}H-$$

Not much is known about aminyl radical spin trapping, but one would expect to find the reactivity of this radical center to be influenced by the same factors, namely resonance and steric hindrance.

For oxygen-centered radicals we know that high reactivity is retained for all alkyl substituted oxyl radicals. Only very little loss in reactivity is experienced in the following series:

$$HO\cdot > CH_3O\cdot > CH_2CH_2O\cdot > (CH_3)_2CHO\cdot > (CH_3)_3CO\cdot$$

Actually only the reactivities of hydroxyl (Sridhar et al., 1986) and *tert*-butoxyl (Janzen and Evans, 1973) radicals are known but it can be assumed that the other examples would fall in between. The lack of steric hindrance as the group attached to the oxygen atom becomes more sterically demanding must be due to the fact that these substitutions come one atom farther removed from the spin center than in the case where carbon carries the unpaired electron.

In the case of phenoxyl radicals it should be pointed out that spin trapping can sometimes take place, but the attachment is on carbon. If a site *ortho* or *para* to the phenoxyl oxygen is unhindered spin trapping can occur here (Terabe et al., 1972):

The case of allyl radicals has already been mentioned. Here the unpaired electron is delocalized over two atoms equally. There are other free radicals where the spin is delocalized over two atoms but the delocalization is unequal. Two cases which have been well studied are peroxyl radicals and hydroxyalkyl radicals:

$$R-\ddot{O}-\dot{O} \qquad \longleftrightarrow \qquad R-\overset{\oplus}{\dot{O}}-\overset{\ominus}{\ddot{O}} \qquad \text{peroxyl radical}$$

$$H-\ddot{O}-\underset{R}{\overset{|}{\dot{C}}} \qquad \longleftrightarrow \qquad H-\overset{\oplus}{\dot{O}}-\underset{R}{\overset{\ominus}{\overset{|}{\ddot{O}}}} \qquad \begin{array}{l}\text{hydroxyalkyl}\\ \text{radical}\end{array}$$

Because of resonance spin delocalization, peroxyl radicals are relatively unreactive. Certainly in comparison with alkoxyl radicals, alkylperoxyl radicals are unreactive and add to spin traps only slowly. Hydroxyalkyl radicals are also less reactive than alkyl radicals and the rate of addition is strongly influenced by steric hindrance (Greenstock and Wiebe, 1982; Madden and Taniguchi, 1991). Thus:

$$HO\dot{C}H_2 \quad > \quad HO\underset{CH_3}{\overset{|}{\dot{C}H}} \quad > \quad HO\underset{CH_3}{\overset{CH_3}{\overset{|}{\underset{|}{C\cdot}}}}$$

A special case of this type is the superoxide anion radical and the hydroperoxyl radical. In both cases the unpaired electron is delocalized over two atoms but in the former example delocalization is equal by symmetry:

$$[\ddot{O}-\dot{O}]^- \qquad \longleftrightarrow \qquad [\dot{O}-\ddot{O}]^-$$

$$H-O-\dot{O} \qquad \longleftrightarrow \qquad H-\overset{+}{\dot{O}}-\ddot{O}$$

Both of these radicals are relatively unreactive, at least in comparison with the hydroxyl radical, and spin trapping is relatively slow. It is believed that the hydroperoxyl radical is more reactive towards spin traps than the superoxide radical anion (Finkelstein et al., 1980).

In a discussion of radicals where the spin is delocalized over two atoms one must include the case of nitric oxide. Here the spin is primarily on the nitrogen atom but a resonance structure can be drawn showing delocalization onto the oxygen atom as well:

$$:\!\overset{\cdot}{N}\!\!=\!\!\overset{\cdot\cdot}{\underset{\cdot\cdot}{O}} \qquad \longleftrightarrow \qquad \overset{\ominus \quad \oplus}{:\!\overset{\cdot\cdot}{N}\!\!=\!\!\overset{\cdot}{\underset{\cdot\cdot}{O}}\!:}$$

This free radical is very unreactive and does not appear to react rapidly with spin traps, if at all. Indeed the reactivity of nitric oxide is best viewed as comparable to dioxygen (triplet ground state) (Janzen et al., 1993a; Wilcox and Janzen, 1993):

$$\cdot\overset{\cdot\cdot}{\underset{\cdot\cdot}{O}}\!-\!\overset{\cdot\cdot}{\underset{\cdot\cdot}{O}}\cdot$$

Some common triatomic radicals and radical ions have become interesting examples for spin trappers. Whereas chlorine dioxide ($\cdot ClO_2$), nitrogen dioxide ($\cdot NO_2$) and sulfur dioxide anion radical ($\cdot SO_2^-$) are all persistent radicals, reactions with spin traps vary greatly. Chlorine dioxide and nitrogen dioxide both lead to oxygenated aminoxyl products (Knechtel et al., 1978; Wilcox and Janzen, unpublished work) while sulfur dioxide radical anion kills aminoxyl radical signals (either by reduction or addition) and does not add to spin traps (E.R. Davis and E.G. Janzen, unpublished work). On the other hand, the carbon dioxide anion radical ($\cdot CO_2^-$) as well as the sulfur trioxide anion radical ($\cdot SO_3^-$) can readily be spin trapped (Knecht and Mason, 1993; Constantin et al., 1994).

In the second horizontal row of the periodic table the following sterically unhindered radicals and atoms are possible:

$H_3Si\cdot$ (silyl), $H_2P\cdot$ (phosphinyl), $HS\cdot$ (thiyl), $Cl\cdot$ (chlorine atom)

In general, the reactivity of these radicals is expected to diminish as one goes down the periodic chart, and so sulfur-centered radicals are less reactive than oxygen-centered radicals, and the chlorine atom is less reactive than the fluorine atom etc. Again steric hindrance is not likely to influence the reactivity of thiyl radicals as much because the substituents are farther removed from the spin center. However, an interesting point here is that aryl group resonance is expected to be less effective for second row elements and thus thiophenoxyl radicals may be as reactive as alkylthiyl radicals in spin trapping reactions (Ito and Matsuda, 1984).

What effect does the reactivity of a radical have on the likelihood of a successful spin trapping experiment? The approach to this question comes from a realization that normally, free radicals have a wide choice of reactions that they can participate in. If the "desired" reaction from the analytical point of view is slow, radicals will find faster avenues to pursue in order to reach stable end products.

Reactions of radicals with other radicals are always faster than reactions of radicals with non-radicals. Since in our definition of spin trapping the spin trap is a diamagnetic molecule, i.e., a non-radical molecule, this route will necessarily be slower than a reaction with another radical. It is fortunate then, that the steady-state concentrations of free radicals is relatively low, or else spin trapping would not work. All radicals would react with each other before they could be spin trapped.

There are other implications of the wide range of free radical reactivity. Frequently the researcher is looking for the effect of a radical on the system but only the initial radical produced is detected. Let us select the hydroxyl radical as an example. Perhaps it is hoped that some new radical resulting from "free radical damage" will be located in an interesting segment of an animal tissue. When the spin trapping experiment is run sometimes only the hydroxyl radical is detected and nothing else. Why? Probably in the presence of a spin trap, the spin trapping reaction between the hydroxyl radical and the spin trap is a much faster reaction than the reaction of the "damage radical" (d-R·) with the spin trap:

$$HO\cdot + ST \longrightarrow HO{-}SA\cdot \qquad [2] \ \text{relatively fast}$$
$$d{-}R\cdot + ST \longrightarrow d{-}R{-}SA\cdot \qquad [3] \ \text{relatively slow}$$

If EPR is being used to monitor this situation then the hope is that sufficient signal is available so that the instrument GAIN can be increased. Often the EPR operator simply displays the strongest spectral pattern on the page so that the peaks stay on scale. If the hydroxyl adduct is the strong spectrum that is all that will be visible. However, if we assume that the relative rates differ by a factor of ten (they could actually differ by as much as ten thousand or a hundred thousand!), then the EPR spectrometer GAIN should be increased by ten, even if the hydroxyl adduct spectrum goes off scale. In this way, perhaps spin adducts of the "damage-radical" species could be detected. With computer-assisted spectrum accumulation perhaps GAINs of up to 100 could be achieved.

EPR work of this type takes some practice and experience. The danger is mis-interpretation of the EPR spectrum at high GAIN. The strongest peaks at high GAIN may be the carbon-13 satellites of the strong spectrum. Although naturally abundant carbon has only 1.1% C-13, equivalent atoms will increase this number. It is imperative to run known spectra of spin adducts under the same conditions at high GAIN to recognize these features in the spectrum.

High GAIN spectra must be obtained from extremely pure spin traps. Any EPR-active impurity brought along with the spin trap which was manageable at normal GAIN settings will become problematic now. Controls on all spin trapping experiments must be run routinely and with some thought as to the process being investigated.

The most commonly encountered oxygen free radical scenario involves the simultaneous or sequential formation of hydroxyl radicals and superoxide/hydroperoxyl radicals. With a commonly used spin trap the rate constant of spin trapping for these two radicals, namely hydroxyl and hydroperoxyl, differ by almost a million times (4.3×10^9 vs. 6.6×10^3) while for hydroxyl and superoxide anion radical the difference is even greater (4.3×10^9 vs. 10) (Janzen and Haire, 1990). There would have to be over 10,000,000 times more superoxide than hydroxyl radical to get the same EPR spin adduct signal!

Choice of reactive spin trap

Spin traps come in all shapes and sizes. In principle any molecule to which a radical can add and stay attached permanently could serve as a spin trap. For this property the most common function needed within the spin trap is a double bond. The radical should add to the double bond irreversibly producing a new radical addition product. If this new radical has the capability of stabilizing the unpaired electron so much the better. A stabilized radical product will increase the intrinsic rate (the rate constant) of spin trapping for all radicals. However, this effect will make the biggest difference with relatively unreactive radicals. Highly reactive radicals, like hydroxyl radicals, will add no matter what happens; less reactive radicals will need all the assistance they can get.

The archetypical radical addition reaction is of course the encounter of a radical with ethylene:

$$R\cdot \;+\; \underset{H}{\overset{H}{\diagdown}}C=C\underset{H}{\overset{H}{\diagup}} \longrightarrow R-\overset{\overset{H}{|}}{\underset{\underset{H}{|}}{C}}-\overset{H}{\underset{H}{C\cdot}} \qquad [4]$$

Radicals add rapidly to the carbon atom ends of the double bond to produce a new radical addition product. However, in the case of ethylene the addition product has no stabilization and remains as reactive (or even more reactive) than $R\cdot$. Thus further radical addition occurs if more ethylene is available:

$$R-CH_2CH_2\cdot \;+\; \underset{H}{\overset{H}{\diagdown}}C=C\underset{H}{\overset{H}{\diagup}} \longrightarrow R-CH_2-CH_2-\overset{\overset{H}{|}}{\underset{\underset{H}{|}}{C}}-\overset{H}{\underset{H}{C\cdot}} \qquad [5]$$

This process is rapidly repeated until all ethylene is used up and the end radical is coupled with another radical (termination). The final product is some kind of polyethylene:

$$R-CH_2-CH_2-(CH_2-CH_2)_n-CH_2-CH_2-R$$

In spin trapping we want the addition to the double bond (reaction 4) to remain fast but ideally all subsequent reactions should be slow. The spin adduct should be intrinsically stable. Also, the spin adduct should be inert to further reactions of radicals or other redox chemicals or biochemicals which might be encountered. This is a tall order.

What choices have we got in selecting a reactive spin trap function? Perhaps one could go with the ethylene molecule itself for starters, and take advantage of the fast radical reaction on the open carbon atom end of ethylene. If so, how do we build stability into the addition product if and when it is formed in the spin trapping reaction? Steric protection seems to be one possibility:

$$R\cdot \;+\; \underset{H}{\overset{H}{>}}C=C\underset{G}{\overset{G}{<}} \longrightarrow \underset{H}{\overset{H}{R-C-}}\underset{G}{\overset{G}{C\cdot}} \qquad [6]$$

If G were an absolutely huge GROUP perhaps enough protection could be provided so that the spin adduct could have some intrinsic stability. However, oxygen (the dioxygen molecule) would still find a way to get in and react with this addition product to form a peroxyl radical. Even this might be acceptable if only the EPR spectrum of a peroxyl radical consisted of a pattern of sharp lines. However, this is not the case and this possibility has not been pursued in an analytical sense.

However, some work has been done on using the *tert*-butyl group as a sterically hindering GROUP on ethylene. Thus 1,1-di-*tert*-butylethylene can be prepared and subjected to radical reactions (Griller and Ingold, 1973):

$$R\cdot \;+\; \underset{H}{\overset{H}{>}}C=C\underset{C(CH_3)_3}{\overset{C(CH_3)_3}{<}} \longrightarrow \underset{H}{\overset{H}{R-C-}}\underset{C(CH_3)_3}{\overset{C(CH_3)_3}{C\cdot}} \qquad [7]$$

If a carbon-carbon double bond is unsuitable for our purposes, can a heteroatom be used? Double bonds are easily formed between carbon and nitrogen, oxygen, or sulfur, but with difficulty between carbon and phosphorus or silicon. The well-known trend in stability is as follows:

$$C=C \sim C=N \sim C=O > C=S > C=P > C=Si$$

If radical addition occurred to the carbon end of any of these double bonds, the resulting radical adduct would not be significantly more stable than the carbon-centered radical product obtained from ethylene. Some examples are known, however, where radical addition can be detected on the heteroatom end.

Thus di-*tert*-butyl thioketone and di-*tert*-butyl selenoketone trap radicals on the heteroatom end (Scaiano and Ingold, 1976) and form new carbon-centered radicals:

$$
R\cdot \;+\; S=C\begin{matrix}\nearrow C(CH_3)_3\\ \searrow C(CH_3)_3\end{matrix} \longrightarrow R-S-\overset{\displaystyle \cdot}{C}\begin{matrix}\nearrow C(CH_3)_3\\ \searrow C(CH_3)_3\end{matrix} \qquad [8]
$$

$$
R\cdot \;+\; Se=C\begin{matrix}\nearrow C(CH_3)_3\\ \searrow C(CH_3)_3\end{matrix} \longrightarrow R-Se-\overset{\displaystyle \cdot}{C}\begin{matrix}\nearrow C(CH_3)_3\\ \searrow C(CH_3)_3\end{matrix} \qquad [9]
$$

Surprisingly, even nitrogen or oxygen can play a role in this type of reaction if the double bond is appropriately substituted (Griller et al., 1974; Krusic et al., 1974):

$$
R\cdot \;+\; \overset{\diagdown}{N}=C\begin{matrix}\nearrow C(CH_3)_3\\ \searrow C(CH_3)_3\end{matrix} \longrightarrow R-\overset{\mid}{N}-\overset{\displaystyle \cdot}{C}\begin{matrix}\nearrow C(CH_3)_3\\ \searrow C(CH_3)_3\end{matrix} \qquad [10]
$$

$$
R\cdot \;+\; O=C\begin{matrix}\nearrow CH(CF_3)_2\\ \searrow CH(CF_3)_2\end{matrix} \longrightarrow R-O-\overset{\displaystyle \cdot}{C}\begin{matrix}\nearrow CH(CF_3)_2\\ \searrow CH(CF_3)_2\end{matrix} \qquad [11]
$$

Surprisingly few double bond scenarios can be imagined that could function as a spin trap. The two most popular of course are the nitrone and the nitroso function:

$$
\overset{\diagdown}{\underset{\diagup}{C}}=\overset{\overset{\textstyle O^-}{\mid}}{\underset{+}{N}}- \qquad\qquad -N=O
$$

The former is simply a variation, albeit an important one, on the iminyl (C=N) function, and the latter is a "cross-product" of the iminyl and carbonyl (C=O) functions. The reason why the nitronyl function should work well seems obvious. Addition can occur on carbon, making a strong bond between the radical (R·) and a carbon atom, and the product is the highly stabilized aminoxyl radical function. What better features could possibly have been chosen for fast radical reactions with an unsaturated bond?

$$R\cdot \; + \; \underset{/}{\overset{\backslash}{C}}=\overset{\overset{O^-}{|}}{\underset{+}{N}}- \;\longrightarrow\; \underset{\underset{H}{/}}{\overset{\backslash}{R}-\overset{\overset{O^-}{|}}{\underset{+}{C}}-\underset{+}{N}}- \;\longleftrightarrow\; \underset{\underset{H}{/}}{\overset{\backslash}{R}-\overset{\overset{O\cdot}{|}}{C}-\overset{\cdot\cdot}{N}}- \qquad [12]$$

The fast addition to the nitroso function by radicals is not nearly that obvious. In general, radical addition to a nitrogen atom in an unsaturated bond rarely if ever happens (e.g., in iminyl groups or unsaturated nitrogen heteroatom cyclic molecules). This reluctance of radical addition to a heteroatom such as nitrogen must be massively overwhelmed by the stability of the product formed in this case, namely the aminoxyl function:

$$R\cdot \; + \; -N=O \;\longrightarrow\; R-\overset{\overset{O\cdot}{|}}{N}- \qquad [13]$$

However, it is clear from studies with sterically hindered nitroso compounds, that the adding radical is somewhat ambivalent about the direction of addition. Attachment to the oxygen atom of the nitroso function can also occur and an alkoxyaminyl radical is formed (Terabe and Konaka, 1973):

$$R\cdot \; + \; -N=O \;\longrightarrow\; R-O-\overset{\cdot}{N}- \qquad [14]$$

There is something unusual about the nitroso double bond. The electronic absorption spectrum is strange for a function with one double bond (nitroso compounds are blue) and nitroso compounds tend to dimerize if unhindered. Nitroso compounds also undergo the electrocyclic "ene" reaction with certain carbon-carbon double bonds and in this sense resemble singlet oxygen. All of this nitroso double bond chemistry indicates a high intrinsic reactivity for the nitroso function which is more than just a difference in polarity between the nitrogen and oxygen atom. The carbonyl bond has a larger difference in polarity between the carbon and oxygen atom than the nitroso function has between the nitrogen and oxygen atom. Perhaps the nitroso function has some singlet diradical character at room temperature, and this is the reason why reaction with a radical is favorable:

$$-N=\overset{\cdot\cdot}{\underset{\cdot\cdot}{O}}: \;\longleftrightarrow\; -\overset{\uparrow\quad\downarrow}{\overset{\cdot}{N}-\overset{\cdot\cdot}{O}}:$$

Before leaving this topic it should be pointed out that connecting additional double bonds to ethylene increases the rate constant of addition by radicals. Thus butadiene is more reactive than ethylene as a trap for radicals (Janzen et al., 1978a):

$$R\cdot \ + \ CH_2{=}CH{-}CH{=}CH_2 \ \longrightarrow \ R{-}\overset{\overset{\displaystyle H}{|}}{\underset{\underset{\displaystyle H}{|}}{C}}{-}\overset{\displaystyle \cdot}{C}H{-}CH{=}CH_2 \quad [15]$$

Extending this point to biological systems brings us to the realization that radical addition to the conjugated butadiene-like function found in polyunsaturated lipids might become more important than bis-allyl hydrogen abstraction. Thus, conjugated hydroperoxides could be quite effective traps for free radicals:

$$R\cdot \ + \ \overset{}{\underset{\underset{\displaystyle OOH}{|}}{CH{=}CH{-}CH{=}CH{-}CH{-}}} \ \longrightarrow \ R{-}\overset{}{\underset{\underset{\displaystyle}{|}}{CH}}{-}\overset{\displaystyle\cdot}{C}H{-}CH{=}CH{-}\overset{}{\underset{\underset{\displaystyle OOH}{|}}{CH{-}}} \quad [16]$$

Choice of reaction conditions

In homogeneous chemical systems the scientist has relatively speaking many choices in the design of a spin trapping experiment. The most obvious are control of concentrations, control of temperature, polarity of solvent, reactivity of solvent, concentration of compatible other reagents etc. In unadulterated biological systems the situation is not so favorable. The system is usually predetermined by "natural factors". Here very little control can be exercised over local conditions such as concentrations of competitive reactants, imposed temperatures, local polarity or effective concentrations of added reagents. Notwithstanding these challenging problems with analytical spin trapping in biological systems, it is still worth looking at the environmental factors which will influence the rate of spin trapping reactions in general.

The most important factor which is under the control of the practitioner in spin trapping is the concentration of the spin trap. Since radical reactions with each other are so fast, the concentration of spin trap must necessarily be high in the immediate environment of free bioradical creation. Low solubility and toxicity however may limit the amount of spin trap that the system can tolerate.

The most desirable spin trapping compound would be one which is soluble in water as well as in a variety of lipid phases. If some solubility in buffer solutions is possible, solutions can be made up in the laboratory and administered either intraperitoneally (I.P.), intravenously (I.V.) or intragastrically (I.G.). All of these methods have been used in laboratory animals such as the rat. The intensity of spin adduct signal in general appears to follow the sequence I.V. > I.P. > I.G. but the right experiments under identical conditions have not actually been performed (L.A. Reinke, personal communication; J.L. Poyer, unpublished work). The spin trap can also be administered in a lipid vehicle like corn oil (McCay et al., 1984) or an emulsifying agent like EMULFOR (Janzen et al., 1990b). Little systematic work has been done to look for the effect of varying the spin trap dose on the EPR spin adduct signal detected in *in vivo* experiments, although some research connected to this point is underway in these laboratories. In a systematic study of *in vitro* spin trapping in rat liver microsomal dispersions (Janzen et al., in press), we have explored the effect of spin trap dose on the resulting spin adduct signal intensity. As expected, initially the yield of spin adduct increases with increase in spin trap concentration but soon this increase disappears and a flat plateau is observed. At high concentrations the yield actually goes down! This effect was also noted by Albano et al. (1982) and interpreted in terms of slight inhibition of the enzyme which is responsible for producing the free radicals the spin trap is sent to detect! Another explanation could involve poisoning the microsomal system or destroying some essential protein/membrane structure. Obviously it is important to use the lowest possible concentration of spin trap if it is supposed to function as a probe of the system under investigation.

The spin trapping approach is actually quite complicated if the most information possible is to be extracted from the system being studied. The initial objective is to know whether free radicals are being produced by the system of interest. How does the concentration of spin trap influence this search? Obviously one has to start somewhere with spin trap concentrations; then the experiment should be repeated varying spin trap concentration both downward and upward.

Three possibilities can be visualized:

1. No EPR spin adduct signal is found.
2. Very weak EPR adduct signal is detected.
3. A rich and intense EPR spin adduct(s) signal is observed.

1. If no EPR spin adduct signal is found the concentration of spin trap should be varied through a wide range of concentrations in spite of this failure. There may be a small window of concentrations of spin trap where the compound can function as a probe but not adulterate the system significantly. In general, this window of opportunity is not known and must be discovered by trial and error.

Also, the time and manner of administration of the spin trap should be varied before the experimenter gives up, because arriving at the right location for spin trapping may take some time for the spin trap. It is known that administering Vitamin E (tocopherol) I.P. is almost ineffective in inhibiting lipid peroxidation in the liver of a rat, presumably because of the logistics of getting the compound to the right place at the right time. So it may also be for spin traps.

Finally, it is highly advisable to follow some other observable which may or may not depend on the occurrence of free radicals to be manifested in the system while at the same time looking for evidence of spin adducts. For example, one could follow the extent of lipid peroxidation in the system of interest as a function of spin trap concentration while free radical generation is manipulated. It is possible that no EPR-detectable spin adducts are recognized but at the same time the spin trap does influence the rate of lipid peroxidation. In fact a concentration of spin trap may exist where no lipid peroxidation is initiated. Clearly the concentration of spin trap is an important variable in this case, although perhaps no EPR-detectable spin adducts are found.

The reason why no EPR-detectable spin adducts are detected may be due to a broad range of reasons. Thus it is not good science to interpret negative EPR-based spin trapping results in a positive way. The absence of EPR spin adduct signals does not mean that no free radicals exist in the system. It is possible that the radicals of interest do not add to the spin trap (too unreactive or too bulky or too low in concentration), or the spin trap concentration in the immediate volume compartment where the radical is produced is too low (poor solubility, slow diffusion into the critical space or improperly oriented), or the spin trap is rapidly metabolized, hydrolyzed, or rearranged by the biological system it is sent to probe, or the spin adducts are too unstable to accumulate to levels detectable by EPR (see later), or the spin adducts undergo further reactions with radicals so that no EPR-active products are formed (double trapping, reduction or oxidation), or the spin adducts themselves are metabolized, hydrolyzed or rearranged by the systems that are present during radical creation. It's a harsh world out there for free radicals *and* for the spin trap probe sent out to find them.

2. If the EPR signal due to spin adducts from a given experiment is very weak a number of factors can be manipulated to help in the analysis. Obviously, varying the concentration of the spin trap through a wide range of soluble levels should be explored and the best concentration selected. Then the volume of sample and the choice of solvent for extraction (e.g., chloroform, benzene, hexane, chloroform/methanol etc.) should be varied. The best time for EPR scanning should be investigated. Usually immediately or as soon as possible is the best time, assuming the solution is homogenous and does not need settling of particulates.

When the chemical and biochemical aspects have been optimized the EPR spectroscopic factors need attention. The complete absence of oxygen from air must be ensured. Any transfer of solutions or needles open to air even for an instant allows enough air to enter the EPR cells so that oxygen broadening occurs. It should be remembered that oxygen in air is a ground state triplet molecule. This means these oxygen molecules have two unpaired electrons and as such are an excellent line-width broadening agent in EPR spectroscopy. Once oxygen broadening has been eliminated, and only then, it is worth the money to employ the use of deuterated spin traps.

The line sharpening and hence the increase in EPR signal intensity available from deuterated spin traps is impressive (Haire and Janzen, 1994). But it cannot be emphasized enough, that deuteration to improve EPR spectra is only useful if and when other factors controlling line width have been eliminated.

If deuteration does not improve the signal to noise characteristics of the spectrum, replacement of ^{14}N with ^{15}N in the spin trap will (Zhang and Janzen, unpublished work). Since ^{15}N has a nuclear spin of 1/2 and ^{14}N has a nuclear spin of 1, an enhancement of 50% is expected (and realized) in signal strength for the same spin trap. Although an increase in signal of 50% does not seem like much, sometimes this is enough to make the experiment worthwhile.

Finally, if the EPR spectrum is stable and the line width reasonably narrow, computer assisted accumulation of the weak spectrum is likely to be useful. If the EPR spectrum is not perfectly stable or other lines are growing in as a function of time, computer spectrum accumulation is not going to be helpful. Not every spectrum can be improved by computer assisted spectrum accumulation. In this sense, EPR spectroscopy is more limited than NMR spectroscopy.

It should also be made clear that EPR spectra must be presented as actual traces, or photographs of actual traces, or photocopies of actual traces, or computer output of actual data. An EPR spectrum (or any other spectrum for that matter) should never be traced by hand and presented for publication. It is more useful to allow some blemishes and embarrassing background signals or noise to be retained in the spectrum than to eliminate them from the record by hand-tracing or by an erasing mechanism. Sometimes the small "extra" peaks become the beginning of a more interesting story than the strong expected spectrum recorded.

In conclusion, when confronted with a very weak signal the main goal is to improve the EPR signal intensity. Hopefully it can be improved enough so that the spectrum can be assigned to a specific adduct reproducibly. It is highly unlikely that by manipulating the system that a whole variety of new spectra will be unearthed. This is because the only detectable spectrum is already very weak. For some reason the method has already reached its detection limit.

3. If the EPR spin adduct spectrum is rich and strong a variety of things are now possible. First, the concentration of the spin trap should again be varied through a wide range of values. This is not simply to increase the signal intensity of the spin adduct but rather at low concentrations to allow the radical normally trapped the freedom to do something else. If at high concentrations of spin trap only the radical of interest is trapped, the experiment only proves that that radical existed in the system. It says nothing about the possibility that that radical can produce any other perhaps interesting radicals. Therefore, at lower concentrations of spin trap the spectra may get more complicated. This is because the number of different spin adducts might increase. Life gets more interesting as it gets more complicated. So also with spin adduct spectra.

In general, spin traps must compete with other local reactants which seek to annihilate radicals. These competitive situations are well known in chemical free radical systems (Pryor, 1966):

$$R\cdot + ST \xrightarrow{k_1} R\text{–}SA\cdot \qquad [17]$$

$$R\cdot + TH \xrightarrow{k_2} R\text{–}TH\cdot \qquad [18]$$

$$R\text{–}TH\cdot + ST \xrightarrow{k_3} R\text{–}TH\text{–}SA\cdot \qquad [19]$$

$$R\cdot + TH \xrightarrow{k_4} T\cdot + RH \qquad [20]$$

$$T\cdot + ST \xrightarrow{k_5} T\text{–}SA\cdot \qquad [21]$$

One could ask what is the ratio of EPR signal due to the various spin adducts in this kind of competitive spin trapping situation? Assuming steady state free radical kinetics it is well documented in chemical systems that these equations are followed (Janzen et al., 1975):

$$\frac{d[R\text{–}SA\cdot]}{d[R\text{–}TH\text{–}SA\cdot]} = \frac{k_1}{k_2}\frac{[ST]}{[TH]} \qquad [22]$$

$$\frac{d[R\text{–}SA\cdot]}{d[T\text{–}SA\cdot]} = \frac{k_1}{k_4}\frac{[ST]}{[TH]} \qquad [23]$$

where TH is some kind of "tissue", R-TH· is a new radical resulting from addition of the radical to some function in the tissue, T· is a radical resulting from hydrogen abstraction from the tissue in question, and R-TH-SA· and T-SA· are spin adducts of "tissue-damage". Clearly as the concentration of spin trap is increased the amount of "normal" spin adduct is increased. However, at lower concentrations the opportunity for "damage-radical" spin adducts to appear is better and so other radical reactions now show their faces. Of course, there must always be enough spin trap present to allow trapping of the secondary radical. Therefore, there will always be a lower limit of spin trap concentration for competitive experiments.

With a strong EPR spin adduct signal it will also be possible to explore competitive experiments with added reagents, e.g., "antioxidants". The same competitive equations can be set up and the same kinetic expressions derived except that with luck the antioxidant radical is detectable by direct EPR methodology (e.g., tocopheroxyl in Vitamin E):

$$R\cdot \ + \ ST \ \xrightarrow{k_1} \ R\text{--}SA\cdot \qquad [17]$$

$$R\cdot \ + \ AH \ \xrightarrow{k_6} \ A\cdot + RH \qquad [24]$$

where AH is a hydrogen donor antioxidant.

$$\text{Then:} \qquad \frac{d[R\text{--}SA\cdot]}{d[A\cdot]} \ = \ \frac{k_1}{k_6} \frac{[ST]}{[AH]} \qquad [25]$$

The EPR signal intensity ratio should respond to changes in the relative concentrations of spin trap and antioxidant. This approach would prove that the free radical trapped is actually free to react with other reagents such as antioxidants in the system.

Choice of spin trap persistence

The spin trap should be persistent in the system under investigation. Only then can the true concentration of the spin trap remain functional for a spin trapping experiment. Something needs to be known about the chemical and biological stability of each spin trap as a function of time, and if free radical generation is to be monitored in a specific location such as an organ of the living system, the rate of build-up of the spin trap concentration and the rate of clearance should be known.

Since spin traps usually have polar functions (and in some cases even ionic groups), it is logical that reaction with nucleophiles is possible. Nucleophilic addition normally responds to concentration in a mass-action manner, but the rate can be enhanced by Lewis acid groups or ions. Thus we can expect some vulnerability to the following nucleophiles under any conditions of spin trapping in biological systems:

$$R\text{--}SH \ > \ R_2NH \ > \ R\text{--}OH > R\text{--}OOH \ > \ H\text{--}OO\text{--}H \ > \ H\text{--}O\text{--}H$$

That metal ions could catalyze the addition of nucleophiles such as acetate to spin traps was recognized in the first paper on spin trapping (Janzen and Blackburn, 1969). To date, no serious problems have been encountered with these possible reactions when the Lewis acid catalyst is at low concentration. However, with ferric iron for example, when the concentration is increased, nucleophilic addition of water seems to be enhanced and oxidation of the molecular addition product results in the EPR spectrum of the hydroxyl adduct (Makino et al., 1990). We are in the process of designing chiral spin traps which should allow a distinction to be made between real free radical trapping and molecular water addition followed by oxidation (N. Sankuratri and Janzen, unpublished work). However, these new spin traps are not ready for the market-place yet.

The logical final result for nucleophilic addition if no oxidation occurs is "hydrolysis" i.e., spin trap alteration to another compound (or other compounds) which do not function as spin traps. Hydrolysis of spin traps is also catalyzed by acidic conditions; hence the unsuitability of administering spin traps orally. Although *in vivo* spin trapping has been shown possible by I.G. administration of spin traps using a corn oil emulsion, the yield of spin adducts is lower than when the I.P. method is used (J.L. Poyer and E.G. Janzen, unpublished work). Hydrolysis of a spin trap could be viewed as a form of "metabolism" but we prefer to think of this process as chemical in nature.

Enzyme-promoted metabolism of spin traps does however seem to occur (Janzen et al., 1994a). Although the metabolic stability of spin traps is only now being investigated it is not surprising that the compounds normally chosen as spin traps should be vulnerable to oxygenating enzymes. Spin traps do have a special and very reactive double bond. Thus oxygen-atom transfers should be possible by enzymes designed to hydroxylate xenobiotics. This is the current mechanism being considered based on the assignment of structure of some spin trap metabolites.

Spin trap distribution in the live animal does not seem to be very specific (Chen et al., 1990). Thus only in adipose tissue is some concentration of spin trap retained. It appears this point is dependent on the lipophilicity of the spin trap. At this time, nothing is known in general about organ-specific spin trap distribution. Further work is needed in this area.

Analysis

Sensitivity
Analysis of the spin trapping products remains the *raison-d'être* of the method. If analysis of the effect of spin trap on the system is not possible there is no point in doing the experiment or reporting the attempt. As stated before a negative result cannot be used to give a positive conclusion.

Ideally the spin trapping product should be a stable adduct detectable *in situ* during the course of the experiment. This ideal situation is still hypothetical for *in vivo* spin trapping experiments but a few laboratories are getting close to obtaining the necessary equipment for such research. A number of approaches have been used to circumvent the inherent problems with this objective. If EPR is considered as the analytical tool for the detection of the spin adduct, the spectrometer must be able to accommodate a live animal, and the sensitivity of the system must be adequate for the detection of spin adducts produced *in vivo*. At this time no-one has achieved this level of performance for an EPR spectrometer, although moving the animal from a zone of free radical generation into the EPR spectrometer and subsequently searching for the spin adduct *in vivo* may be possible (H.H. Halpern, personal communication). Failing this, the only other approach is to remove body fluids from the living animal (e.g., blood, bile, liver perfusate, urine) or sacrificing the animal and homogenizing the tissue of interest. In either case the fluid or homogenate can be viewed directly by EPR or organic extractions can be performed. Since spin adducts seem more stable in non-polar organic solvents, extraction of body fluids or homogenized tissues has been the most successful route in most experiments. This points out the fact that there are actually very few detectable free radicals formed *in vivo* in the animals investigated. Organic solvent extraction serves to concentrate the few spin adduct molecules produced to the point where EPR analysis is possible.

Of course another interpretation is possible. Namely that lots of free radicals are formed but most are quenched by some other "natural" process and that spin traps compete very poorly with these processes. For example, frequently the ascorbyl radical is seen after a bout of oxidative stress in a variety of systems. Perhaps this is an indicator of the effectiveness of ascorbic acid (Vitamin C) in quenching a variety of water soluble radicals (or "repair" of tocopheroxyl radicals previously recruited to quench lipid soluble radicals). In any event EPR still appears to be the most sensitive analytical tool for the detection of spin adducts which are themselves radicals. This situation could change in the future depending on the development of better mass spectrometric or chromatographic techniques.

What about spin adducts which are not themselves free radicals and hence go undetected by EPR? It might be quite common that the spin adduct is changed in some way so that the radical character is lost, but the signature left by the free radical trapped could still remain. If this is the case the yield of spin adducts would have been underestimated as compared to the actual flux of free radicals produced.

The tools for addressing this possibility must be at least or perhaps even more sensitive than EPR. The only possibilities which come to mind are mass spectroscopy (MS) or/and high pressure liquid chromatography with electrochemical detection (EC-HPLC). Although MS has a reputation for very high sensitivity even with small amounts of sample, analytical MS has acquired this recognition through gas chromatographic MS with known rock-stable compounds, e.g., polychlorinated biphenyls (PCBs) and dioxins. The same sensitivity is not possible with room temperature HPLC methodology. In fact in the few cases explored, the HPLC-MS sensitivity is barely comparable to EPR (Iwahashi et al., 1993). This itself is no mean feat and should provide a companion methodology with EPR for compounds that are not free radicals. Somewhat the same situation exists for EC-HPLC. For known functions such as hydroxylated aromatic compounds EC-HPLC can be very sensitive when known compounds are monitored. Of course, HPLC does not give structural information and suffers from lack of information for unknown compounds.

Concentrated efforts combining MS with EC-HPLC could provide possible structural information for assignment but this exercise is not trivial.

It should be noted that at best NMR cannot reach the sensitivity of the above methods although
much more structural information can be obtained if suitable concentrations can be reached.

Provided we have adequate sensitivity in the analytical method of choice, what are the features of
the spin adduct most useful for the spin trap practitioner? Three characteristics are important:
stability of spin adduct, uniqueness of parameter, and lack of artefacts.

Stability

Infinite stability of all spin adducts was assumed in much of the early work done on spin trapping
in biological systems and in live animals. Unfortunately, it was only much later that some knowledge about this problem was acquired and it continues to be a question of paramount importance
at this time.

One can consider two kinds of stability: intrinsic stability and extrinsic stability. If a spin adduct
is intrinsically unstable this compound spontaneously decomposes and is itself a molecule with a
short life-time. The products of the decomposition may not be unique and thus no permanent
mark is left by the presence of the radical which was trapped. An example might be the hydroxyl
radical adduct of a spin trap. Upon decomposition the same products result as would be available
from hydrolysis. In such cases, it seems that polar environments are most damaging to the stability of these spin adducts. Thus sometimes rapid removal by organic solvent extraction allows
capturing of the spin adduct and preserving it long enough for analytical manipulation. Nevertheless it is our goal to synthesize spin traps for ultimate intrinsic stability of the spin adducts. It
is difficult enough to protect the spin adducts from external destructive forces without having to
worry about spontaneous self-destruction.

The lack of extrinsic stability in spin adducts could be due to a variety of factors. Mostly one
considers reduction or oxidation to be the main difficulty, although double spin trapping could
also happen. Spin adducts do seem to be subject to reduction to the hydroxylamine in spite of the
protection built into the molecule by having an N-*tert*-butyl group attached.

$$R_2NO\cdot \quad \xrightarrow{\text{[H]}} \quad R_2NOH \qquad \qquad [26]$$

It is still not clear what the actual reducing agent is in biological systems, and until this is known
it is difficult to prepare for this eventuality by synthesizing a prevention factor into the spin trap.
If prevention is not possible some spin trappers use oxidation as a cure. Thus the use of oxygen
or a mild oxidizing agent such as potassium ferricyanide in the last step before EPR analysis
presumably oxidizes all the hydroxylamine to the aminoxyl spin adduct. However, one should be

cautious about this step. If the spin adduct has a β-hydrogen, this carbon-hydrogen bond can also very easily be oxidized by these reagents. One should only consider the fact that a key step in preparing some nitrones is the copper (II) acetate oxidation in air of the precursor hydroxylamine with a β-hydrogen. Spin adduct hydroxylamines with a β-hydrogen are not the same as spin label hydroxylamines. Spin labels have no β-hydrogens. Thus oxidation of the hydroxylamines of spin labels is a safe procedure whereas oxidation of the hydroxylamine of a spin adduct with a β-hydrogen may not be advisable.

Oxidation of aminoxyls is also possible, as has been shown by electrochemical methods (Stronks et al., 1984).

$$R_2N{-}O{\cdot} \quad \xrightarrow{[O]} \quad R_2\overset{\oplus}{N}{=}O \qquad\qquad [27]$$

In typical spin trapping experiments this result would seem to sound the death knell of the spin adduct because either the β-hydrogen or some other group would dissociate itself from the spin adduct cation (nitroxonium ion) and cleavage or hydrolytic degradation would occur. Again, for spin labels this process might be tolerated because the aminoxyl has a cyclic structure, and hydride reduction may occur fast enough so that the hydroxylamine is formed. Until the mechanism of oxidation is known it is difficult to imagine what features to build into the spin trap so as to avoid oxidation of the spin adduct.

Ideally one would want the aminoxyl function totally protected from outside assault while at the same time leaving the site of free radical addition completely open for outside attack. Attempts to come up with the perfect molecule to serve as a spin trap under these conditions are still underway.

Another form of spin adduct decay is further reaction with more radicals. Radical-radical reactions could involve self-reactions (disproportionation), oxidation, reduction, double radical trapping or mixed double radical trapping. Although all of these possibilities have been experienced in biological systems it is not very likely that local radical concentrations would get high enough for serious problems of this type to be encountered. Nevertheless when MS or HPLC is used as an analytical tool one should be on the lookout for such products. They may also indicate encounters of spin traps with free radicals.

Unique parameters

Ideally it would be nice to have each spin adduct endowed with a unique set of parameters for every situation encountered. This would mean unique within EPR, MS and HPLC methodologies. Such spin traps do not exist although this does not stop us from trying to find them.

$$2RCHNR \xrightarrow{\quad\quad} RC{=}NR + RCHNR \qquad [28]$$

where the first RCHNR has $\overset{O\cdot}{|}$ group, products $RC{=}NR$ has $\overset{O^-}{|}$ and $+$ under N, and $RCHNR$ has $\overset{OH}{|}$

$$\overset{O\cdot}{\underset{|}{RCHNR}} + R\cdot \xrightarrow{\quad\quad} \overset{O^-}{\underset{|}{RC{=}\underset{+}{N}R}} + RH \qquad [29]$$

$$\overset{O\cdot}{\underset{|}{RCHNR}} + R\dot{C}HOH \xrightarrow{\quad\quad} \overset{OH}{\underset{|}{RCHNR}} + RCHO \qquad [30]$$

$$\overset{O\cdot}{\underset{|}{RCHNR}} + R\cdot \xrightarrow{\quad\quad} \overset{OR}{\underset{|}{RCHNR}} \qquad [31]$$

$$\overset{O\cdot}{\underset{|}{RCHNR}} + R'\cdot \xrightarrow{\quad\quad} \overset{OR'}{\underset{|}{RCHNR}} \qquad [32]$$

In EPR it is essential to have at least two parameters in order to try and assign a given spin adduct but even this is not enough. Although a large database of hyperfine splitting constants (hfsc's) of spin adducts is being accumulated based on essentially two EPR parameters, the aminoxyl nitrogen and the β-hydrogen hfsc (N and β-H hfsc), much overlap exists and assignments on the basis of two parameters alone is a risky business. This is because the range of values displayed for some families of spin adducts is relatively small and the effect of solvent can be as large (or larger) than the effect of different groups which were the radicals (radical addends). In an attempt to improve on this problem 100% ^{13}C-labeled nitrones have been synthesized so that the α-^{13}C-hfsc is also available as a third parameter for spin adduct structural assignment (Haire et al., 1988; Janzen et al., 1994b). With three parameters the chances of coming up with a more correct assignment for a spin adduct are much better. Moreover, the range of α-^{13}C-hfsc's surprisingly is much larger than the range for β-H hfsc's, and so this parameter may become more useful than the β-H hfsc in a final analysis of structure and assignment.

Unfortunately, however, all EPR parameters we depend on for making assignments of spin adducts are primarily dependent on each other. When the N-hfsc decreases for some reason the β-H and α-^{13}C-hfsc go down as well. For example, as the electronegativity of the radical addend group goes up all the EPR parameters, N-, β-H- and α-^{13}C-hfsc's go down. The explanation appears to be that for the normal aminoxyl function the unpaired electron is delocalized over the two aminoxyl atoms, N and O, approximately equally. When the electronegativity of the neighboring group attached to the nitrogen atom goes up, the spin on nitrogen goes down and moves to the oxygen atom. No hyperfine splitting influence comes from spin on the oxygen atom and thus the result is as if the spin effect on the hyperfine splitting of nitrogen, β-hydrogen and α-carbon-13 has decreased. However no-one has made spin traps from a nitrone with 100% ^{17}O-labeled nitronyl oxygen to test this hypothesis.

Of course if the entire structure assignment of the spin adduct rested on MS and/or chromatography one would not need hyperfine splitting parameter uniqueness. Perhaps in the future spin adducts will only need to report their presence by EPR and their structural identity will be disclosed by parent ion mass and fragmentation patterns.

Artefacts

It would be nice if life had no artefacts. Or would it? Maybe it would be boring. Spin trapping, as any other analytical technique, has its share of artefacts.

Artefacts can be considered in terms of different types. The most nauseating are those artefacts which come from impure reagents. Spin traps appear to be the most difficult of all compounds to obtain absolutely pure. Of the thousands of standard synthetic techniques used by practicing organic chemists, most create undesirable side products. Everyone knows this and lives with it. Organic chemicals with 98% purity are considered good commercial grade products. However, consider 2% EPR-active impurity. In an *in vitro* spin trapping experiment utilizing 0.01 M spin trap the 2% EPR impurity will produce a solution with a 2×10^{-4} M EPR-active impurity spectrum. This is completely intolerable when the actual spin adduct signal of interest is perhaps 1×10^{-6} or 1×10^{-7} M in concentration. This fact is not appreciated by typical synthetic organic chemists who do not do EPR. The synthetic methods in the older literature have no regard for the EPR purity of nitrones they have synthesized for other purposes. Today we must rethink each and every synthetic step for a preparation of pure nitrones, since it is very difficult to purify a nitrone after all steps have been executed. In fact in some cases purification makes the situation worse (Janzen et al., 1993b). Pure spin traps attract impurities. They must be protected from light, water, air, heat and environmental pollutants. Although the mechanism of impurity formation is not known, experience teaches that these precautions do help in keeping spin traps EPR-clean.

Artefacts can also come from the system itself under investigation. By this is meant that sometimes EPR-active species are detected which are formed "spontaneously" without the intervention of free radicals as such. For example, what if the spin trap preparation contained an impurity which was the hydroxylamine of some undesired compound which had come along in the preparation? Then simple exposure to oxygen in air would produce an EPR-active aminoxyl compound with its own spectrum. Of course oxidizing agents in the solution more aggressive than oxygen would do the same. This process would constitute a kind of artefact creation not attributable to a fault of the spin trapping methodology. Indeed it is likely that many wrong conclusions have been drawn and subsequently published because of impure commercial sources of spin traps. If a 98% purity spin trap will sell, why try and provide a purer product? The EPR background is not something that is checked by big commercial houses who sell spin traps.

Another form of artefact may come from dormant compounds in the system under investigation. For example if a polyunsaturated fatty acid lipid undergoes lipid peroxidation while standing around, it is possible to imagine that reactive hydroperoxides will accumulate in the oil waiting for something to happen. In fact it is well known that old ether bottles are explosive for this reason. Reactive hydroperoxides may react spontaneously with spin traps to give strong EPR signals; e.g., *meta*-chloroperbenzoic acid (Janzen et al., 1992b). Some artefacts may have this kind of origin. Interesting, but not the desired result. Again this is not the fault of spin trapping methodology. If one selects a sensitive technique for research, one must be prepared to accept some responsibility for understanding the process one is dealing with.

The influence a spin trap might have on a key component in the system which is needed to affect free radical generation might be considered an artefact. Thus enzyme inhibition might be one of these influences. For example, let us consider the possibility that an enzyme is an obligatory precursor of a possible free radical generating process, but the spin trap somehow inhibits the enzyme function:

$$Enz + \quad Substrate \quad \longrightarrow \quad R\cdot \qquad\qquad\qquad [33]$$
$$R\cdot + \quad ST \quad \longrightarrow \quad R\!-\!SA\cdot \qquad\qquad\qquad [34]$$
$$Enz + \quad ST \quad \longrightarrow \quad Enz\!-\!ST \text{ (stable complex)} \quad [35]$$

Then the spin trapping experiment would report no free radical generation but in this case it would not be due to the inability of the spin trap to detect the radicals produced. Rather the spin trap exhibits a special inhibitory effect on the system which has nothing to do with the actual spin trapping process. On the other hand, free radical management schemes of this type might be very desirable for oxidative stress therapy, since it would be expected to take much less spin trap to inhibit an enzyme than it would to run down all the radicals produced throughout the generating volume of the system. Notwithstanding this possibility it should be emphasized that in this case the fault does not rest on the shoulders of spin trapping. Spin traps like humans come with their own idiosyncrasies and faults. None are exempt.

Another potential "artefact" of the system is toxicity. If the animal is subjected to a dose of spin trap so that other physiological effects take control, the presence or absence of free radicals as evidenced by the detection of spin adducts should be interpreted with caution. The lethal doses of various spin traps administered I.P. have now been estimated albeit with very few animals (rats) (Janzen et al., submitted). The observed tolerance varies considerably. However, the structural factors most important in determining toxicity are not understood. It appears cyclic nitrones are more toxic than non-cyclic for the same lipophilicity. More detailed inspection of damage to organs caused by the spin traps is currently under study.

Final overview

How do currently available spin traps stack up against the increasingly stringent demands of the spin trapping community? It goes without saying that spin traps must be commercially available before the populace will make use of them. Most spin trappers are not synthetic organic chemists.

PBN

Spin trapping history began with PBN. This was the acronym used to describe C-phenyl N-*tert*-butyl nitrone. It seemed appropriate at the time to write the nitrone function with carbon on the left and nitrogen on the right, so that attack of the radical was a straight-ahead process. If so, then logically one would name the groups in the same order in which they are attached to the nitrone:

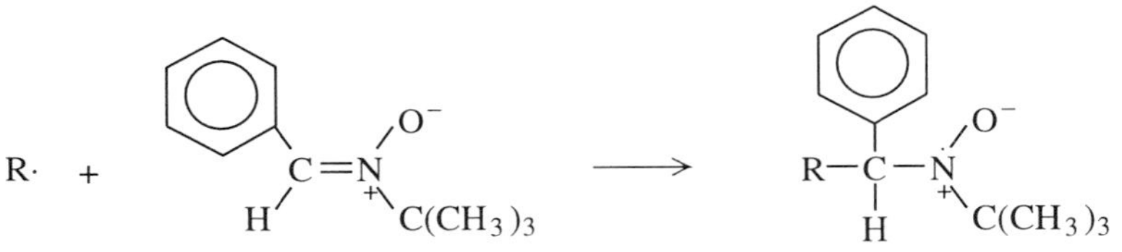

[36]

C-phenyl N- *tert* -butyl nitrone
PBN

However, nomenclature is not logical (or rather various nomenclatures dance to the tune of different nomenclaturologists). The IUPAC name should be with the groups listed in alphabetical order, namely N-*tert*-butyl C-phenyl nitrone, ignoring the non-alphabetic order of the atoms to which they are attached. However, fortunately (or unfortunately) PBN is now in the literature and BPN will never replace it!

PBN is an absolutely amazing molecule if all of its attributes can be believed. Space is too limited here to tell all of the stories concerning its prowess, but a review recently written on biological spin trapping (DeGray and Mason, 1994) can be consulted. It is perhaps useful to list those PBN derivatives and PBN-type nitrones which have become popular as spin traps, including those which are isotopically labeled for better EPR and MS performance. These compounds are now available from OMRF-STS (825 N.E. 13th, Oklahoma City, OK, USA 73104, phone (405) 271-7570):

PBN-d_9 (perdeuterated *tert*-butyl)

PBN-d_5 (perdeuterated phenyl)

PBN-d_{14} (perdeuterated phenyl and *tert*-butyl)

PBN-^{15}N (nitronyl nitrogen 100% ^{15}N)

As discussed above deuteration improves the line-width of the spin adduct EPR spectrum if every other line broadening factor has been eliminated. Deuteration of PBN is also excellent for keeping track of the spin adduct molecules in mass spectroscopy since modern equipment provides such very complicated spectra with a multitude of distracting peaks. The α-^{13}C hfsc can be diagnostic for assigning EPR spectra and ^{15}N-PBN provides 50% higher sensitivity for the same concentration of spin adducts (Y.K. Zhang and E.G. Janzen, unpublished work).

As a spin trap, PBN is extremely versatile. Almost all reactive radicals add to PBN. Over 30 different types of radical adducts have been studied and this is only a beginning (Haire et al., 1988). The rate constants for addition to PBN vary greatly however. For small reactive radicals, the rate constant is fast, but sterically-challenged radicals have considerable difficulty. The solubility of PBN is just about right. Its solubility in water is adequate (approximately 0.1 M saturated) but the partition coefficient with respect to 1-octanol is about 20:1 in favor of 1-octanol. In a lipid-bilayer mimic the nitrone function resides on a time-average basis near the polar region (Haire et al., 1989) although exchange is quite fast. The toxicity of PBN is relatively low. About 250 mg/100 g rat body weight kills the rat, but levels of PBN about 100 x less are still adequate for some spin trapping experiments (J.L. Poyer and E.G. Janzen, unpublished work). The extent of EPR artefactual problems is not too serious since PBN can be recrystallized with care to produce beautiful (but not easily water-soluble) crystals of high purity.

The major problem with PBN is the lack of stability of the spin adducts in aqueous solutions (for some radical addends) or in metabolically active biological systems. It appears that polar solvents such as water (or other solvents used in HPLC) enhance spontaneous decomposition of the spin adducts. In the case of the hydroxyl radical adduct of PBN we know α-carbon bond cleavage occurs (Janzen et al., 1992a) in aqueous solutions, and we assume this process also happens in other spin adducts:

$$\begin{matrix} HO \\ \quad\diagdown \quad \overset{O\cdot}{|} \\ H-C-N-C(CH_3)_3 \\ C_6H_5 \diagup \end{matrix} \quad\longrightarrow\quad C_6H_5CHO \quad + \quad \overset{\overset{O\cdot}{|}}{H-N-C(CH_3)_3} \qquad [37]$$

However, this process takes time and a quick organic solvent extraction can succeed in capturing some of the undecomposed spin adduct (Baker et al., 1994).

An interesting new result has recently been obtained in our laboratory where hydroxy-PBN (not to be confused with hydroxyl-PBN which is the name for the EPR-active spin adduct) with the hydroxy group attached to the phenyl ring is formed by rat liver microsomal metabolism of CCl4. It may be that when the hydroxyl radical approaches PBN it is somewhat ambivalent about its point of final attachment. Perhaps it initially nests in the aromatic-nitrone π-system and skitters around for a while before settling down to form a covalent bond to a carbon. In this process aromatic carbons may be almost as good as the nitronyl carbon and some hydroxyl radicals end up attached to the phenyl ring. This observation may lead to an analytical methodology similar to the use of salicylate or phenylalanine for detection of hydroxyl radicals (Powel, 1994). The advantage of this possibility is that a more permanent record of hydroxyl radical formation might be written, and since PBN probably samples a different region of the biological system perhaps true hydroxyl radical production in polar regions of actual membranes could be monitored.

4-POBN

In an attempt to improve on the capabilities of PBN we synthesized the pyridine-N-oxide analogue of PBN. Again we drew the molecule as follows and named the groups attached to the nitrone in a left to right manner:

$$[38]$$

C-(4-pyridinyl-N-oxide) N-*tert*-butyl nitrone
4-POBN

The acronym for C-(4-pyridinyl-N-oxide) N-*tert*-butyl nitrone was suggested as 4-PyOBN or 4-POBN or POBN for short. However, again like PBN, this is not an alphabetically correct name. The number position of 4-POBN should be retained however, since 3-POBN and 2-POBN have also been studied (Janzen et al., 1992a, Janzen et al., 1978b).

The virtue of 4-POBN is that the hydroxyl radical adduct has a distinguishing feature in the EPR spectrum. The hydroxyl hydrogen hfs can be resolved at very low levels of microwave power and modulation amplitude and slow scan (sweep width), provided oxygen or nitric oxide or other line-width enhancing reagents (like ferric iron) have been removed. The additional γ-H-hfs provides definitive proof of the assignment of the spin adduct structure.

However, few spin trappers have taken advantage of this feature. Other characteristics have been found to be more helpful such as longevity of spin adducts. Perhaps the net electron-withdrawing character of the pyridine-N-oxide function helps to stabilize the spin adduct since the transition state for the cleavage step probably has significant positive polar character. If this advantage is general we plan to explore more members of this family as spin traps.

DMPO

Probably the most popular spin trap for the search of hydroxyl radicals is 5,5-dimethylpyrroline-N-oxide. In the first paper demonstrating the general utility of this nitrone for spin trapping we christened this compound DMPO (Janzen and Liu, 1973). Now with more experience in the synthesis of pyrroline-N-oxide nitrones a more generic name is suggested, such as M_2PO (where M=methyl). Then M_3PO and M_4PO would have the structures shown below:

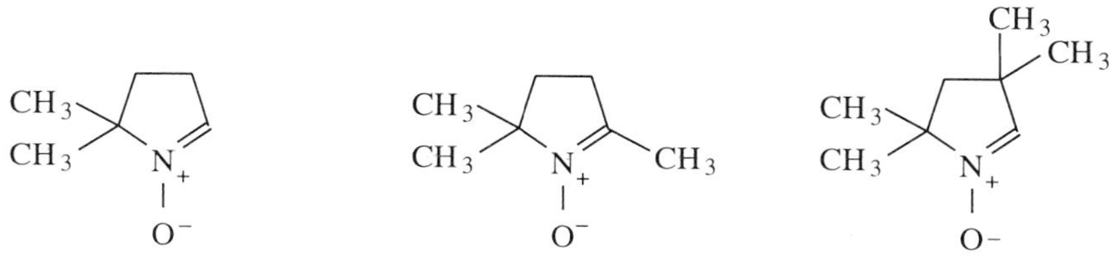

5,5-dimethylpyrroline-N-oxide
DMPO or M_2PO

2,5,5-trimethylpyrroline-N-oxide
M_3PO

3,3,5,5-tetramethylpyrroline-N-oxide
M_4PO

DMPO has almost become a household word in connection with the search for hydroxyl radicals. Probably its wide popularity for this purpose comes from the ease of recognizing the 4-line

EPR spectrum produced in aqueous solutions. Because in water there is accidental overlap of the ^{14}N triplet and the β-H hfs the typical 6-line spectrum reduces to only 4 lines with an intensity ratio of approximately 1:2:2:1.

However, DMPO is a molecule much too reactive to be left to its own devices. Artefactual sources of the hydroxyl adduct 4-line pattern abound (Tomasi and Iannone, 1993) and competitive spin trapping experiments had to be set up to sort this out. The argument was that if truly free hydroxyl radicals were trapped they should also be capable of abstracting the most reactive hydrogen from compounds such as ethanol. Then the 1-hydroxyethyl radical spin adduct spectrum should result simultaneously with the hydroxyl radical spectrum. The ratio of these spin adduct spectra should respond to the relative concentration variations in a logical way (see earlier equations). This clever method is now being implemented routinely and confirms the existence of true free hydroxyl radicals as evidenced by spin trapping.

Additional features of DMPO as a spin trap are also quite remarkable. The hydroperoxyl /superoxide adduct gives a distinctive spectrum due to γ-H hfs but unfortunately this adduct decays with production of the hydroxyl adduct spectrum in some systems. Also the rate constant of addition is relatively slow for hydroperoxyl/superoxide although probably still faster than for other spin traps. This characteristic may be improved by substitution of electron-withdrawing groups on DMPO but nucleophilic addition of water, amines or sulfides could become a problem.

The spin trapping chemistry of other 5-membered ring nitrones such as 2-phenyl DMPO (Janzen et al., 1994b; Janzen and Zhang, 1993) and 3-imidazoline-N-oxides (Klauschenz et al., 1993) is currently under investigation by various groups. Their usefulness in general for biological systems needs to be investigated and elaborated before common utility is known.

MGD

A new "unconventional" type of spin trap has recently become recognized for the specific purpose of detecting nitric oxide. Since nitric oxide is too stable to be trapped by nitrones, coordination to a metal ion center is necessary in order to capture this molecule. Surprisingly, dithiocarbamate complexes of iron (II) coordinate with nitric oxide to produce a new complex which is EPR-active at room temperature (Mordvintcev et al., 1991). Very few iron complexes are EPR-active in the usual sense; i.e., most give either very broad or infinitely broad EPR lines. In the case of N-methylglucamine dithiocarbamate, MGD, a triplet spectrum with N-hfsc = 12.6 Gauss is obtained from ^{14}N nitric oxide with a line width of 4.0 Gauss. This kind of EPR spectrum is quite manageable and serves quite nicely as a tool for the detection of nitric oxide (Komarov et al., 1993):

Conclusion

It has been said that spin trapping has not proved anything that was not already known in free radical biology. It all depends on how one defines "known". When enough people believe in something it becomes "known" or accepted. This is called a religion. However, proposing, or suggesting, or postulating, or believing, or dreaming, that free radicals exist in the system, or even are responsible for some damage in the system, does not necessarily make it so. Free radical biology must provide proof for its precepts or else it is not a practicing science, only an art form. If free radicals are produced, prove it; then also prove that the free radicals are obligatory intermediates for the "free radical damage" presumably identified by the observer. Then, and only then, can scientific progress be made.

Acknowledgement
Spin trapping has been generously supported over the years by the Oklahoma Medical Research Foundation, The University of Guelph and the University of Georgia (where it all began). The current sources of funding are the NIH Center for Research Resources (Grant # RR05517) and the Natural Sciences and Engineering Research Council of Canada. This paper could not have been prepared without the excellent assistance and encouragement of Michelle Evans and Dr. Yashige Kotake.
The wonderful hospitality of the Yamagata Conference organizers and generous financial support from sponsors of the meeting will be fondly remembered by my wife and myself. Grateful acknowledgement for all of the above is hereby offered to all those who make this work possible.

References

Albano, E.F., Lott, A.K., Slater, T.F., Stier, A., Symons, M.C.R. and Tomasi, A. (1982) *Biochem. J.* 204: 593–603.
Baker, J.E., Konorer, E.A., Tse, S.Y.H., Joseph, S. and Kalyanaraman, B. (1994) *Free Rad. Res.* 20: 145–163.
Chen, G., Bray, T.M., Janzen, E.G. and McCay, P.B. (1990) *Free Rad. Res. Comm.* 9: 317–323.
Constantin, D., Mehrotra, K., Jernstrom, B., Tomasi, P. and Moldeus, P. (1994) *Pharmacol. Toxicol.* 74: 136–140.
Davis, E.R. and Janzen, E.G. (1974) The sulfur dioxide anion radical can be detected by EPR directly: Janzen, E.G. (1972) *J. Phys. Chem.* 76: 157–162.
DeGray, J.A. and Mason, P.B. (1994) Biological Spin Trapping: A Specialist Periodical Report. The Royal Society of Chemistry, *Electron Spin Resonance* 14: 246–303.
Finkelstein, E., Rosen, G.M. and Rauckman, E.J. (1980) *J. Am. Chem. Soc.* 102: 4994–4999.
Greenstock, C.L. and Wiebe, R.H. (1982) *Can. J. Chem.* 60: 1560–1564.
Griller, D. and Ingold, K.U. (1973) *J. Am. Chem. Soc.* 95: 6459–6466.
Griller, D., Kaba, R.A. and Ingold, K.U. (1974) *J. Am. Chem. Soc.* 96: 6202.

Haire, D.L., Oehler, U.M., Krygsman, P.H. and Janzen, E.G. (1988) *J. Org. Chem.* 53: 4535–4542.
Haire, D.L., Janzen, E.G., Coulter, G.A., Hilborn, J.W., Towner, R.A., Krygsman, P.H. and Stronks, H.J. (1989) *J. Org. Chem.* 54: 2915–2920.
Haire, D.L. and Matsuda, M. (1994) *Mag. Res. Chem.* 32: 151–157.
Ito, O. and Matsuda, M. (1984) *Bull. Chem. Soc. Japan* 57: 1745–1749.
Iwahashi, H., Parker, C., Mason, R. and Tomer, K. (1993) *Free Rad. Biol. Med.* 15: 337–342.
Janzen, E.G. and Blackburn, B.J. (1969) *J. Am. Chem. Soc.* 91: 4481–4490.
Janzen, E.G. and Evans, C.A. (1973) *J. Am. Chem. Soc.* 95: 8205–8206.
Janzen, E.G. and Liu, J.I.-P. (1973) *J. Mag. Res.* 9: 510–512.
Janzen, E.G., Nutter, D.E. and Evans, C.A. (1975) *J. Phys. Chem.* 79: 1983–1984.
Janzen, E.G., Evans, C.A. and Davis, E.R. (1978a) *In*: W.A. Pryor (ed.): *American Chemical Society Symposium Series* 69: 433.
Janzen, E.G., Wang, Y.Y. and Shetty, R.V. (1978b) *J. Am. Chem. Soc.* 100: 2923.
Janzen, E.G. and Haire, D.L. (1990) *In*: D.D. Tanner *Advances in Free Radical Chemistry*, JAI Press Inc, Greenwich, Conn., USA, pp 253–295.
Janzen, E.G., Krygsman, P.H., Lindsay, D.A. and Haire, D.L. (1990a) *J. Am. Chem. Soc.* 112: 8279–8284.
Janzen, E.G., Towner, R.A. and Yamashiro, S. (1990b) *Free Rad. Res. Comm.* 9: 325–335.
Janzen, E.G., Kotake, Y. and Hinton, R.D. (1992a) *Free Rad. Biol. Med.* 12: 169–173.
Janzen, E.G., Lin, C.R. and Hinton, R.D. (1992b) *J. Org. Chem.* 57: 1633–1635.
Janzen, E.G. and Zhang, Y.K. (1993) *J. Mag. Res.* 101B: 91–93.
Janzen, E.G., Wilcox, A.L. and Manoharan, V. (1993a) *J. Org. Chem.* 58: 3597–3599.
Janzen, E.G., Zhang, Y.K. and Arimura, M. (1993b) *Chemistry Letters* 3: 497–500.
Janzen, E.G., Chen, G., Bray, T.M., Lindsay, D.A., Lloyd, S. and McCay, P.B. (1994a) *In*: R.P. Mason (ed.): *Free Radicals in the Environment Medicine and Toxicology: Critical Aspects and Current Highlights*, pp 327–344.
Janzen, E.G., Zhang, Y.K. and Haire, D.L. (1994b) *J. Am. Chem. Soc.* 116: 3738–3743.
Janzen, E.G., Poyer, J.L., West, M.S., Crossley, C. and McCay, P.B. (1994c) *J. Biochem. Biophys. Meth.* 29: 189–205.
Janzen, E.G., Poyer, J.L., Schaefer, C.F., Downs, P.E. and DuBose, C.M. *J. Biochem. Biophys. Methods*; *submitted*.
Kalyanaraman, B., Mottley, C. and Mason, R.P. (1984) *J. Biochem. Biophys. Meth.* 9: 27–31.
Klauschenz, E., Haseloff, R.F., Volodarskii, L.B. and Blasig, I.E. (1993) *Free Rad. Res.* 20: 103–111.
Knecht, K.T. and Mason, R.P. (1993) *Arch. Biochem. Biophys.* 303: 185–194.
Knechtel, J.R., Janzen, E.G. and Davis, E.R. (1978) *J. Am. Chem. Soc.* 100: 2923.
Komarov, A., Mattson, D., Jones, M.M., Singh, P.K. and Lai, C.-S. (1993) *Biochem. Biophys. Res. Comm.* 195: 1191–1198.
Krusic, P.J., Chen, K.S., Meakin, P. and Kochi, J. (1974) *J. Phys. Chem.* 78: 2036–2047.
Madden, K.P. and Taniguchi, H. (1991) *J. Am. Chem. Soc.* 113: 5541–5547.
Makino, K., Hagiwara, T., Hagi, A., Nishi, M. and Murakami, A. (1990) *Biochem. Biophys. Res. Commun.* 172: 1073–1080.
McCay, P.B., Lai, E.K., Poyer, J.L., DuBose, C.M. and Janzen, E.G. (1984) *J. Biol. Chem.* 259: 2135–2143.
Mordvintcev, P., Mulsch, R., Busse, R. and Vanin, A.F. (1991) *Anal. Biochem.* 199: 142–146.
Powel, S.R., (1994) *Free Rad. Res.* 21: 355–370.
Pryor, W.A. (1966) *Free Radicals*, McGraw-Hill, New York, NY, USA, pp 311–319.
Scaiano, J.C. and Ingold, K.U. (1976) *J. Am. Chem. Soc.* 98: 4727–4732.
Sridhar, R., Beaumont, P.C. and Powers, E.L. (1986) *J. Radioanalyt. Nucl. Chem.* 101: 227–237.
Stronks, H.J., Janzen, E.G. and Weber, J.R. (1984) *Analytical Letters* 17: 321–328.
Terabe, S., Kuruma, K. and Konaka, R. (1972) *Chem. Lett.* 2: 115–118.
Terabe, S. and Konaka, R. (1973) *J. Chem. Soc. Perkin Trans.* 2: 369–374.
Tomasi, A. and Iannone, A. (1993) *EMR of Paramagnetic Molecules*, Plenum Press, New York, USA, pp 353–384.
Wilcox, A.L. and Janzen, E.G. (1993) *J. Chem. Soc. Chem. Commun.* 18: 1377–1378.

Bioradicals Detected by ESR Spectroscopy
H. Ohya-Nishiguchi & L. Packer (eds)
© 1995 Birkhäuser Verlag Basel/Switzerland

History of spin trapping in Japan

R. Konaka

Institute for Life Support Technology, Yamagata Technopolis Foundation, Numagi, Yamagata 990, Japan

Summary. Since the spin trapping technique was first investigated by several research groups around the world in the late 1960s, many studies have been reported in a diversity of fields. In this review, the history of spin trapping in Japan i.e., development of spin traps, general studies of spin trapping, development of the spin trapping technique and application of spin trapping to various fields is described.

Early stages of spin trapping in Japan

The notion that the addition reaction of radicals to nitroso or nitrone compounds could have analytical utility for the detection and identification of free radicals was independently proposed in 1968 and 1969 by five groups in different parts of the world: Forshult et al. (1969) in Sweden, Chalfont et al. (1968) in England, Leaver and Ramsey (1969) in Australia, Terabe and Konaka (1968, 1969) in Japan, and Janzen and Blackburn (1968, 1969) in USA.

Prior to this, in 1967; Iwamura and Inamoto (1967) reported that the 2-cyanopropyl radical adds to the carbon atom of phenyl *tert*-butyl nitrone (PBN) to produce a stable nitroxide, and a cyclic nitrone, 5,5-dimethyl-1-pyrroline N-oxide (DMPO), exhibited the same radical addition reaction to produce a stable cyclic nitroxide.

Janzen and Blackburn (1968, 1969) paid attention to this study and in 1968 developed a general means of detecting short-lived free radicals using these nitrones. They named the technique of detecting and identifying short-lived free radicals by addition to an unsaturated function to produce an ESR detectable radical, "Spin Trapping".

In another approach, Maruyama et al. (1964) observed the ESR spectra of diphenylnitroxides during the photo-decomposition of several nitrosobenzene derivatives in tetrahydrofuran.

Terabe and Konaka (1968, 1969) applied the ESR technique using nitrosobenzene as a scavenger to detect and identify free radical intermediates during the oxidation of various kinds of hydrocarbons with nickel peroxide (Ni-PO).

$$RH \xrightarrow{Ni-PO} R \cdot \xrightarrow{PhNO} \underset{O\cdot}{\overset{}{\langle\!\!\langle \quad \rangle\!\!\rangle}}\!\!-\overset{}{\underset{|}{N}}\!\!-R$$

They determined the ESR parameters of the nitroxides produced, and showed this technique to be a general means applicable to the detection and identification of short-lived free radicals. This work was initially presented at the 7th ESR symposium in Japan in 1968, and preliminarily published in 1969. A more detailed paper on this study is also available (Terabe and Konaka, 1972).

Development of spin traps

Because of the complexity of spectra, the disadvantage of nitrosobenzene as a spin trap is obvious.Therefore, Konaka's group subsequently pursued better spin traps than nitrosobenzene itself. They synthesized several kinds of nitrosobenzene derivative and evaluated the utility of these compounds for spin trapping. Among these derivatives, 2,4,6-tri-*tert*-butyl- nitrosobenzene (TBN) (Terabe and Konaka, 1971,1973 a) and nitrosodurene (ND) (Terabe et al., 1973) were particularly useful.

TBN functions as an ambient spin trap. Thus, primary alkyl radicals add to form nitroxides in the normal way, but with bulky molecules like tertiary alkyl radicals, addition occurs at the oxygen with alkoxyaminyl radicals being formed. Secondary alkyl radicals give mixtures of both species (Fig. 1).

ND, which produces relatively simple spectra of spin adducts, has three significant advantages over 2-methyl-2-nitrosopropane (MNP). First, the scatter of hfscs of β-hydrogen is wide, and therefore individual spectra are easiliy distinguishable from each other. Second, ND is more reac-

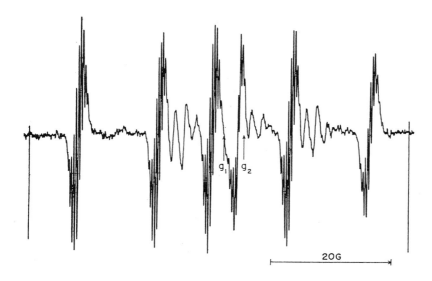

Figure 1. ESR spectrum of a mixture of 2,4,6-tri-*tert*-butylphenyl isopropyl nitroxide and N-isopropoxy-2,4,6-tri-*tert*-butylanilino radical in benzene solution at 2.5 h after spin trapping of the isopropyl radical by use of TBN.

tive toward radical addition. Finally, it is not sensitive to visible light. An example of the spectrum of a spin adduct is shown in Figure 2.

Doba et al. (1977a) sythesized pentamethylnitrosobenzene (PMNB) and evaluated its usefulness for spin trapping. PMNB was found to show similar properties to ND.

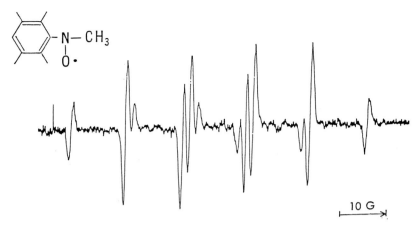

Figure 2. ESR spectrum of the spin adduct of methyl radical produced in photolysis of di-*t*-butyl peroxide with ND in benzene. g = 2.0060.

Although ND is an excellent spin trap, it cannot be used in aqueous systems due to its insolubility in water. Konaka and Sakata (1982) synthesized sodium 2,4-dimethyl-3-nitrosobenzenesulfonate (DMNS), which is a deformation compound of ND, as well as its perdeuterio compound. They found them efficient spin traps in aqueous solutions. An example of such a spin adduct spectrum is shown in Figure 3.

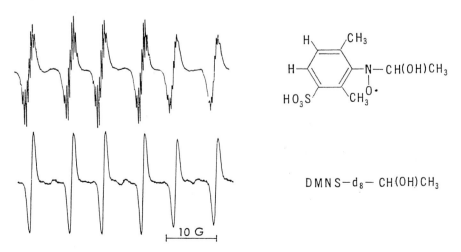

Figure 3. ESR spectrum of the α-hydroxyethyl spin adducts of DMNS (upper) and DMNS-d$_8$ (lower) obtained from ethanol by photolysis of an aqueous H$_2$O$_2$ solution at room temperature.

Because ND is an effective spin trap, duryl nitrones such as methyl-N-durylnitrone (MDN) and
N-durylnitrone (DN) can be expected to be useful nitrone-type spin traps rather like PBN.

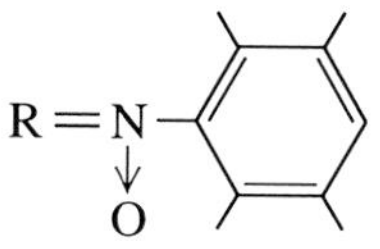

$$R=N-\underset{O}{\overset{\downarrow}{\Big|}}\bigcirc$$

R : CH_2 (DN)

R : CH_3CH (MDN)

Konaka et al. (1982) synthesized these nitrones and evaluated their availability for spin trapping.
   They found that the spin adducts of the *tert*-butyl peroxy radical and the *tert*-butoxy radical
gave spectra showing considerably different β-hydrogen hfscs. The spin adducts of the hydro-
peroxy and hydroxyl radicals also showed similar behavior.

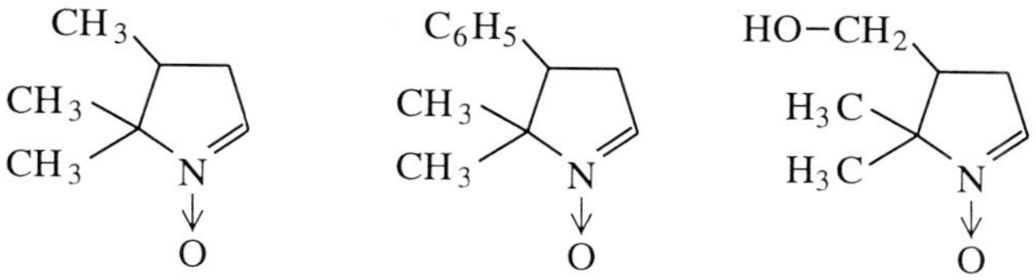

4MDMPO   4PDMPO   4HMDMPO

Further, Konaka et al. (1993, 1995) reported that spin traps substituted with some groups at the
4-position of DMPO were comparable to DMPO itself regarding their abilities as spin traps and
their physical properties such as melting points and partition coefficients. In conclusion,
4MDMPO, 4PDMPO and 4HMDMPO were recommended as useful spin traps in addition to
DMPO.
   Recently, polymer-supported spin traps, poly[N-(*p*-vinylbenzylidene)-*tert*-butylamine N-oxide]
(polyPBN), and its copolymers with styrene or methyl methacrylate (copolyPBNs) were synthesized, and their spin trapping abilities were investigated by Ren et al. (1994). They found that
both polyPBN and copolyPBNs in the solid state could capture carbon- and oxygen-centered
radicals in the heterogeneous phase (solid-liquid system).

Poly PBN

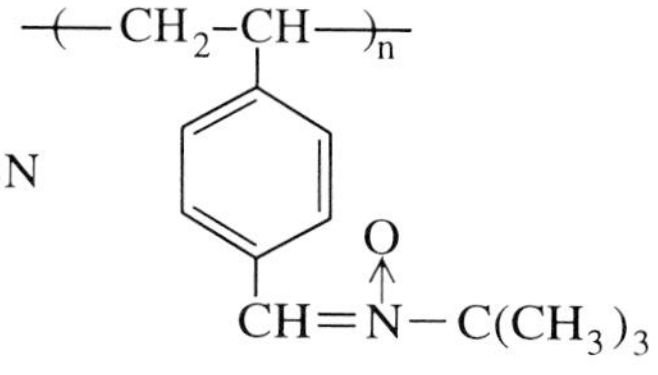

$$\left(\!\!-CH_2-CH-\right)_n$$

$CH=\overset{O}{\overset{\uparrow}{N}}-C(CH_3)_3$

General studies of spin trapping

Terabe and Konaka (1973b) explored the hyperfine splitting constants of β-fluorine atoms in aryl perfluoroalkyl nitroxides, $ArN(R_f)O\cdot$, with the spin trapping method. A discussion of the conformational dependence of β-fluorine splitting constants showed that the $\cos^2\theta$ low is also applicable to β-fluorine splitting constants for a series of nitroxides.

$$R_fI \quad + \quad A_rN{=}O \quad \xrightarrow{h\nu} \quad A_r\text{-}\underset{\underset{O\cdot}{|}}{N}\text{-}R_f$$

I	A_r : 2,4,6-$Bu^t{}_3C_6H_2$	R_f :	
II	A_r : C_6H_5	a)	CF_3
III	A_r : C_6F_5	b)	CF_2CF_3
IV	A_r : 2,3,5,6-$(CH_3)_4C_6H$	c)	$CF(CF_3)_2$

Doba et al. (1977b; Doba and Yoshida, 1982) measured the rate constants of trapping the *tert*-butyl radical with several spin traps, and concluded that the spin trapping rate constants of the nitroso compounds varied from 2.3×10^5 to 2.0×10^8 mol^{-1}dm^{3}s^{-1} or more toward the *tert*-butyl radical in benzene at 299 K, and PBN was less efficient in trapping than the nitroso compounds. Furthermore, they determined the trapping rate constants of commonly used spin traps for the cyclohexyl radical.

Makino et al. (1981) reported a fundamental study of aqueous solutions of MNP as a spin trap using product analysis. The reaction mechanism for the light-irradiated aqueous solution is summarized in the following scheme:

$$\begin{aligned}
\text{dimer MNP} \quad &\longrightarrow \quad tBuN(OH)\text{-}N{=}O \ + \ tBuOH \ + \ (CH_3)_2C{=}CH_2 \\
\downarrow \qquad\qquad& \\
\text{monomer MNP} \quad &\longrightarrow \quad tBuNO_2 \\
| \qquad\qquad& \\
&\longrightarrow \quad tBu\cdot + \ NO\cdot \\
\text{monomer MNP}& \\
tBu\cdot \quad &\longrightarrow \quad tBu_2N\text{-}O\cdot
\end{aligned}$$

Makino et al. (1991) resolved in detail the reaction mechanism of DMPO on the Fenton system. In conclusion, 1) the DMPO-OH adduct is generated by the nucleophilic attack of H_2O on the DMPO-Fe(III) complex, and 2) the DMPO-OH adduct produces additional nitroxides under the oxidative conditions, so that it causes the appearance of the background spectra during spin trapping on the Fenton system.

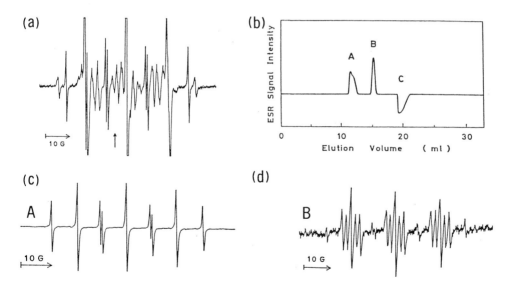

Development of the spin trapping technique

Hatano and his colleagues developed the ESR-HPLC combined technique which is a useful and general method for the analysis of mixture spectra by spin trapping. Initially, Kominami et al. (1976) studied short-lived radicals in the γ-irradiated aqueous solution of uridine-5'-monophosphate by the ESR-HPLC method. Since then many studies involving spin trapping of the short-

Figure 4. (a) ESR spectrum of an aqueous solution of glycine with MNP observed just after γ irradiation (at pH 6.0). During chromatography the magnetic field was fixed at the position indicated by the vertical arrow. (b) Chromatogram of the γ-irradiated aqueous solution of glycine with MNP drawn by ESR detection. (c) ESR spectrum of the separated spin adduct obtained from the fractions giving peak A in Figure (b). (d) ESR spectrum of the separated spin adduct obtained from the fractions giving peak B in Figure (b).

lived radicals formed in γ-irradiated aqueous solutions of amino acids, peptides and nucleic bases using this method were reported by Hatano's group. For an example, Moriya et al. (1980) reported the analysis of spin trapping with MNP upon γ-irradiation of an aqueous solution of glycine (Fig. 4).

Hiramatsu and Kohno (1987) established the superoxide detection method in the hypoxanthine and xanthine oxidase system with spin trapping by use of DMPO, and they developed the method for analyzing superoxide dismutase activity using this method. Since then these methods have been widely used for biological studies and medicine in Japan (Fig. 5).

Application of spin trapping

Since 1972, many researchers in Japan have been applying the spin trapping technique to various fields. Several of the papers are described below.

Organic chemistry

Suehiro et al. (1976) reported that aroyloxycyclohexadienyl radicals, the key intermediates in homolytic aromatic aroyloxylation, were detected and identified during the decomposition of bis-aroyl peroxides in benzene-d_6 as spin adducts of ND.

The spin trapping technique using PBN was applied to the radiation chemistry of gaseous mixtures of carbon monoxide and hydrogen irradiated with electron beams (Nagai et al., 1978). ESR spectra of the spin adducts provided evidence for the formation of H atoms and methyl radicals.

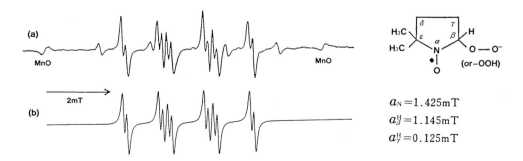

$a_N = 1.425 \text{mT}$

$a_\beta^H = 1.145 \text{mT}$

$a_\gamma^H = 0.125 \text{mT}$

Figure 5. (a) ESR spectrum of DMPO-O_2^- spin adduct. (b) Simulated ESR spectrum of Figure (a).

Niki et al. (1983) studied the spin trapping of *tert*-butylperoxy and tetralylperoxy radicals by PBN and MDN. They reported that the spin adducts of peroxy radicals by MDN could be clearly distinguished from that of an alkoxy radical. The nitrone adducts of oxygen radicals were found to be reasonably stable at room temperature in the dark, but readily decayed in ordinary room light.

Okazaki et al. (1985) applied the spin trapping techniqe using PBN and DMNS during the photoreduction of naphthoquinone in a SDS micellar solution in an external magnetic field. The effects of the magnetic field on the photochemical reaction were successfully interpreted in terms of the radical pair model, in which the intersystem crossing rate between triplet and singlet radical pairs is influenced by an external magnetic field.

Polymer chemistry

Otsu and his colleagues published many papers on the application of spin tapping to polymer chemistry. For example, Sato et al. (1973) investigated the initiation mechanism of radical polymerization resulting from the initiator system of triethylboron and oxygen by means of the spin trapping technique using MNP, PBN and 3,5-di-*t*-butyl-4-hydroxyphenyl-N-*t*-butylnitrone (BHPBN). They observed both ethoxy and ethyl radical adducts in every case, and concluded that the ethoxy radical was produced initially and the ethyl radical was formed through the subsequent reaction of the ethoxy radical with triethylboron. As another example, Sato and Otsu (1977) reported the investigation of the initiation reaction of various vinyl monomers with the *tert*-butoxyl radical, using the spn trapping technique with MNP to detect and identify the intermediate radical species, and to evaluate the monomer reactivities toward the *tert*-butoxyl radical.

Ouchi et al. (1984) studied the initiation mechanism in the radical polymerization of vinyl monomers by polyethylene glycol (PEG-300) in aqueous solution by means of the spin trapping technique using DMNS-d_8, and concluded that the initiating radical species were generated by hydrogen atom transfer from the monomer adsorbed at the ether group of PEG-300 to the free monomer.

Biology and medicine

Recently, strong interest has been turned on the intermediates of free radical species in biological systems. Since 1977 the spin trapping technique has been noted as a useful method for evaluating the behavior of bio-radicals. Since the 10th Conference of Magnetic Resonance in Medicine was held in 1978, many researchers in Japan have been applying the spin trapping technique in the biological and medical fields, and now they are using it to advance bio-radical research.

Fujita (1987) observed radicals in the rat brain mitochondria under various conditions by spin trapping with DMPO, and described the usefulness of this method and presented problems to be resolved.

Naito et al. (1989) reported the semi-quantitative measurement of superoxide radical in polymorphonuclear leukocytes by spin trapping with DMPO and discussed the effects of various stimulatives.

Mitsuta et al. (1990) evaluated the SOD-like activity, which is defined as the 50% inhibitory dose (ID_{50}) for superoxide radical generation by the competition method, for several biological substances by means of spin trapping using DMPO.

Tanigawa et al. (1990) observed superoxide radicals in a solution of fresh main stream cigarette smoke by applying the trapping technique using DMPO. They also detected the thiyl radical in dithiothreitol-induced protein gastroenteropathy using spin trapping with DMPO.

Ueta and Ogura (1990) applied spin trapping using DMPO to determine the generation of superoxide radicals in submitochondrial particles prepared from ischemic heart, and found it to be higher in the particles obtained from the ischemic region than in those from non-ischemic regions.

Shiroto et al. (1991) examined the effects of irrigation solvents and hyaluronic acid on free radical production in jaw cyst cavities *in vitro* using the spin trapping technique. They found that the addition of solcoseryl decreased the DMPO-O_2^- relative signal intensity, while the DMPO-OH signal was decreased by the addition of hyaluronic acid and glucose.

Sonoda et al. (1993) detected free radicals in tissue samples of the rabbit gastric mucosa by spin trapping with PBN, but only during reperfusion and the milk-intake period. They concluded that gastric rupture could be initiated by the action of oxygen radicals followed by lipid peroxidation and promoted by stomach acid.

Togashi et al. (1994) reported that superoxide and hydroxyl radicals were produced after reperfusion of isolated and perfused rat livers subjected to global ischemia after showing that these species were trapped by DMPO, and they also concluded that the production of these radicals occurs at an early stage of reperfusion in ischemic liver and that the amount of production is closely related to the duration of ischemia.

In recent years, many substances have been evaluated for free radical scavenging activity or antioxidant properties by means of the spin trapping technique, e.g., tannins and related polyphenols by Hatano et al. (1989), some Chinese medicines by Wang et al. (1991), Liu et al. (1992), and Takahashi et al. (1992), dipyridamole by Hiramatsu et al. (1992), and rebamipide by Naito et al. (1992).

References

Chalfont, G.R., Perkins, M.J. and Horsfield, A. (1968) A probe for homolytic reactions in solution. II. The polymerization of styrene. *J. Am. Chem. Soc.* 90: 7141–7142.

Doba, T., Ichikawa, T. and Yoshida, H. (1977a) Pentamethylnitrosobenzene as a spin trapping agent. *Bull. Chem. Soc. Japan.* 50: 3124–3126.

Doba, T., Ichikawa, T. and Yoshida, H. (1977b) Kinetic studies of spin trapping reaction. I. The trapping of the *t*-butyl radical generation by the photodissociation of 2-methyl-2-nitrosopropane by several spin-trapping agents. *Bull. Chem. Soc. Japan.* 50: 3158–3163.

Doba, T. and Yoshida, H. (1982) Kinetic studies of spin trapping reaction. III. Rate constants for spin trapping of the cyclohexexyl radical. *Bull. Chem. Soc. Japan.* 55: 1753–1755.

Forshult, S., Lagercrantz, C. and Torssell, K. (1969) Use of nitroso compounds as scavengers for the study of short-lived free radicals in organic reactions. *Acta Chem. Scand.* 23: 522–530.

Fujita, Y. (1987) Radicals observed in the rat brain mitochondria by spin trapping method. *Free Radicals in Clinical Medicine* 2: 39–49.

Hatano, T., Edamatsu, R., Hiramatsu, M., Mori, A., Fujita, Y., Yasuhara, T., Yoshida, T. and Okuda, T. (1989) Effects of the interaction of tannins with co-existing substances on superoxide anion radical, and on DPPH radical. *Chem. Pharm. Bull.* 37: 2016–2021.

Hiramatsu, M. and Kohno, M. (1987) Determination of superoxide dismutase activity by electron spin resonance spectrometry using the spin trap method. *JEOL NEWS* 23 A: 7–9.

Hiramatsu, M., Edamatsu, R., Kadowaki, D., Okamura, Y., Kanakura, K. and Mori, A. (1992) Dipyridamole quenched free radicals and inhibited lipid peroxidation. *Magnetic Resonance in Medicine* 3: 129–133.

Iwamura, M. and Inamoto, N. (1967) Novel formation of nitroxide radicals by radical addition to nitrones. *Bull. Chem. Soc. Japan* 40: 702–703.

Janzen, E.G. and Blackburn, B.J. (1968) Detection and identification of short-lived free radicals by an electron spin resonance trapping technique. *J. Am. Chem. Soc.* 90: 5909–5910.

Janzen, E.G. and Blackburn, B.J. (1969) Detection and identification of short-lived free radicals by electron spin resonance trapping techniques (Spin Trapping). Photolysis of organolead, -tin, and -mercury compounds. *J. Am. Chem. Soc.* 91: 4481–4490.

Kominami, S., Rokushika, S. and Hatano, H. (1976) Studies of short-lived radicals in the γ-irradiated aqueous solution of uridine-5'-monophosphate by the spin trapping method and the liquid chromatography. *Int. J. Radiat. Biol.* 30: 525–534.

Konaka, R. and Sakata, S. (1982) Spin trapping by use of a water-soluble nitroso compound. *Chemistry Letters* 411–414.

Konaka, R.,Terabe, S., Mizuta, T. and Sakata, S. (1982) Spin trapping by use of nitrosodurene and its derivatives. *Can. J. Chem.* 60: 1532–1541.

Konaka, R., Abe, M., Noda, H. and Kohno, M. (1993) Studies on development of new spin traps. Synthesis and evaluation of DMPO-type spin traps. *Magnetic Resonance in Medicine* 4: 17–20.

Konaka, R., Kawai, M., Noda, H., Kohno, M. and Niwa, R. (1995) Synthesis and evaluation of DMPO-type spin traps. *Free Rad. Res.* 23: 5–25.

Leaver, I.H. and Ramsay, G.C. (1969) Trapping of radical intermediates in the photoreduction of benzophenone. *Tetrahedron* 25: 5669–5675.

Liu, J., Wang, X., Hiramatsu, M. and Mori, A. (1992) Free radical scavenging activity of Kanglaojianshenye. A possible mechanism for its antiaging effect. *Magnetic Resonance in Medicine* 3: 134–137.

Makino, K., Suzuki, N., Moriya, F., Rokushika, S. and Hatano, H. (1981) A fundamental study on aqueous solutions of 2-methyl-2-nitrosopropane as a spin trap. *Radiation Research* 86: 294–310.

Makino, K., Hagiwara, T., Hagi, A., Takeuchi, T. and Murakami, A. (1991) Reaction in DMPO spin trapping induced by Fe ions. *Magnetic Resonance in Medicine* 2: 128–132.

Maruyama, K., Tanikaga, R. and Goto, R. (1964) The appearance of free radicals during the photo-decomposition of nitrosobenzenes. *Bull. Chem. Soc. Japan* 37: 1893–1894.

Mitsuta, K., Mizuta, Y., Kohno, M., Hiramatsu, M. and Mori, A. (1990) The application of ESR spin-trapping technique to the evaluation of SOD-like activity of biological substances. *Bull. Chem. Soc. Japan* 63: 187–191.

Moriya, F., Makino, K., Suzuki, N., Rokushika, S. and Hatano, H. (1980) Studies on spin-trapped radicals in γ-irradiated aqueous solutions of glycine and L-alanine by high-performance liquid chromatography and ESR spectroscopy. *J. Phys. Chem.* 84: 3085–3090.

Nagai, S., Matsuda, K. and Hatada, M. (1978) Application of the spin trapping technique to the study of radiation effects on gaseous mixture of carbon monoxide and hydrogen. *J. Phys. Chem.* 82: 322–325.

Naito, Y., Yoshikawa, T., Tanigawa, T., Oyamada, Y., Ueda, S., Takemura, T., Tainaka, K., Morita, Y., Sugino, S. and Kondo, M. (1989) In: H. Nishikawa and T. Yoshikawa (eds): *ESR and Free Radicals*, First Edition, NIHON IGAKUKAN Inc., Tokyo, pp 115–120.

Naito, Y., Yoshikawa, T., Kokura, S., Tomii, T., Tsujigiwa, M., Takahashi, S., Ichikawa, H., Takano, H., Yasuda, M., Tasaki, N., Tanigawa, T. and Kondo, M. (1992) Free radical scavenging by rebamipide: An ESR assay. *Magnetic Resonance in Medicine* 3: 145–148.

Niki, E., Yokoi, S., Tuschiya, J. and Kamiya, Y. (1983) Spin trapping of peroxy radicals by phenyl-N-(*tert*-butyl)nitrone and methyl-N-durylnitrone. *J. Am. Chem. Soc.* 105: 1498–1503.

Okazaki, M., Sakata, S., Konaka, R. and Shiga, T. (1985) Application of spin trapping to probe the radical pair model in magnetic-field-dependent photoreduction of naphthoquinone in SDS micellar solution. *J. Am. Chem. Soc.* 107: 7214–7216.

Ouchi, T., Hosaka, Y. and Imoto, M. (1984) Vinyl Polymerization. Part 422. Initiation mechanism of uncatalyzed polymerization by polyethyleneglycol. *J. Polymer Sci.* 22: 1507–1514.

Ren, J., Sakakibara, K. and Hirota, M. (1994) Synthesis and spin-trapping properties of poly[N-(p-vinylbenzylidene)-tert.-butylamine N-oxide] and its copolymers as a solid spin trap. *Reactive Polymers* 22: 107–114.

Sato, T., Hibino, K., Fukumura, N. and Otsu, T. (1973) A study of the initiator system of trialkylboron and oxygen by the spin trapping technique. *Chem. Ind.* 4: 745–746.

Sato, T. and Otsu, T. (1977) Application of spin trapping technique to radical polymerization, 14. Initiation reaction of vinyl monomers with the *tert*-butoxyl radical and evaluation of monomer reactivities. *Makromol. Chem.* 178: 1941–1950.

Shiroto, M., Watanabe, S., Kimura, H. and Suzuki, M. (1991) Effects of irrigation solvent and components of jaw cyst fluid on free radical generation system. *Magnetic Resonance in Medicine* 2: 147–151.

Sonoda, M., Matsuki, M., Asakuno, G., Satomi, A., Ishida, K. and Sakagishi, Y. (1993) Free radical production during ischemia-reperfusion injury in rabbit gastric mucosa. *Magnetic Resonance in Medicine* 4: 107–110.

Suehiro, T., Kamimori, M., Sakuragi, H., Yoshida, M. and Tokumaru, A. (1976) Spin trapping of benzoyloxycyclohexadienyl radicals by 2,3,5,6-tetramethylnitrosobenzene (nitrosodurene) in the decomposition of dibenzoyl peroxide in benzene. *Bull. Chem. Soc. Japan* 49: 2594–2595.

Takahashi, S., Yoshikawa, T., Ichikawa, H., Naito, Y.,Tanigawa, T., Kokura, S., Tomii, T. and Kondo, M. (1992) Studies on superoxide scavenging activity of crude drugs. *Magnetic Resonance in Medicine* 3: 149–152.

Tanigawa, T., Yoshikawa, T. and Kondo, M. (1990) Application of electron spin resonance to clinical medicine. *Free Radicals in Clinical Medicine* 5: 37–40.

Terabe, S. and Konaka, R. (1968) ESR of radicals produced on the reaction of nitroso-compounds. *7th ESR Symposium (Sapporo)* Abstract pp 44.

Terabe, S. and Konaka, R. (1969) Electron spin resonance studies on oxidation with Nickel peroxide. Spin trapping of free-radical intermediates. *J. Am. Chem. Soc.* 91: 5655–5657.

Terabe, S. and Konaka, R. (1971) Spin trapping of short-lived free radicals by use of 2,4,6-tri-*tert*-butylnitrosobenzene. *J. Am. Chem. Soc.* 93: 4306–4307.

Terabe, S. and Konaka, R. (1972) Spin trapping by use of nitroso-compounds. Part IV. Electron spin resonance studies on oxidation with nickel peroxide. *J. Chem. Soc. Perkin Trans. II.* 2163–2172.

Terabe, S. and Konaka, R. (1973a) Spin trapping by use of nitroso-compounds. Part V. 2,4,6-tri-t-butylnitrosobenzene: a new type of spin-trapping reagent. *J. Chem. Soc. Perkin Trans. II.* 369–374.

Terabe, S. and Konaka, R. (1973b) Spin trapping by use of nitroso-compounds. VII. β-fluorine splitting constants in nitroxides. *Bull. Chem. Soc. Japan* 46: 825–829.

Terabe, S., Kuruma, K. and Konaka, R. (1973) Spin trapping by use of nitroso-compounds. Part VI. Nitrosodurene and other nitrosobenzene derivatives. *J. Chem. Soc. Perkin Trans. II.* 1252–1258.

Togashi, H. Shinzawa, H., Yong, H., Takahashi, T., Noda, H., Oikawa, K. and Kamada, H. (1994) Ascorbic acid radical, superoxide, and hydroxyl radical are detected in reperfusion injury of rat liver using electron spin resonance spectroscopy. *Arch. Biochem. Biophys.* 308: 1–7.

Ueta, H. and Ogura, R. (1990) Detection of superoxide anion generated from submitochondrial particles prepared from ischemic heart by electron spin resonance spin trapping. *Magnetic Resonance in Medicine* 1: 87–91.

Wang, G., Nomoto, K., Nakamura, K., Kumano, K. and Sakai, T. (1991) Mimicking of superoxide dismutase by extracts from the traditional chinese medicines analysed by ESR. *Magnetic Resonance in Medicine* 2: 215–218.

Bioradicals Detected by ESR Spectroscopy
H. Ohya-Nishiguchi & L. Packer (eds)
© 1995 Birkhäuser Verlag Basel/Switzerland

In vivo ESR observation of bioradical metabolites in living animals

H. Fujii, J. Koscielniak and L.J. Berliner

Department of Chemistry, The Ohio State University, Laboratory of In vivo Electron Spin Resonance Spectroscopy, Columbus, Ohio 43210, USA

Summary. L-band ESR spectroscopy with a loop-gap resonator demonstrated the first *in vivo* detection of a "bioradical", generated from the metabolism of nitrosobenzene in live mice. A broad three-line ESR spectrum ($a_N = 11.6$ G) was detected in the buttocks or stomach region of a mouse after intramuscular or intraperitoneal injection of 0.2 mmol/kg nitrosobenzene. The signal intensity reached a maximum at 20 to 30 min and remained constant well beyond twelve hours. When muscle tissue was doped with nitrosobenzene and excised within 5 min, a similar three-line spectrum was also detected at X-band, which was preceded by the rapid growth and subsequent decay of a spectrum identical to that of the phenylhydronitroxide radical. A model system containing nitrosobenzene and unsaturated fatty acids yielded an identical three-line spectrum which came from the radical adduct of nitrosobenzene across double bonds. These results suggest that one of the first possible targets of nitroso compounds *in vivo* may be regions of polyunsaturated fatty acid clusters in fat or membranes. From the success in detecting "bioradicals" generated by pollutants *in vivo*, two- and three-dimensional visualization of localized bioradicals seems quite feasible.

Introduction

Historically, the detection of free radicals in biological systems has been fraught with problems: e.g, the tissue excision protocol and the need to freeze quickly to 77 K may be accompanied by ESR signal artefacts (Hutchison et al., 1971). In order to correlate the presence of some paramagnetic intermediate with a metabolic or disease state, the first *in vivo* ESR study was reported by Feldman et al. (1975) who implanted a helix antenna into the regenerated liver of a live rat and observed exogenous nitroxide radicals in the hepatic circulation at X-band (9.5 GHz). The unacceptability of dielectric microwave heating at 9.5 GHz and the geometric size constraints at this frequency prompted us to develop alternative strategies at much lower frequencies, especially L-band and lower radio frequencies (Berliner and Fujii, 1985).

The use of *in vivo* L-band ESR has expanded rapidly since the mid-1980s, beginning with monitoring and imaging of nitroxides in plant samples (Berliner and Fujii, 1985). This was followed by work on live animals, such as the monitoring of nitroxide distributions in melanoma tumors (Berliner et al., 1987; Berliner and Fujii, 1992) and nitroxide pharmacokinetics in rats or organs (Berliner and Wan, 1989; Bacic et al., 1989; Utsumi et al., 1990). Several groups have reported the development and application of specialized low field ESR spectrometers for animal studies (Ono et al., 1986; Ishida et al., 1989; Halpern and Bowman, 1991; Chzhan et al., 1992).

More recently several low field imaging studies of whole mice or organs have been reported (Ishida et al., 1992; Kuppusamy et al., 1994). However, the absolute sensitivity reported for most L-band spectrometers to date has been poor for most realistic *in vivo* experiments, thus most applications have used nitroxide spin labels as the vehicle for observing ESR signals in animals. Unfortunately, the introduction of exogenous nitroxide spin labels at concentrations of 10^{-3} M and above raises several concerns about long term toxicity or other side effects. Until recently, there was only one previous demonstration reporting the direct measurement of naturally occurring free radicals *in vivo*, namely the observation of melanin radicals in a melanoma tumor (Lukiewicz and Lukiewicz, 1984). Presently, the L-band bridge/loop gap resonator assembly at Ohio State University (Koscielniak and Berliner, 1994) has achieved the highest reported sensitivity, which is currently about 0.5 µM 2,2,5,5-tetramethylpyrrolidine-1-oxyl-3-carboxamide (CTPO). With such remarkable sensitivity, it should be feasible and desirable to observe several other "intrinsic bioradicals" which are directly involved in free radical processes *in vivo* instead of the more common protocol which requires following external imaging agents or probes in order to obtain decent signal to noise ratios.

Nitrosobenzene is an important environmental pollutant which is widespread in the environment. Furthermore, its directly related higher and lower oxidation states, nitrobenzene, phenylhydroxylamine and aniline, are all interconvertible to nitrosobenzene through hepatic and other metabolic pathways *in vivo* according to the report of Mason's group (Maples et al., 1990). While a wealth of *ex vivo* tissue extraction studies and model chemical systems have been exploited in the past, a direct *in vivo* study has not been possible until recently (Fujii et al., 1994a). In this

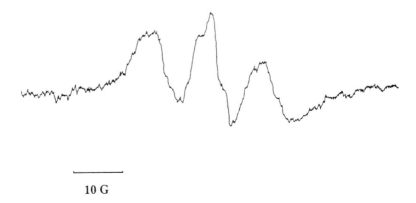

10 G

Figure 1. L-band ESR spectrum of an anesthetized mouse at room temperature 20 min after administration of nitrosobenzene. A 50 mM (0.1 ml) nitrosobenzene suspension in DMPC liposomes or ethanol was injected i.p. or i.m. as a 0.2 mmol/kg dose into 25 g balb/c male mice. The anesthetized mouse was monitored in the modified loop-gap resonator/L-band bridge apparatus described above (Koscielniak and Berliner, 1994). Spectrometer conditions were: frequency, 1.128 GHz; microwave power, 10 mW; modulation, 1.5 gauss.

chapter, we review the first detection of "bioradicals" generated in metabolic interconversions ofnitrosobenzene in living mice.

In vivo mouse experiment

Figure 1 depicts the resultant ESR spectrum of the buttock muscle tissue region measured at 1.128 GHz which reached a maximum in 20–30 min and remained stable at an unusually high intensity for several hours to days. The signal intensity, measured at only 10 mW applied power, reflects about 50 µM radical. As a control to check whether the injection or other artefact was responsible for this spectrum, the mouse was injected with ethanol or liposome suspension alone where no ESR signals were detected.

In vitro muscle experiment

A buttock muscle tissue sample was excised 5 min after injection of 0.1 ml ethanolic 50 mM nitrosobenzene into a mouse, and monitored by X-band ESR. A broad 11.6 ± 0.5 G hyperfine

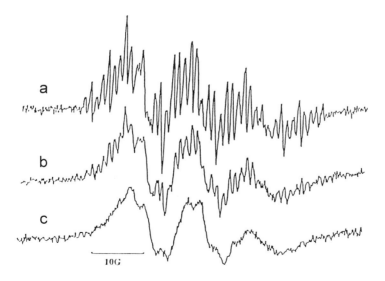

Figure 2. X-band ESR spectra of excised mouse buttock tissue 5 min after *in vivo* intramuscular administration of 0.1 ml 50 mM nitrosobenzene into the mouse buttocks: (a) 20 min, (b) 80 min, and (c) 200 min after injection. From Fujii et al. (1994a) with permission.

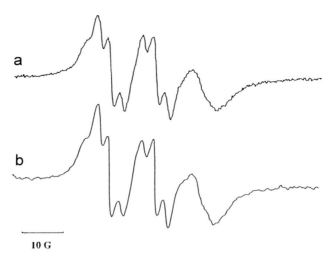

10 G

Figure 3. X-band ESR spectra of nitrosobenzene in buttock tissue or a lipid model system: (a) Excised muscle tissue 200 min after injection (50 mW microwave power); (b) Reaction product obtained 120 min after mixing 0.2 ml olive oil with 10 μl of 50 mM nitrosobenzene.

spectrum was eventually obtained which was stable over at least 4 h. As shown in Figure 2, first a complex hyperfine pattern appeared (Fig. 2a), which slowly converted to a broad three-line spectrum over 85 to 130 min (Fig. 2b). By 150 min the spectrum had entirely converted to the broad three-line pattern (Fig. 2c), which was very similar to that observed in Figure 1 at L-band. The initial multiline spectral pattern detected in the first stages (Fig. 2a) of the experiment was identical to that of a phenylhydronitroxide radical obtained from mixtures of nitrosobenzene and reducing agents such as NAD(P)H or ascorbic acid, both commonly found in biological systems (Takahashi et al., 1988). Some physico-chemical properties of this radical are discussed later in this chapter.

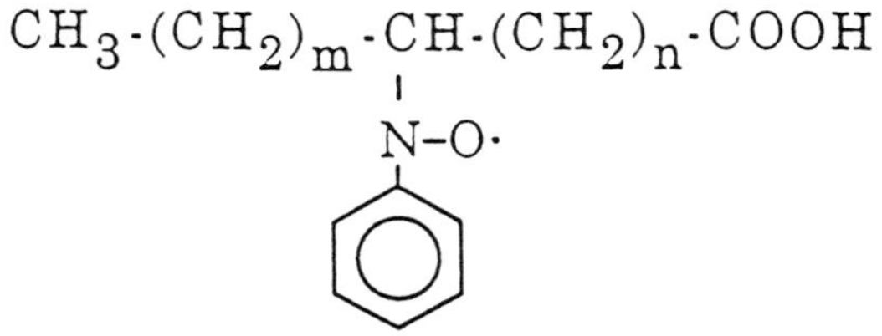

$$CH_3 \cdot (CH_2)_m \cdot CH \cdot (CH_2)_n \cdot COOH$$

$$N-O \cdot$$

Figure 4. Proposed structure of the nitrosobenzene radical adduct.

In order to identify and characterize further the bioradical associated with this broad three-line spectrum, both the whole mouse and excised tissue (Fig. 2c) spectra were compared at different microwave powers. At relatively high microwave power, e.g., 50 mW, several additional hyperfine components became visible (Fig. 3a), which appeared to be identical to published spectra from model systems containing nitrosobenzene and oleic acid. For comparison, the ESR spectrum resulting 120 min after mixing nitrosobenzene with olive oil is shown in Figure 3b. Remarkably, this ESR spectrum (Fig. 3b) was essentially identical to that of excised muscle tissue (Fig. 2c), strongly suggesting that the broad three-line spectrum was due to a bioradical which is the adduct of nitrosobenzene with unsaturated fatty acids.

The same broad three-line spectrum was also obtained at L-band for the nitrosobenzene/olive oil model system and for a sample of porcine animal fat injected with nitrosobenzene, respectively. Notably, both samples yielded essentially identical line shapes, which were also quite similar to the whole mouse three-line spectrum in Figure 1 (data not shown).

Several previous reports have shown that model reactions between nitrosobenzene and polyunsaturated fatty acids proceed by a pseudo Diels-Alder mechanism yielding a free radical adduct (Sullivan, 1966; Iwashita et al., 1991). The proposed chemical structure of this nitrosobenzene radical adduct, observed both *in vivo* and in the model lipid system, is depicted in Figure 4.

Nitrosobenzene reactions *in vitro*

In vitro nitrosobenzene chemistry was examined in more detail, particularly after evidence for the transient production of phenylhydronitroxide radicals was found in excised tissue, red blood cells and red blood cell hemolysates (Fujii et al., 1994b). Interestingly, there exists evidence for both aerobic and anaerobic pathways of nitrosobenzene reduction/oxidation in addition to the lipid radical adduct pathway described earlier. The most significant observation was evidence for the generation of superoxide radical anion as a direct result of oxidation of phenylhydronitroxide radicals by molecular oxygen back to nitrosobenzene. The scheme below (Fig. 5) summarizes the related aerobic and anaerobic pathways of nitrosobenzene metabolism *in vitro*: Nitrosobenzene (Phe-NO) is reduced to the phenylhydronitroxide radical (Phe-NHO·) *via* a one-electron pathway. Under aerobic conditions molecular oxygen reoxidizes the radical back to diamagnetic nitrosobenzene. Thus a continuous cycle occurs with a steady state generation of superoxide, while the equilibrium lies predominantly towards the phenylhydronitroxide species. The generated Phe-NHO· and O lasted for a relatively long period (up to 60 min) until most of the oxygen in the system was consumed by Phe-NHO·. In the alternative case where a two-electron reduction of nitrosobenzene to Phe-NH(OH) occurred, a one-electron oxidation may occur concurrently, generating superoxide, and resulting in the production of phenylhydronitroxide radicals. Under anaerobic conditions, where the oxidation reaction is essentially nonexistent, a two-electron reduction can occur driving the equilibrium to the diamagnetic hydroxylamine, consistent with the lack of an ESR signal in oxygen depleted solutions of nitrosobenzene and various reducing agents. Lastly, the bottom pathway in Figure 5 represents the nitroxide radical adduct reaction across double bonds of unsaturated lipids in fatty tissue with Phe-NO, accompanied by an interim generation and decay of Phe-NHO·. At the present time, the physiological implication of these nitroxide radical adducts is unknown; however, this product was shown in model systems to be

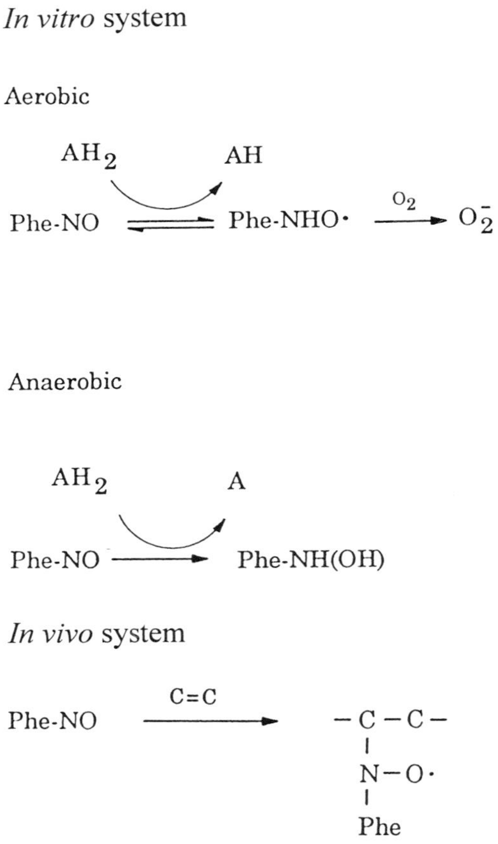

In vitro system

Aerobic

$$AH_2 \qquad AH$$

$$Phe\text{-}NO \rightleftharpoons Phe\text{-}NHO\cdot \xrightarrow{O_2} O_2^-$$

Anaerobic

$$AH_2 \qquad A$$

$$Phe\text{-}NO \longrightarrow Phe\text{-}NH(OH)$$

In vivo system

$$Phe\text{-}NO \xrightarrow{C=C} \begin{array}{c} -C-C- \\ | \\ N-O\cdot \\ | \\ Phe \end{array}$$

Figure 5. Schematic representation of the *in vitro* oxidation/reduction equilibria of nitrosobenzene under aerobic and anaerobic conditions, and the *in vivo* metabolic fate of nitrosobenzene. Abbreviations used are: Phe-NH(OH), phenylhydroxylamine; Phe-NHO·, phenylhydronitroxide radical; Phe-NO, nitrosobenzene; AH2, reductant (such as NAD(P)H).

capable of hydrogen abstraction from fatty acids, generating a carbon-centered lipid peroxidation *in vivo* system leading to radicals (Stier et al., 1982).

Conclusions

The metabolic physiology of aromatic nitroso compounds is broad and plentiful. In particular, the *in vivo* fate of these potent xenobiotics is of intense interest as a result of their potential entry into humans and animals by many environmental pathways. The application of high sensitivity localized *in vivo* ESR spectroscopy has now been proven as a valuable and necessary technology for non-invasive monitoring of the metabolic fate of xenobiotics in living systems (Knecht and Mason, 1993; Fujii et al., 1994a). With success in detecting bioradicals generated *in vivo*, more elaborate two- and three-dimensional visualization (imaging) of localized bioradicals should be feasible in the next few years (Eaton et al., 1991).

Acknowledgements
This work was supported in part by grants from the Ministry of Education, Science and Culture of Japan (HF), and the U.S. National Institutes of Health (LJB).

References

Bacic, G., Nilges, M.J., Magin, R.L. and Swartz, H.M. (1989) *In vivo* localized ESR spectroscopy reflecting metabolism. *Magn. Reson. Med.* 10: 266–272.
Berliner, L.J. and Fujii, H. (1985) Magnetic resonance imaging of biological specimens by electron paramagnetic resonance of nitroxide spin labels. *Science* 227: 517–519.
Berliner, L.J., Fujii, H., Wan, X. and Lukiewicz, S.J. (1987) Feasibility study of imaging a living murine tumour by electron paramagnetic resonance. *Magn. Reson. Med.* 4: 380–384.
Berliner, L.J. and Wan, X. (1989) *In vivo* pharmacokinetics by electron paramagnetic resonance spectroscopy. *Magn. Reson. Med.* 9: 430–434.
Berliner, L.J. and Fujii, H. (1992) Some applications of ESR to *in vivo* animal studies and EPR imaging. *In*: L.J. Berliner and J. Reuben (eds): *Biological Magnetic Resonance*. Vol. 11, Plenum Press, New York, pp 307–319.
Chzhan, M., Shteynbuk, M., Kuppusamy, P. and Zweier, J.L. (1992) An optimized L-band ceramic resonator for EPR imaging of biological samples. *J. Magn. Reson.* 105: 49–53.
Eaton, G.R., Eaton, S.S. and Ohno, K. (1991) *EPR imaging and in vivo EPR*, CRC Press, Boca Raton, FL.
Fujii, H., Koscielniak, J. and Berliner, L.J. (1994a) *In vivo* ESR studies of the metabolic fate of nitrosobenzene in the mouse. *Magn. Reson. Med.* 31: 77–80.
Fujii, H., Koscielniak, J., Kakinuma, K. and Berliner, L.J. (1994b) Biological and model reduction reactions of aromatic nitroso compounds: Evidence for the involvement of superoxide anions. *Free Radical Res. Commun.* 21: 235–243.
Feldman, A., Wildman, E., Bartolini, G. and Piette, L.H. (1975) *In vivo* electron spin resonance in rats. *Phys. Med. Biol.* 20: 602–612.
Halpern, H.J. and Bowman, M.K. (1991) Low-frequency EPR spectrometers: MHz Range. *In:* G.R. Eaton, S.S. Eaton and K. Ohno (eds): *EPR Imaging and In vivo EPR*, CRC press, Boca Raton, FL, pp 45–63.
Hutchison, J.M., Foster, M.A. and Mallard, J.R. (1971) Description of anomalous ESR signals from normal rabbit liver. *Phys. Med. Biol.* 16: 655–661.

Ishida, S., Kumashiro, H., Tsuchihashi, N., Ogata, T., Ono, M., Kamada, H. and Yoshida, E. (1989) *In vivo* analysis of nitroxide radicals injected into small animals by L-band ESR technique. *Phys. Med. Biol.* 34: 1317–1323.

Ishida, S., Matsumoto, S., Yokoyama, H., Mori, N., Kumashiro, H., Tsuchihashi, N., Ogata, T., Yamada, M., Ono, M., Kitajima, T., Kamada, H. and Yoshida, E. (1992) An ESR- CT imaging of the head of a living rat receiving an administration of a nitroxide radical. *Magn. Reson. Imag.* 10:109–114.

Iwashita, H., Parker, C.E., Mason, R.P. and Tomer, K.B. (1991) Radical adducts of nitrosobenzene and 2-methyl-2-nitrosopropane acid with 12,13-epoxylinoleic acid radical, 12,13-epoxylinolenic and 14,15-epoxyarachidonic acid radical. *Biochem. J.* 276: 447–453.

Knecht, K.T. and Mason, R.P. (1993) *In vivo* spin trapping of xenobiotic free radical metabolites. *Arch. Biochem. Biophys.* 303: 185–194.

Koscielniak, J. and Berliner, L.J. (1994) Dual diode detector for homodyne EPR microwave bridges. *Rev. Sci. Instr.* 65: 2227–2230.

Kuppusamy, P., Chzhan, M., Vij, K., Shteynbuk, M., Lefer, D.J., Giannella, E. and Zweier, J.L. (1994) Three-dimensional spectral-spatial EPR imaging of free radicals in the heart: A technique for imaging tissue metabolism and oxygenation. *Proc. Natl. Acad. Sci. USA* 91: 3388–3392.

Lukiewicz, S.J. and Lukiewicz, S.G. (1984) *In vivo* ESR spectroscopy of large biological objects. *Magn. Reson. Med.* 1: 297–298.

Maples, R., Eyer, P. and Mason, R.P. (1990) Aniline-, pheylhydroxylamine-, nitrosobenzene- and nitrobenzene-induced hemoglobin thiyl free radical formation *in vivo* and *in vitro*. *Mol. Pharmacol.* 37: 311–318.

Ono, M., Ogata, T., Hsieh, K.-C., Suzuki, M., Yoshida, E. and Kamada, H. (1986) L-band ESR spectrometer using a loop-gap resonator for *in vivo* animals. *Chem. Lett.* 1986: 491–494.

Stier, A., Clauss, R., Lucke, A. and Retiz, I. (1982) Radicals in carcinogenesis by aromatic amines. *In:* D.C.H. McBrien and T.F. Slater (eds): *Free Radicals, Lipid Peroxidation and Cancer*, Academic Press, London, pp 327–343.

Sullivan, A.B. (1966) Electron spin resonance studies of a stable arylnitroso-olefin adduct free radical. *J. Org. Chem.* 31: 2811–2817.

Takahashi, N., Fischer, V., Schreiber, J. and Mason, R.P. (1988) An ESR study of nonenzymatic reactions of nitroso compounds with biological reducing agents. *Free. Rad. Res. Commun.* 4: 351–358.

Utsumi, H., Muto, E., Masuda, S. and Hamada, A. (1990) *In vivo* ESR measurement of free radicals in whole mice. *Biochem. Biophys. Res. Commun.* 172: 1342–1348.

Bioradicals Detected by ESR Spectroscopy
H. Ohya-Nishiguchi & L. Packer (eds)
© 1995 Birkhäuser Verlag Basel/Switzerland

Dithiocarbamate spin traps for *in vivo* detection of nitric oxide production in mice

C.-S. Lai and A.M. Komarov

Biophysics Research Institute, Medical College of Wisconsin, 8701 Watertown Plank Road, Milwaukee, Wisconsin 53226, USA

Summary. This chapter covers the use of dithiocarbamate derivatives chelating ferrous iron (Fe) as spin trap reagents for detection of nitric oxide (·NO) produced *in vivo* in small laboratory animals by electron paramagnetic resonance (EPR) spectroscopy. The principle of this method is based on the high affinity of ·NO toward dithiocarbamate-Fe complexes, forming stable nitrosyl-Fe-dithiocarbamate complexes whose characteristic three-line spectrum with $g_{iso} = 2.04$ can readily be detected by EPR spectroscopy. Experiments in our laboratory showed that among several dithiocarbamate derivatives, N-methyl-D-glucamine dithiocarbamate, which is highly water-soluble and relatively non-toxic, is the most promising derivative for studying the production of ·NO *in vivo* in physiological as well as pathophysiological conditions where ·NO is overproduced.

Introduction

Vanin and coworkers were the first group to use the diethyldithiocarbamate-Fe(II) complex, [(DETC)$_2$/Fe], as an ·NO spin trap reagent for direct EPR detection of ·NO in aqueous solution, cells and tissues (Kubrina et al., 1992). Incubating cells or tissues with DETC results in the accumulation of the intracellular iron in the form of the [(DETC)$_2$/Fe] complex in hydrophobic environments, where the complex traps the ·NO produced in these samples as the [(DETC)$_2$/Fe-NO] complex. The latter complex exhibits the characteristic three-line spectrum of a nitrosyl-Fe-dithiocarbamate derivative. The major drawback of the use of the [(DETC)$_2$/Fe] complex as an ·NO spin trap reagent is its low water-solubility, limiting its usefulness for *in vivo* real time measurement of ·NO production in living animals.

In our opinion, an ideal spin trap reagent for *in vivo* real time measurement of ·NO production in living animals should fulfill the following criteria: (i) it should have a high affinity toward ·NO either in the presence or absence of oxygen and the resulting spin adduct should exhibit sharp EPR lines at ambient temperature; (ii) it should be a low molecular weight compound with moderate hydrophilicity, thus permitting it to penetrate the cell membrane freely and trap cellular ·NO, as well as to release the stable spin adduct readily into the circulation for detection; (iii) it should be relatively non-toxic, therefore not interfering with normal cellular functions, and (iv) the resulting spin adduct should be relatively stable in tissues as well as in the circulation. Utilizing these criteria as guidelines, we have examined several dithiocarbamate derivatives including iminodiacetic acid dithiocarbamate, sarcosinedithiocarbamate, di(hydroxyethyl)dithiocarbamate, diethyldithiocarbamate (DETC) and N-methyl-D-glucamine dithiocarbamate (MGD), for their ·NO binding properties. Among these dithiocarbamates, MGD was found to be the best derivative for *in vivo* use (Komarov et al., 1993; Lai and Komarov, 1994).

Materials and methods

ICR mice (female, 25–30 g) were supplied by Harlan-Sprague-Dawley (Indianapolis, IN). Measurement of the effect of dithiocarbamates on superoxide dismutase (SOD) activity was carried out as follows: After mixing SOD (150 units/ml; Sigma Chemical, St. Louis, MO) with dithiocarbamates (5 mM each) in phosphate buffered saline (PBS), pH 7.4, the reaction mixture was incubated at 4°C for two hours. SOD without dithiocarbamates was used as a control. Both experimental samples and controls were dialyzed overnight against PBS. SOD activities were determined by using the SOD-525 kit from Bioxytech (Nebot et al., 1993). Dithiocarbamate derivatives including iminodiacetic acid dithiocarbamate trisodium salt, sarcosine dithiocarbamate

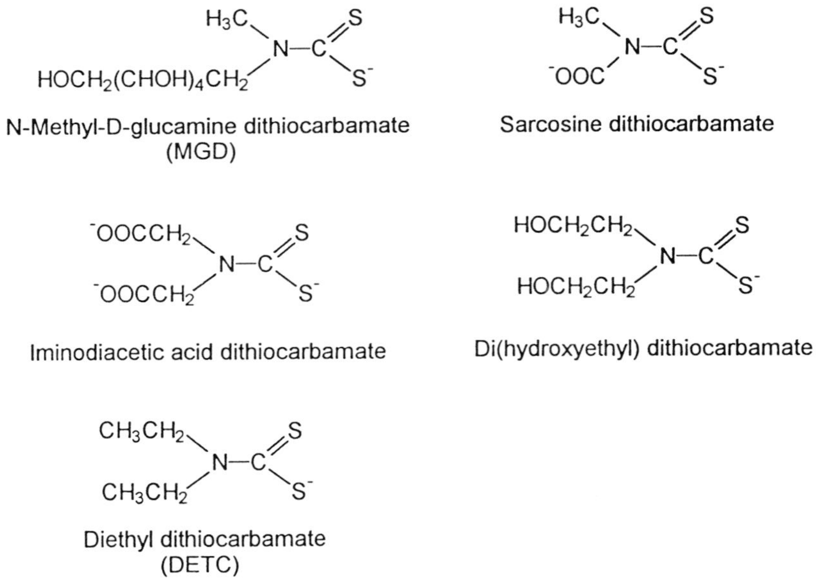

Figure 1. Chemical structures of dithiocarbamate derivatives used in this study.

disodium salt and di(hydroxylethyl) dithiocarbamate sodium salt (see Fig. 1 for chemical structures) were gifts of Dr. Mark M. Jones of Vanderbilt University in Tennessee. N-methyl-D-glucamine dithiocarbamate was synthesized by the method of Shinobu et al. (1984). Other reagent-grade chemicals, materials and EPR instrumentation used in this study have been described elsewhere (Komarov et al., 1993; Lai and Komarov, 1994).

Results and discussion

Conventionally, the term spin trapping refers to the use of diamagnetic nitrone or nitroso compounds that react with a reactive free radical to generate a more stable radical adduct. In this study, the term is extended to include the stabilization of a reactive free radical through the interaction with a metal-chelator complex without the formation of a new covalent bond. Although the chemistry of ·NO interactions with nonheme iron complexes has been extensively studied (Henry et al., 1991 for a review), the field of the dithiocarbamate spin trapping technique is young and many aspects of it remain to be explored. Here we use MGD as an example to describe this technique in general and our recent *in vivo* studies in particular.

The basis of the dithiocarbamate spin trapping method

The principle of this method is based on the observation that MGD chelates with ferrous iron as a two-to-one $[(MGD)_2/Fe]$ complex, which interacts strongly with ·NO, forming a stable and water-soluble $[(MGD)_2/Fe-NO]$ complex in aqueous solution. The latter complex gives rise to a sharp three-line spectrum with $g_{iso} = 2.04$ characteristic of a nitrosyl-Fe-dithiocarbamate complex (Komarov et al., 1993). Ferrous iron (d^6) is preferred over ferric iron (d^5) as an ·NO spin trap reagent inasmuch as the former has one additional d-orbit electron which contributes to its higher affinity toward neutral gaseous ligands such as ·NO (Tsai, 1994). Although this ·NO-specific spin trapping method is rather straightforward, there are several precautionary notes that need to be addressed:

(i) Like other dithiocarbamate derivatives, MGD in aqueous solution is not stable, especially if the pH falls below 7 (Martens et al., 1993). This hydrogen ion-catalyzed decomposition reaction produces volatile carbon disulfide, a known toxic compound. The powder form of MGD therefore should be stored in a desiccator at 4°C and the working solution of MGD in water or saline solution has to be freshly prepared and used within a few hours. The stock solution of ferrous sulfate (100 mM) in water is prepared by adding ferrous sulfate to deoxygenated water, and after mixing the solution is continuously purged with a stream of nitrogen gas for an additional 30 min. The ferrous sulfate solution can then be aliquoted into small quantities and stored at −20°C for several months. The spin trap stock solution consisting of the [(MGD)$_2$/Fe] complex is freshly prepared just prior to use by adding a ferrous sulfate solution into an MGD solution to give a ratio of 5-to-1 (MGD-to-Fe), which yields the optimal trapping efficiency. The reason why excess MGD is needed is discussed below. The ·NO trapping efficiencies of various dithiocar- bamates followed the order: MGD > sarcosine dithiocarbamate > iminodiacetic acid dithio- carbamate >> di(hydroxyethyl) dithiocarbamate > DETC in aqueous solution at ambient temperature.

(ii) Even though the [(MGD)$_2$/Fe] complex has a high affinity toward ·NO, the reaction between the [(MGD)$_2$/Fe] complex and ·NO in solution appeared to take about 30 min to reach completion as judged from the time-dependent increase in the signal intensity of the

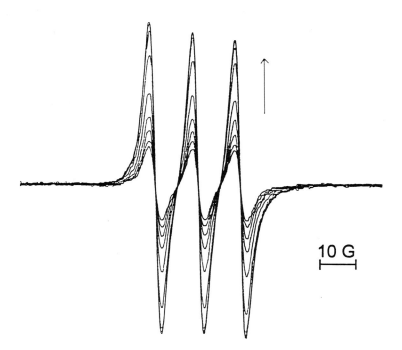

Figure 2. Time-dependent changes in the X-band EPR spectra of the [(MGD)$_2$/Fe-NO] complex in an air-saturated solution at 22°C. An aliquot (10 μl) of the MGD solution (100 mM in water) and an aliquot of (20 μl) of ferrous sulfate solution (10 mM in water) were added to 160 μl of Hepes (50 mM) buffered solution, pH 7.4. The reaction was initiated by adding 10 μl of ·NO saturation solution, and the reaction mixture was transferred immediately into a quartz flat cell for EPR measurement at 22°C. The superimposed spectra were obtained by repetitive scanning using a four minute scan time. Other instrumental settings included 200 G-scan range, 2 G-modulation amplitude, and 60 mW microwave power. The arrow indicates the direction of signal increases.

[(MGD)$_2$/Fe-NO] complex as shown in Figure 2. Purging the reaction mixture containing MGD and Fe(II) with nitrogen gas prior to mixing with ·NO solution shortened the time to reach the maximum, but did not seem to affect appreciably the final signal intensity, suggesting that the observed increase in the signal intensity in Figure 2 is not due to the presence of dissolved oxygen in the reaction mixture. It is worth noting that the addition of an ·NO solution (colorless) into an MGD solution (colorless) yielded a yellowish color solution, a result indicating the formation of an S-nitroso-MGD derivative (data not shown). Since small S-nitrosothiol derivatives are known to undergo slow decomposition in aqueous solution (Mathews and Kerr, 1993), especially in the presence of transition metals, the S-nitroso-MGD derivative formed in the reaction mixture between the [(MGD)$_2$/Fe] complex and ·NO might also decompose slowly with time to produce free ·NO, which in turn is trapped by the [(MGD)$_2$/Fe] complex. This could account for the observed time-dependent increase in the signal intensity (Fig. 2). Additionally, that excess MGD is required to achieve the optimal trapping efficiency as described above may at least in part relate to its transient stabilization of ·NO in the form of the S-nitroso-MGD derivative, preventing ·NO from reaction with dissolved oxygen or escape into the atmosphere. Figure 3 shows a linear correlation between EPR signal intensities of the [(MGD)$_2$/Fe-NO] complex and various ·NO concentrations used. The spectra were recorded 30 min after the mixing of various ·NO concentrations with a fixed concentration of the [(MGD)$_2$/Fe] complex to ensure the signal intensity reached the maximum. Another important implication for the results presented here is that the trapping of ·NO by the [(MGD)$_2$/Fe] solution in air-saturated solution appears not to be

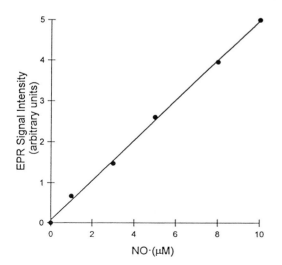

Figure 3. Plot of the signal intensities of the [(MGD)$_2$/Fe-NO] complex as a function of ·NO concentrations at 22°C. The experimental conditions were the same as described in Figure 2, except for the various ·NO concentrations added to the reaction mixture. The reaction mixture transferred into a quartz flat cell was kept at room temperature for 30 min prior to EPR measurement at 22°C.

influenced by dissolved oxygen, therefore this ·NO spin trapping method may be carried out either in the presence or the absence of dissolved oxygen. However, at higher concentrations of dissolved oxygen, it is conceivable that the reaction between oxygen and ·NO yielding nitrite and nitrate may compete effectively against the trapping reaction between the [(MGD)$_2$/Fe] complex and ·NO in aqueous solution. The rates of reaction are not known at present.

(iii) Dithiocarbamates such as DETC are known to be potent inhibitors of superoxide dismutase (SOD) enzymes (Heikkila et al., 1976). In this study, we have compared SOD-inhibitory activities of MGD and DETC, and found that MGD and DETC reduced SOD activites by $25 \pm 4\%$ and $44 \pm 3\%$, respectively, suggesting that MGD is only about half as potent as DETC with respect to the inhibitory action against SOD. In addition, DETC has been shown to be a potent immunomodulator, possibly effecting maturation of T-lymphocytes, activating natural killer cells and prolonging immunological memory (Renoux and Renoux, 1980). Lombardi et al. (1991) studied the immunomodulatory potential of several dithiocarbamates including MGD and DETC in mice, and found that MGD expressed weak *in vivo* immunostimulatory potential when compared with the effect of DETC.

In vivo *spin trapping of ·NO produced in septic shock mice*

The overproduction of ·NO has been implicated as the basis for septic shock, a disease resulting from bacterial infection that is manifested by hypotension, insufficient organ perfusion, multi-organ failure and eventually death (St. John and Dorinsky, 1993).

In a recent paper, we have demonstrated the *in vivo* real time measurement of the overproduction of ·NO in the circulation of conscious, lipopolysaccharide (LPS)-treated mice (Lai and Komarov, 1994). After LPS challenge, the mice were injected subcutaneously with the [(MGD)$_2$/Fe] complex at either 0, 2, 4, or 6 h, respectively. The *in vivo* EPR signal was recorded at two hours after injection of the spin trapping reagent, i.e., at 2, 4, 6, or 8 h after LPS. While no signal was detected at 2 or 4 h (Fig. 4a or 4b), a three-line EPR spectrum of the [(MGD)$_2$/Fe-NO] complex was observed at 6 h (Fig. 4c). At 8 h after LPS challenge, the signal intensity was further increased (Fig. 4d). When N-monomethyl-L-arginine (NMMA), an inhibitor of nitric oxide synthase enzymes, was injected at 6 h after LPS administration, the signal intensity was reduced by more than one-half (Fig. 4e), confirming that the ·NO trapped in the form of the [(MGD)$_2$/Fe-NO] complex was produced *via* the arginine-·NO synthase pathway. Several important aspects of this dithiocarbamate spin trap technique for *in vivo* ·NO studies are discussed below.

(i) The mouse was chosen for this *in vivo* experiment because the tip of the mouse tail (2–3 mm) fits well into the diameter of the resonator (4 mm), thus allowing the direct detection of *in*

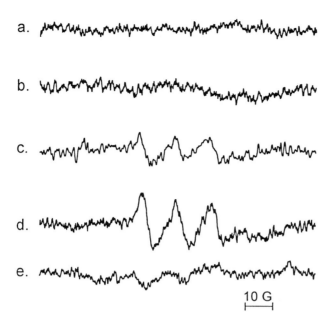

a.

b.

c.

d.

e.

10 G

Figure 4. 3.5 GHz EPR spectra of the [(MGD$_2$/Fe-NO] complex in the circulation of the mouse tail at various times after LPS administration. After intravenous injection of LPS (6 mg; E. coli 026:B6) into the lateral tail vein, an aliquot of (0.4 ml) of the [(MGD)$_2$/Fe] complex in water was injected subcutaneously at (a) 0 h, (b) 2 h, (c) 4 h, and (d) 6 h. (e) Before the injection of the [(MGD)$_2$/Fe] complex at 6h as in (d), an aliquot of NMMA in saline (50 mg/kg) was injected intraperitoneally. The concentrations of MGD and FeSO$_4$ in the injected solution were 83.3 mM and 16.7 mM, respectively. All spectra were recorded two hours after the injection of the [(MGD)$_2$/Fe] complex. Each of the spectra presented here was an average of nine 30-second scans (Lai and Komarov, 1994).

vivo ·NO production in the circulation. In addition, animals were conscious throughout experiments, thus reducing problems generally associated with the administration of anesthetic agents, including affects on the blood circulation, especially in the tail region (Waynforth and Flecknell, 1992).

(ii) The LD$_{50}$ of the [(MGD)$_2$/Fe] complex in mice is still not known. In our laboratory, we have injected subcutaneously four separate aliquots (0.4 ml each) of the [(MGD)$_2$/Fe] solution (326 mg/kg of MGD and 34 mg/kg of FeSO$_4$) into normal mice over six hours, which amounted to about 4% of the mouse body weight, and observed no ill effects; all three mice survived. Previously, Shinobu et al. (1984) showed that MGD is not toxic to mice at doses of up to 2.5 g/kg. Nevertheless, it is recommended here that the molar ratio of MGD-to-Fe for *in vivo* use also be kept at five-to-one (MGD-to-Fe) to alleviate potential oxidative damage generally associated with free ferrous iron. Using the Fenton reaction as a chemical model system for hydroxyl radical production and DMPO as a spin trap reagent, we found that ferrous iron chelated by MGD as the [(MGD)$_2$/Fe] complex substantially reduces its reactivity toward hydrogen peroxide to generate ·OH radical, compared to free ferrous iron. However, by virtue of its chemical reactivity, it is likely that the [(MGD)$_2$/Fe] complex may participate in free radical reactions either *in vitro* or *in vivo*.

(iii) Another significant question is whether the [(MGD)$_2$/Fe] complex traps the ·NO near its *in vivo* sites of production and/or action. This question is important in the light of potential future imaging applications using the [(MGD)$_2$/Fe] complex as a contrast agent to identify *in vivo* loci where ·NO is produced. We showed previously that after subcutaneous injection of the [(MGD)$_2$/Fe] complex into LPS-treated mice, the [(MGD)$_2$/Fe-NO] complex was detected in various isolated tissues, for example, the liver tissue (Lai and Komarov, 1994). Interestingly, when the preformed [(MGD)$_2$/Fe-NO] complex was injected intravenously into mice, the [(MGD)$_2$/Fe-NO] signal was not detected in the liver. This argues that the [(MGD)$_2$/Fe-NO] signal detected in the liver tissue in our previous experiments is due to the reaction between the [(MGD)$_2$/Fe] complex and the ·NO produced in hepatocytes, rather than to the [(MGD)$_2$/Fe-NO] complex formed elsewhere and then taken up by hepatocytes. Therefore, at least for liver tissue, the distribution of the [(MGD)$_2$/Fe-NO] complex seems to correlate with the site of ·NO production and action.

Conclusion

In this chapter, we have emphasized water-soluble dithiocarbamate-Fe complexes for *in vivo* EPR use. However, there may be situations where lipid-soluble dithiocarbamate-Fe complexes will be more suitable for monitoring the ·NO production in lipophilic compartments. Because of the ease of its derivatization, a series of ·NO-specific dithiocarbamate spin traps with different hydrophilicities may be made for functional MRI or EPR imaging of different ·NO-producing tissues. Although S-band (3.5 GHz) EPR spectroscopy is the frequency of choice for *in vivo* real time measurement of ·NO production in the mouse circulation, low-frequency (0.2–1.5 GHz) EPR spectroscopy should permit the study of ·NO production in various organs or tissues in small living animals.

Acknowledgement
We thank Dr. Mark M. Jones of Vanderbilt University in Nashville, Tennessee for his generous gifts of several dithiocarbamates used in this work. This study was supported in part by NIH grant R01008.

References

Heikkila, R.E., Cabbat, F.S. and Cohen, G. (1976) *In vivo* inhibition of superoxide dismutase in mice by diethyl-dithiocarbamate. *J. Biol. Chem.* 251: 2182–2185.

Henry, Y., Ducrocq, C., Drapier, J.-C., Servent, D., Pellat, C. and Guissani, A. (1991) Nitric oxide, A biological effector-electron paramagnetic resonance detection of nitrosyl-iron-protein complexes in whole cells. *Eur. Biophys. J.* 20: 1–15.

Komarov, A., Mattson, D., Jones, M.M., Singh, P.K. and Lai, C.-S. (1993) *In vivo* spin trapping of nitric oxide in mice. *Biochem. Biophys. Res. Commun.* 195: 1191–1198.

Kubrina, L.N., Caldwell, W.S., Mordvintcev, P.I., Malenkova, I.V. and Vanin, A.F. (1992) EPR evidence for nitric oxide production from guanidino nitrogens of L-arginine in animal tissues *in vivo. Biochim. Biophys. Acta* 1099: 233–237.

Lai, C.-S. and Komarov, A.M. (1994) Spin trapping of nitric oxide produced *in vivo* in septic-shock mice. *FEBS Lett.* 345: 120–124.

Lombardi, P., Fournier, M., Bernier, J., Mansour, S., Neveu, P. and Krzystyniak, K. (1991) Evaluation of the immunomodulatory potential of diethyl dithiocarbamate derivatives. *Int. J. Immunopharmac.* 13: 1073–1084.

Martens, T., Langevin-Bermond, D. and Fleury, M.B. (1993) Ditiocarb: decomposition in aqueous solution and effect of the volatile product on its pharmacological use. *J. Pharmaceut. Sci.* 82: 379–383.

Mathews, W.R. and Kerr, S.W. (1993) Biological activity of S-nitrosothiols: the role of nitric oxide. *J. Pharmacol. Experiment. Therapeut.* 267: 1529–1537.

Nebot, C., Moutet, M., Huet, P., Xu, J.Z., Yadan, J.-C. and Chaudiere, J. (1993) Spectrophotometric assay of superoxide dismutase activity based on the activated autoxidation of a tetracyclic catechol. *Anal. Biochem.* 214: 442–451.

Renoux, G. and Renoux, M. (1980) The effects of sodium diethyldithiocarbamate, azathioprine, cyclophosphamide, or hydrocortisone acetate administered alone or in association for 4 weeks on the immune response of BALB/c mice. *Clin. Immun. Immunopath.* 15: 23–32.

Shinobu, L.A., Jones, S.G. and Jones, M.M. (1984) Sodium N-methyl-D-glucamine dithiocarbamate and cadmium intoxication. *Acta Pharmacol. Toxicol.* 54: 189–194.

St. John, R.C. and Dorinsky, P.M. (1993) Immunologic therapy for ARDC, septic shock, and multiple-organ failure *Chest* 103: 932–943.

Tsai, A.-L. (1994) How does NO activate hemoproteins? *FEBS Lett.* 341: 141–145.

Waynforth, H.B. and Flecknell, P.A. (1992) *Experimental and Surgical Technique in the Rat.* Academic Press, New York.

Bioradicals Detected by ESR Spectroscopy
H. Ohya-Nishiguchi & L. Packer (eds)
© 1995 Birkhäuser Verlag Basel/Switzerland

Recent developments in spin labeling

D. Marsh

Max-Planck-Institut für biophysikalische Chemie, Abteilung Spektroskopie, D-37077 Göttingen, Germany

Introduction

Stable bioradical reporters can be generated by attaching a nitroxide spin label group covalently to biomolecules. The conventional electron spin resonance (ESR) spectra are sensitive to rotational motions of the spin label in the $10^{-11} - 10^{-8}$ s time regime (determined by the T_2-relaxation time), and the saturation transfer ESR spectra are sensitive to molecular rotation on the $10^{-8} - 10^{-3}$ s timescale (determined by the T_1-relaxation time). Applications of conventional spin label ESR include the study of lipid mobility in biological membranes, particularly of lipid-protein interactions, and of conformational changes in proteins. Applications of saturation transfer ESR (STESR) spectroscopy are confined almost exclusively to spin label studies, and include the determination of rotational diffusion rates (and hence the state of oligomeric association) of integral proteins in membranes, and of supramolecular assemblies of biomolecules.

Recent developments have been made in progressive saturation ESR and in saturation transfer ESR, for the study of slow exchange processes and of proximity relations to paramagnetic relaxation agents. These new spin label applications will be concentrated on in the following sections, with review also of some recent developments in the more established methods that were mentioned in the previous paragraph. The newer applications open up the possibility to study such processes as the local translational diffusion of integral proteins in membranes, the exchange of lipids associated with membrane proteins, domain formation in lipid membranes, and the mode of insertion and assembly of proteins in membranes. This work with stable bioradical reporters therefore contributes to our knowledge of the structure and dynamics of biological assemblies at the molecular level, and hence to systems for life-support and their technology.

Conventional spin label spectroscopy and lipid-protein interactions

The angular anisotropy in the ESR spectra of nitroxide bioradical labels is sensitive to rotational motions that take place on a timescale which matches very well that for the motions of the lipid molecules in biological membranes. Conventional ESR spectroscopy has therefore been able to contribute materially to the study of membrane structure and dynamics (Knowles and Marsh, 1991). An important example is the application to the characterization of lipid interactions with integral membrane proteins (see e.g., Marsh, 1985, 1989), which are a crucial feature of biomembrane assembly. The lipids that are associated directly with the intramembranous surface of integral proteins are restricted in their rotational motions relative to those in fluid lipid bilayers, and have rotational rates that lie in the slow motional regime of conventional spin label ESR spectroscopy. This lipid population therefore may be distinguished readily in the conventional ESR spectra of lipids that are labelled at a position close to the terminal methyl end of their chains. Quantification by ESR difference spectroscopy then allows determination of both the stoichiometry and specificity of the lipid-protein interaction.

The equilibrium exchange between labelled lipids, L^*, at probe amounts and unlabelled lipids, L, in association with an integral protein, P, can be depicted by:

$$P \cdot L_{N_b} + L^* \xleftrightarrow{\ K_r\ } P \cdot L_{N_b-1} L^* + L$$

If f_b is the fraction of spin labelled lipid that is motionally restricted by the protein and n_t is the lipid/protein mole ratio in the membrane, the equation for equilibrium lipid-protein association is given by (Marsh, 1985):

$$\frac{1-f_b}{f_b} = \frac{1}{K_r}\left(\frac{n_t}{N_b}-1\right) \qquad [1]$$

where N_b is the number of lipid sites on the protein, and K_r is the association constant for the labelled lipid relative to that for the unlabelled host membrane lipid. In this way, the stoichiometries, N_b, and selectivities, K_r, of the lipid interactions with a wide variety of membrane proteins have been determined by ESR spectroscopy (Knowles and Marsh, 1991; Marsh, 1995).

In general, the lipid/protein stoichiometry of association sites correlates with estimates of the size of the intramembranous perimeter of the protein, or protein oligomer, studied. The pattern of lipid specificity is characteristic of the particular protein and is susceptible to changes in ionic strength, to pH titration, and to covalent modification of the protein side chains. Specifically for the myelin proteolipid protein, it has been found that 35 contiguous amino acid residues of the principal polar loop, which are deleted in the DM-20 isoform, are responsible for the major part of the lipid selectivity of this protein (Horváth et al., 1990a). Additionally, the extent of lipid specificity was found to depend on the secondary structure of the membrane-incorporated M-13 bacteriophage coat protein, being much stronger for the β-sheet conformation than for the α-helical conformation of this protein (Peelen et al., 1992). From a functional point of view, it has been demonstrated that aminated local anaesthetics, which are non-competitive blockers of the nicotinic acetylcholine receptor ion channel, are able to substitute competitively for phospholipids at the lipid-protein interface (Horváth et al., 1990b). Other functional aspects of the lipid specificity for integral membrane proteins have been reviewed by Marsh (1987). Also, it has been demonstrated by delipidation experiments, or by competition with non-activating detergents, that full occupation of the first association shell of lipids is necessary for activity of the protein (see e.g., Knowles et al., 1979).

More recently, the method has been extended to use measurements of the lipid-protein interactions as a means to study the intramembranous assembly of integral proteins (Marsh, 1993a). The impetus for this new development comes from the large amount of data recently made available on the primary sequences of membrane proteins from the methods of modern molecular biology. This wealth of information has not been accompanied by a corresponding increase in structural information on integral proteins, and it is a major challenge to devise physical methods for this purpose. The philosophy is illustrated in Figure 1 by a molecular

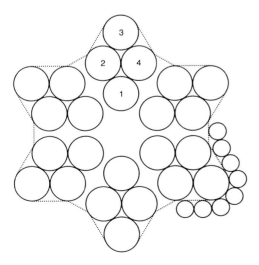

Figure 1. Schematic scale diagram of the helix packing in the hexameric assembly of the 16-kDa proteolipid protein (Holzenburg et al., 1993). The helices are represented as cylinders of 1 nm diameter, where each 4-helix bundle represents a protein monomer. The dashed lines represent the hydrophobic surface of the hexamer which is covered by lipid chains (shaded circles) of 0.48 nm diameter.

model that has been proposed for a class of proteolipid membrane channels which include those of the vacuolar ATPases (Holzenburg et al., 1993). The number of lipid molecules that can be accommodated around this structure is approximately seven per monomer, as opposed to approximately 18 lipids that could be packed around an isolated four-helix bundle. It is clear that the stoichiometry of the lipid-protein association is related very directly to the pattern of assembly of the protein monomers in the membrane. Also, any selectivity of interaction with negatively charged lipids is likely to reflect the presence of charged amino acid residues that are located close to the phospholipid headgroups in the protein structure (cf. above for the DM-20 proteolipid isoform).

In general, the number of phospholipids associated at the surface of an integral protein that has a helical sandwich structure will be given by: $N_b \approx 10 + 2n_\alpha$, where n_α is the number of transmembrane helices (Marsh, 1993a). For rhodopsin, the seven-helix visual receptor protein, this number is expected to be: $N_b \approx 24$ lipids/protein, in reasonable agreement with that obtained experimentally by spin label ESR measurements on native rod outer segment disc membranes (Watts et al., 1979; Pates et al., 1985). On reconstituting rhodopsin in a lipid whose chainlength exceeds that of the hydrophobic span of this integral membrane protein, however, the number of lipids contacting the protein surface is reduced drastically. This is because the protein undergoes aggregation in these membranes – but not in those of lipids with a good hydrophobic match to the protein (Ryba and Marsh, 1992). By contrast, removal of the extramembranous portions of the Na,K-ATPase by proteolytic digestion does not change the number of lipids that are motionally restricted by the protein (Esmann et al., 1994b). This indicates that the structural assembly of the intramembranous sections of the Na,K-ATPase is preserved on trypsinization, which is consistent with their retention of the ability to occlude K⁺ ions.

The method also shows great promise for studying the assembly of small peptides that correspond to the transmembrane sections of integral proteins. The peptide of sequence: *KE*AL**Y** I**LM**V**LG****FF**G**FF**T**LG**I**ML**S**Y**I*R* (charged residues in italics; hydrophobic residues in bold) corresponds to the single apolar span of a small 130-residue protein (I_{sK}) that induces slowly activating voltage-gated K⁺ channels (Takumi et al., 1988). Whereas the apolar stretch of 23 residues is sufficient to span the membrane in an α-helical conformation, the peptide incorporated in membranes of a phospholipid with saturated chains is in a β-sheet structure. The number of lipids that are motionally restricted by the intramembranous peptide assembly is correspondingly reduced compared with that predicted for an α-helix, and additionally a pronounced lipid specificity is observed which locates the terminal charged residues of the peptide in a region close to the lipid headgroups (Horváth et al., 1995).

Progressive saturation ESR and relaxation agents

Continuous wave (CW) saturation of the conventional ESR spectrum, with increasing microwave power, has recently found increasing application in spin label studies. This has come from a recognition of the usefulness of spin-lattice relaxation measurements for determining weak exchange interactions that do not give rise to appreciable linebroadening (i.e., contribute negligibly to spin-spin relaxation). The exchange interactions are related directly to the accessibility of the species that induces the relaxation to the group that is spin labelled. Hence, by suitable choice of the relaxation agents, direct information can be obtained on the location and structure of the labelled species. Such studies are especially useful in combination with site-directed spin labelling in which single cysteine residues are introduced at specific locations in the protein sequence for covalent labelling with a nitroxide derivative (Todd et al., 1989; Altenbach et al., 1990).

For these reasons, developments have been made recently in the analysis of progressive CW saturation experiments. The saturation of the integrated intensity, S, of the ESR spectrum (i.e., the second integral of the conventional first-derivative spectrum) is given by (Marsh, 1994):

$$S(H_1) = \frac{S_o . H_1}{\sqrt{1 + \gamma^2 H_1^2 T_1^{eff} T_2^{eff}}} \qquad [2]$$

where H_1 is the microwave magnetic field at the sample, γ is the electron gyromagnetic ratio, and T_1^{eff} and T_2^{eff} are the effective spin-lattice and spin-spin relaxation times, respectively. It has been shown that use of the integrated absorption intensity circumvents many of the difficulties that are associated with the well-known problems arising from inhomogeneous broadening in the analysis of progressive saturation experiments (Páli et al., 1993a). The saturation of the integrated ESR intensity is insensitive to inhomogeneous broadening, which is a result that extends also to unoriented powder lineshapes, provided that the anisotropy of the relaxation times can be neglected or reasonably represented by an effective average. The integrated intensity also has the advantage that it is directly additive in multicomponent systems and the corresponding saturation curves can be analysed to detect and quantitate a second component, even if it is not evident in the conventional lineshape, provided that the difference in spin-lattice relaxation times is appreciable (Páli et al., 1993a).

The location of a group bearing a bioradical spin label can be determined from its differential accessibility to polar and apolar paramagnetic relaxation agents. In general, the enhancement in spin-lattice relaxation rate that is induced by spin-spin interaction with a fast-relaxing paramagnetic species is linearly proportional to the concentration, c, of the relaxant:

$$\frac{1}{T_1^{eff}} = \frac{1}{T_1^o} + k_{RL} \cdot c \qquad\qquad [3]$$

where k_{RL} depends on the diffusion coefficient and cross-section for collision of the relaxant, in the case of Heisenberg spin exchange, and T_1^o is the spin-lattice relaxation time in the absence of relaxant. For the systems of interest, the effect on the spin-spin relaxation rate will be negligible. Therefore, the $T_1^{eff}T_2^{eff}$ product that is determined from the saturation curve (cf. Eq. 2) will reflect directly the spin-lattice relaxation enhancement by the relaxation agent.

Paramagnetic relaxation agents frequently used are paramagnetic ions which are confined to an aqueous environment, and molecular oxygen which is preferentially concentrated in fluid hydrophobic environments. The accessibility parameters (essentially values of k_{RL}) for both these types of reagents, that are deduced from the relaxation enhancements of spin labelled lipid molecules in lipid membranes, are given in Figure 2. These values were established by progressive saturation experiments (Snel and Marsh, 1993) and indicate rather clearly the degree of

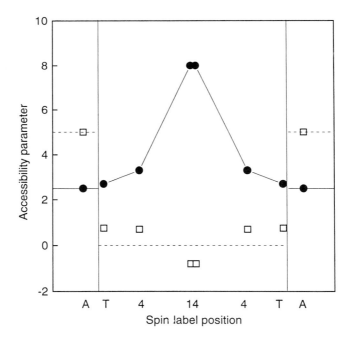

Figure 2. Dependence on the spin label location in lipid bilayer membranes of the accessibility to oxygen (●, solid line) and to chromium oxalate (□, dashed line). T is a spin label in the lipid headgroup; 4 and 14 are spin labels on the respective C-atoms of the lipid chains; and A is a spin label in water. The vertical lines have a separation of 4 nm and indicate the surface of the membrane (Snel and Marsh, 1993).

discrimination that can be obtained between different environments. The paramagnetic ions are restricted to the aqueous phase, and there is a profile of increasing oxygen concentration on entering the hydrophobic core of the membrane. Such data serve to calibrate the location of other labelled groups, e.g., on membrane proteins, from corresponding measurements of the accessibility. The method is most powerful when combined with site-directed spin labelling, the results of which have been reviewed recently (Marsh, 1994). In particular, a sequence periodicity characteristic of an α-helical secondary structure has been found for the accessibility to oxygen of residues that were labelled step-wise throughout one of the transmembrane segments of bacteriorhodopsin (Altenbach et al., 1990).

Alleviation of saturation of a spin labelled species by Heisenberg exchange with a second spin labelled species also may be used to define their mutual accessibilities, in double-labelling experiments. This method has been developed recently to determine the extent of penetration of a spin labelled precursor protein into lipid membranes that additionally contained lipids which were spin labelled at different positions (Snel et al., 1993; Snel and Marsh, 1994). The linear additivity of the integrated ESR intensities (cf. Marsh, 1993b) was exploited in CW progressive saturation experiments. Spin exchange, and hence mutual accessibility, could be detected by comparing the saturation behaviour measured for the doubly-labelled system with that predicted for no exchange, from the results of saturation experiments on the singly labelled systems. The frequencies of spin exchange between the two labels could also be determined, by using a method of analysis that is presented later in connection with STESR studies (cf. Eq. 13, below). For spin labelled apocytochrome c bound to negatively charged lipid membranes, the Heisenberg exchange with spin labelled lipids was found to depend sensitively on the location of the label in the lipid molecule. This positional dependence was used to define the degree of penetration of the N-terminal section of apocytochrome c into the membrane. The penetration step is likely to be a controlling factor in the translocation of this precursor protein across the outer mitochondrial membrane. In contrast, the mature protein, cytochrome c, was found to be located at the membrane surface and not to penetrate the membrane appreciably, which is consistent with its electron transport function in the mitochondrial respiratory chain.

Rather similar CW progressive saturation experiments have been used to determine the exchange rates of lipids at the intramembranous surface of an integral protein from nerve myelin (Horváth et al., 1993a). The method is similar to that for determining the Heisenberg spin exchange between different labelled species, and is discussed in detail later for corresponding measurements with STESR spectroscopy.

Saturation transfer ESR and rotational diffusion

Saturation transfer ESR (STESR) spectroscopy is used normally for determining slow rotational diffusion rates that lie in the region of the T_1-relaxation rate of the spin label, i.e., in the microsecond time regime (Thomas et al., 1976). Such measurements are not accessible to conventional spin label ESR spectroscopy, the sensitivity of which (as seen above) is restricted to rotational motion in the nanosecond time regime. Determination of slow rotational rates is particularly useful in studying both spin labelled supramolecular aggregates in solution and spin labelled integral proteins in membranes. The overall rotational mobility in biomolecular assemblies is of interest because it is related directly to the size of the rotating species, i.e., to the state of molecular association, and also to the asymmetry in shape of the assembly.

For rotational diffusion measurements by STESR, as in conventional ESR, advantage is taken of the angular anisotropy of the spin label hyperfine splittings. Unlike the situation in conventional spin label ESR, the STESR spectrum invariably extends over the full range of the spectral anisotropy, because motional averaging is absent in this time regime. Regions are defined in the different hyperfine manifolds where the sensitivity to alleviation of saturation by rotational motion is greatest, and the spectral lineheights in these regions are normalized to those at the stationary extrema in the manifolds to give diagnostic lineheight ratios, P, that are conventionally designated L''/L, C'/C and H''/H for the $m_I = +1$, 0 and -1 manifolds, respectively (Thomas et al., 1976). Rotational correlation times, τ_R, are then obtained from these diagnostic spectral parameters by using standard calibration systems. Recent developments in the understanding of the effects of exchange processes on the STESR spectral intensities have led to a very simple formulation of these calibrations (Marsh and Horváth, 1992a; Marsh, 1992a). The rotational correlation times are given by:

$$\tau_R = \frac{k}{P_o - P} - b \qquad [4]$$

where k, b and P_o (which is the value of P in the absence of rotational motion) are constants that have been fitted to the calibrations. The values of these calibration constants can be found in Marsh (1992b).

A recent example of the application of STESR to rotational diffusion measurements – that also introduces some new principles of analysis – is afforded by a study of the effects of poly(ethylene glycol) (PEG) on the aggregation state of the Na,K-ATPase ion pump, which is an ubiquitous integral membrane protein (Esmann et al., 1994a). The problem is of interest with respect to the use of PEG as a potent agent for induction of membrane fusion (e.g., for hybridoma production in monoclonal antibody technology), and also because of the fundamental insight involved into hydration forces between proteins and the use of PEG as a precipitant in protein purification. Additionally, the study involved consideration of the effects of aqueous viscosity on the rotational diffusion rates of integral membrane proteins, and this led to a method for determining the size of the extramembranous part of the protein. These aspects of the analysis are now presented immediately below.

In order to account for the dependence on both the aqueous and membrane viscosities, a model is considered (see Fig. 3) in which different sections of an integral protein, of volume V_i, are immersed in media of different viscosities, η_i. The diffusion coefficient for uniaxial rotation about the membrane normal is given by:

$$D_{R//} = \frac{kT}{f_{R//}}$$

[5]

where $f_{R//}$ is the rotational frictional coefficient. Because the frictional torques exerted on the separate sections of the protein are additive, so are the individual contributions to the overall

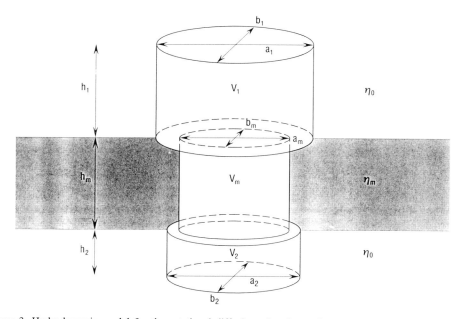

Figure 3. Hydrodynamic model for the rotational diffusion of an integral protein that consists of a central cylindrical section of volume V_m embedded in the membrane of viscosity η_m and two extramembranous sections of total volume $V_o = V_1 + V_2$ in an aqueous medium of viscosity η_0. The heights, h_i, and cross-sectional dimensions, a_i and b_i, of the different sections are indicated (Esmann et al., 1994a).

frictional coefficient:

$$f_{R//} = \sum_i \frac{f^o_{R//,i}}{F_{R//,i}} \qquad [6]$$

where the frictional coefficients of right circular cylinders with volumes equal to those of the different sections of the protein are given by:

$$f^o_{R//,i} = 4\eta_i V_i \qquad [7]$$

and the corresponding shape factors are:

$$F_{R//,i} = 2\frac{a_i / b_i}{1 + (a_i / b_i)^2} \qquad [8]$$

where a_i and b_i are the semi-axes of the elliptical cross-sections of the protein (see Fig. 3). Thus from Equations 5–7, the rotational correlation time of the protein, which is defined as: $\tau_{R//} = 1/6D_{R//}$, is given by:

$$\tau_{R//} = \frac{2}{3kT} \sum_i \frac{\eta_i V_i}{F_{R//,i}} \qquad [9]$$

which specifies the dependence of the measured correlation time on the size and shape of the various parts of the protein.

Specifically, the dependence of the measured rotational correlation time on the aqueous viscosity, η_0, is given by:

$$\tau_{R//} = \tau_{R,memb} + \frac{2\eta_o}{3kT} \sum_j{}' \frac{V_j}{F_{R//,j}} \qquad [10]$$

where the summation now extends only over the extramembranous sections of the protein, and $\tau_{R,memb}$ is the rotational correlation time for a vanishingly small viscosity of the aqueous phase compared with that of the membrane. The latter is determined by the intramembranous dimensions of the protein and by the membrane viscosity, and is given by:

$$\tau_{R,memb} = \frac{2\pi\eta_m h_m}{3kT} \frac{a_m b_m}{F_{R//,m}}$$

[11]

where typically $\eta_m \approx 2-5$ P. Therefore, this equation allows determination of the overall intra-membranous dimensions and aggregation state of the protein, and the previous equation allows determination of the relative size of the extramembranous sections of the protein from the viscosity dependence.

By making STESR measurements on membranes suspended in aqueous buffer alone (i.e., with $\eta_0 = 1$ cP), the results obtained from application of Equation 11 have been compared with estimates of the overall size of the intramembranous section of the protein protomer (Esmann et al., 1987). From this it was concluded that the Na,K-ATPase enzyme is present in the membrane as a diprotomer or higher oligomer. It should be noted that the rotational correlation time deduced by calibration of the experimental spectra depends on the orientation of the spin label axes relative to the rotation axis (Marsh and Horváth, 1989). However, statistically, an orientation that is close to perpendicular is the most likely and gives rise to the maximum sensitivity to rotational diffusion. The dependence of the correlation time on the external viscosity, η_0, was found to be linear for membranes suspended in various aqueous glycerol solutions, in accordance with Equation 10, and from this dependence it was concluded that approximately 50–70% of the protein is external to the membrane (Esmann et al., 1994a). This extramembranous section bears the ATP-binding site, and the results on its relative size obtained from these hydrodynamic measurements were found to be consistent with low-resolution structural data.

In contrast to the experiments with glycerol, the rotational correlation times of the protein in membranes suspended in PEG were found to increase non-linearly with the aqueous viscosity (Esmann et al., 1994a). The values of the correlation times in PEG solutions were also much higher than those in aqueous glycerol of the same viscosity, indicating that PEG induced aggregation of the protein in the plane of the membrane. The value reached at 50% PEG in the aqueous phase corresponded to a degree of aggregation of the proteins between two and five, depending on whether it was assumed that the ethylene glycol polymer was excluded from the membrane surface region.

Another example of the use of STESR spectroscopy to study protein aggregation processes in membranes is afforded by the spin labelled visual receptor protein, rhodopsin, that was purified and reconstituted into membranes composed of a single lipid whose chainlength was varied (Ryba and Marsh, 1992). It was found that the protein was monomeric (as it is in the natural membrane) in disaturated phosphatidylcholines of intermediate chainlengths (C-14:0 or C-15:0), but was aggregated in those of considerably shorter or considerably longer chainlengths (C-12:0

or C-18:0). This result emphasises the importance of matching the length of the lipid fatty acid chains to the hydrophobic span of the integral protein, for a proper integration and function of the protein in the membrane. This has potential significance both for the technology of membrane reconstitution and for the influence of dietary factors on membrane function.

A related problem concerns the state of assembly of the transmembrane sections of integral proteins that have been simplified by proteolytic digestion of their extramembranous domains. If these simplified systems are to be a useful tool for yielding reliable information on the intact protein, the state of intramembranous assembly must remain unchanged on proteolysis. Whether this is so can be checked by measuring the rotational correlation time of the intramembranous assembly. For spin labelled membranous Na,K-ATPase that was trypsinized in the presence of Rb^+ and retained the ability to occlude K^+ ions, it was found that the rotational correlation time remained unchanged relative to that of the native enzyme (Esmann et al., 1994b). In contrast, for the enzyme trypsinized in the presence of Na^+, which underwent more extensive proteolysis and had lost the ability to occlude K^+ ions, it was found that the intramembranous aggregation state of the native protein was no longer retained. The STESR method is therefore also of great value in studying membrane proteins that have been tailored enzymatically.

Saturation transfer ESR and exchange processes

The standard application of STESR spectroscopy described above relies on the anisotropy in the spin label spectrum. More recently, different applications of the STESR method have been developed that rely directly on the sensitivity to T_1-relaxation, but concentrate on the overall integrated intensity of the STESR spectrum rather than on its lineshape (Marsh, 1992b). The motivation for such developments is the study of a variety of slow exchange processes in biological systems – particularly in membranes – the rates of which are comparable to the spin-lattice relaxation of the nitroxide spin label, i.e., which lie in the MHz range (Marsh, 1993b). The rates both of physical exchange between different molecular sites and of Heisenberg spin exchange between labels may be measured. The latter are related directly to the bimolecular collision rates and therefore yield information both on the relative accessibility of the labelled species and on their mutual rates of translational motion. This new class of experiments is therefore capable of yielding significant information on both molecular structure and dynamics, in membranes and in other biological assemblies.

It has been demonstrated from such experiments that the integrated intensity, I_{ST}, of the STESR spectrum is approximately proportional to the nitroxide spin-lattice relaxation time, T_1^{eff} (Páli et al., 1992, 1993b). For the kth component in the STESR spectrum, the integrated intensity can therefore be expressed as:

$$I_{ST,k} = \left(\frac{I^o_{ST,k}}{T^o_{1,k}}\right) T^{eff}_{1\,k}$$

[12]

where $I^0{}_{ST,k}$ and $T^0{}_{1,k}$ are the values of $I_{ST,k}$ and $T_{1,k}$, respectively, in the absence of relaxation enhancement. This linear relationship between the STESR intensity and the effective T_1-relaxation time has also been suggested from the results of theoretical simulations (Thomas et al., 1976). It will be seen that the sensitivity of I_{ST} to T_1 is greater than that in the progressive saturation of the integrated intensity of the conventional CW ESR spectrum. The method therefore offers significant advantages for measurements with strongly immobilized spin labels, which have intense STESR spectra.

The effective spin-lattice relaxation time, $T^{eff}_{1,b}$, of a spin label at site b (e.g., a bound label) that is undergoing exchange at a rate τ^{-1}_{ex} with a label at site f (e.g., a free label) is given by (Horváth et al., 1993a):

$$\frac{T^o_{1,b}}{T^{eff}_{1b}} = 1 + \frac{f_f T^o_{1,b}\tau^{-1}_{ex}}{1 + f_b T^o_{1,f}\tau^{-1}_{ex}}$$

[13]

where f_b and f_f are the fractional populations at sites b and f, respectively ($f_b + f_f = 1$). For Heisenberg spin exchange between the two labels, τ^{-1}_{ex} is the spin exchange frequency that is proportional to the total spin label concentration and to the bimolecular collision rate constant (Marsh, 1992a). For two-site physical exchange, the intrinsic rates, τ^{-1}_k, for leaving the sites k are related to the exchange rate by the condition for detailed balance: $\tau^{-1}_k = (1 - f_k)\tau^{-1}_{ex}$, with $k = b, f$ (Marsh, 1992a).

The two-site exchange method has been used to determine the intrinsic rates of exchange of spin labelled lipids at the intramembranous surface of integral proteins (see Fig. 4). The lipid exchange rates at the hydrophobic interface with the proteolipid protein from nerve myelin are found to be in the region of 10^6 s^{-1}, which is significantly slower than the translational diffusion rates of lipids in fluid membranes, and they are found to reflect the thermodynamic selectivity of different lipid species for the proteolipid protein (Horváth et al., 1993a). The method therefore is applicable to the study of the dynamics of lipid-protein interactions and could be used, for instance, to investigate the specificity of association of spin labelled local anaesthetics with the acetylcholine receptor (Horváth et al., 1990b). In gel-phase lipids, whose chains are highly

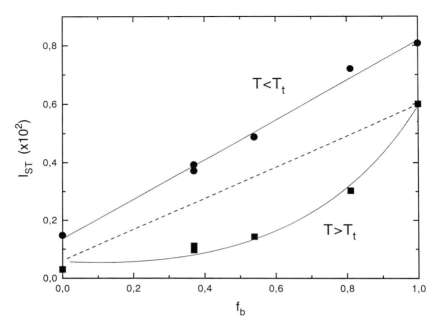

Figure 4. Dependence of the integrated intensity, I_{ST}, of the STESR spectra from different spin labelled lipids on the fraction, f_b, of lipid associated with the myelin proteolipid protein in reconstituted membranes of fixed lipid/protein ratio. Squares correspond to measurements at a temperature below the lipid chain-melting transition, and circles to measurements at a temperature above this transition. In the former case, the exchange rate is immeasurably slow, $\tau_f^{-1} \approx 0$ (solid line), and in the latter case, negative departures from linearity (dashed line) correspond to alleviation of saturation by exchange of lipids from fluid bilayer regions of the membrane at a rate $\tau_f^{-1} = 2.88 / T_{1,b}^0$ (solid line) (Horváth et al., 1993a).

ordered, the lipid exchange rates at the interface of the myelin proteolipid protein are much slower than the spin label T_1-relaxation rate and could not be measured. However, the method could still be used to study the stoichiometry and specificity of the lipid-protein interaction (which was not possible by conventional spin label ESR spectroscopy), because the spin label T_1-relaxation rates still differed between the two environments (Horváth et al., 1993b). This again extends the range of versatility of the spin label method to the study of more immobilized systems and more condensed states of biomolecular organisation.

Heisenberg spin exchange frequencies between different sites have been determined by progressive saturation methods, in a study of the relative locations of spin labelled lipids and proteins in membranes. The exchange frequencies were obtained by using Equation 13, together with estimates of f_b from quantitative difference spectroscopy. This work (Snel et al., 1993; Snel and Marsh, 1994) was aimed at elucidating the import mechanism of a precursor protein and has been discussed more fully above in the section on progressive saturation techniques.

Heisenberg spin exchange interactions between spin label species of the same type can be used to determine bimolecular collision rates. For spin labels in identical environments and with the same T_1-relaxation times, T_1^o, the effective spin-lattice relaxation time in the presence of spin exchange, is given by (Marsh, 1992a):

$$T_{1,k}^{eff} = T_1^o \frac{1 + Z_k T_1^o \tau_{ex}^{-1}}{1 + T_1^o \tau_{ex}^{-1}} \qquad [14]$$

where Z_k is the fractional population or degeneracy at the resonance position k. For weak exchange, the relaxation rate is increased by an amount equal to the exchange frequency τ_{ex}^{-1}, just as in a true relaxation process. However, for fast exchange, T_1^{eff} reaches a limiting value of $Z_k T_1^o$ that represents the maximum extent of redistribution of saturation throughout the entire population of spin label states. This approach has been used for determining the translational diffusion rates of spin labelled proteins, where the spin-exchange frequencies generally are low. Viability of the method was demonstrated from the viscosity and temperature dependence of the translational diffusion coefficients of spin labelled human serum albumin determined in homogeneous solution (Khramtsov and Marsh, 1991). Local translational diffusion rates were determined by STESR for the Na,K-ATPase, a large integral protein, in reconstituted membranes, from the spin-concentration dependence of the exchange frequencies. The diffusion coefficients were found to lie in the range expected for proteins of this size in homogeneous fluid membranes ($\sim 2 \times 10^{-8}$ cm$^2 \cdot$s^{-1}), but were much larger than the long-range diffusion coefficients determined by other means in whole cell systems (Esmann and Marsh, 1992). This indicates that there are considerable barriers to long-range diffusion, possibly associated with cytoskeletal networks, in the cellular systems. Slow translational diffusion of spin labelled lipids in gel-phase bilayer membranes, where the lipid chains are ordered, has similarly been studied by saturation transfer ESR (Marsh and Horváth, 1992b). Additionally, the method has been used to determine local concentrations of spin labelled lipids in heterogeneous systems, in a study of the kinetics of nucleation and growth of lipid domains in model membranes (Páli et al., 1993b).

Saturation transfer ESR and paramagnetic relaxation

Fast relaxing paramagnetic ions are a valuable bioradical species that can be used in conjunction with slower relaxing spin label bioradicals to determine mutual accessibilities and intermolecular distances. In this spin label-spin probe method, which has been used previously in progressive saturation studies (Likhtenstein et al., 1986; Hyde et al., 1979), the paramagnetic ion is detected by its effect on the spin-lattice relaxation of the spin label. The relaxation takes place *via* the dipole-dipole interaction between the two bioradicals and therefore is very sensitively dependent on their distance apart. For a three-dimensional distribution of the paramagnetic ion, the enhancement in relaxation rate of the spin label is inversely proportional to the third power of the distance, R, of their closest approach. For a two-dimensional surface distribution of the paramagnetic ion, the enhancement is proportional to the inverse fourth power of R. In terms of Equation 3, the dependence of the paramagnetic relaxation enhancement on concentration of the paramagnetic ion is given by (Páli et al., 1992):

$$k_{RL} = f^o \int \frac{d\tau}{r^6} = \frac{f_m}{R^m} \qquad [15]$$

where f^o and f_m are constants that contain all other parameters relating to the dipole-dipole relaxation, and the exponent m is determined by the dimensionality of the paramagnetic ion distribution (Marsh, 1992b).

Recently, the new saturation transfer ESR methods have been used to determine the paramagnetic relaxation enhancement by aqueous Ni^{2+} ions of spin labelled lipids in gel-phase bilayer membranes (Páli et al., 1992). Because the lipid hydrocarbon chain geometry is known in these ordered gel phases, the experiments serve not only as a verification of the method but also as a calibration, by yielding the value for the scaling constant f_m (see Fig. 5). By using the C—C bond length as the reference, values of the membrane and lipid headgroup thicknesses were obtained that are in good agreement with the results from measurements by x-ray diffraction. Precision of up to ≈ 0.1 nm can be obtained by the method, for relatively close separations, and interdipole distances of up to ≤ 5 nm are accessible at high paramagnetic ion concentrations. Very clear differences in the profile with chain position of the relaxation enhancement were found between normal gel-phase bilayers and those in which the lipid chains are interdigitated, yielding detailed information on the nature of the interdigitated membrane structure.

Conclusion

The nitroxide spin label is a versatile stable bioradical reporter in biological systems. Coupled with a variety of different ESR methods, spin labels have a wide range of applications for investigating structure and dynamics in biological systems, and for derived technologies. The examples

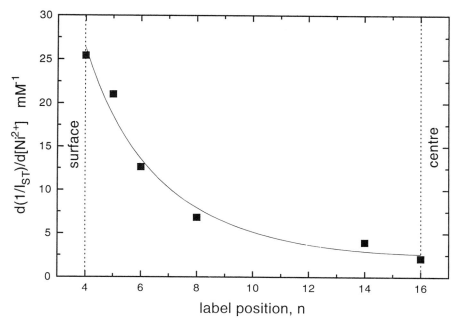

Figure 5. Dependence of the relaxation enhancement by aqueous Ni^{2+} ions on the nitroxide position, n, in the fatty acid chain of spin labelled phospholipids in gel-phase lipid bilayers, as recorded by the intensity of the STESR spectrum. The full line is a least squares fit to a $1/R^3$ dependence, allowing for the presence of paramagnetic ions on both sides of the membrane (Páli et al., 1992).

reviewed in this chapter, in particular the new developments, contribute materially to our knowledge of structure and function in biological membranes at the molecular level, and indicate the continuing expansion in spin label methodology. It is to be anticipated that a major role will be played by saturation techniques and relaxation measurements in future developments. The potential most definitely exists for exploiting these advances in work on life support technologies.

References

Altenbach, C., Marti, T., Khorana, H.G. and Hubbell, W.L. (1990) Transmembrane protein structure: Spin labeling of bacteriorhodopsin. *Science* 248: 1088–1092.

Esmann, M., Horváth, L.I. and Marsh, D. (1987) Saturation-transfer electron spin resonance studies on the mobility of spin labeled sodium and potassium ion activated adenosinetriphosphatase in membranes from *Squalus acanthias*. *Biochemistry* 26: 8675–8683.

Esmann, M. and Marsh, D. (1992) Local translocational diffusion rates of membranous Na+,K+-ATPase measured by saturation transfer ESR spectroscopy. *Proc. Natl. Acad. Sci. USA* 89: 7606–7609.

Esmann, M., Hideg, K. and Marsh, D. (1994a) Influence of poly(ethylene glycol) and aqueous viscosity on the rotational diffusion of membranous Na,K-ATPase. *Biochemistry* 33: 3693–3697.

Esmann, M., Karlish, S.J.D., Sottrup-Jensen, L. and Marsh, D. (1994b) Structural integrity of the membrane domains in extensively trypsinized Na,K-ATPase from shark rectal glands. *Biochemistry* 33: 8044–8050.

Holzenburg, A., Jones, P.C., Franklin, T., Páli, T., Heimburg, T., Marsh, D., Findlay, J.B.C. and Finbow, M.E. (1993) Evidence for a common structure for a class of membrane channels. *Eur. J. Biochem.* 213: 21–30.

Horváth, L.I., Brophy, P.J. and Marsh, D. (1990a) Influence of polar residue deletions on lipid-protein interactions with the myelin proteolipid protein. Spin label ESR studies with DM-20/lipid recombinants. *Biochemistry* 29: 2635–2638.

Horváth, L.I., Arias, H.R., Hankovszky, H.O., Hideg, K., Barrantes, F.J. and Marsh, D. (1990b) Association of spin labeled local anaesthetics at the hydrophobic surface of acetylcholine receptor in native membranes from *Torpedo marmorata*. *Biochemistry* 29: 8707–8713.

Horváth, L.I., Brophy, P.J. and Marsh, D. (1993a) Exchange rates at the lipid-protein interface of the myelin proteolipid protein determined by saturation transfer electron spin resonance and continuous wave saturation studies. *Biophys. J.* 64: 622–631.

Horváth, L.I., Brophy, P.J. and Marsh, D. (1993b) Spin label saturation transfer EPR determinations of the stoichiometry and selectivity of lipid-protein interactions in the gel phase. *Biochim. Biophys. Acta* 1147: 277–280.

Horváth, L.I., Heimburg, T., Kovachev, P., Findlay, J.B.C., Hideg, K. and Marsh, D. (1995) Integration of a K⁺ channel-associated peptide in a lipid bilayer: Conformation, lipid-protein interactions and rotational diffusion. *Biochemistry* 34: 3893–3898.

Hyde, J.S., Swartz, H.M. and Antholine, W.E. (1979) The spin-probe-spin label method. *In*: L.J. Berliner (ed.): *Spin Labeling. Theory and Applications*. Vol. II, Academic Press, New York, pp 71–113.

Khramtsov, V.V. and Marsh, D. (1991) Measurement of the local translational diffusion rates of proteins by saturation transfer EPR spectroscopy. *Biochim. Biophys. Acta* 1068: 257–260.

Knowles, P.F., Watts, A. and Marsh, D. (1979) Spin label studies of lipid immobilization in dimyristoylphosphatidylcholine-substituted cytochrome oxidase. *Biochemistry* 18: 4480–4487.

Knowles, P.F. and Marsh, D. (1991) Magnetic resonance of membranes. *Biochem. J.* 274: 625–641.

Likhtenstein, G.I., Kulikov, A.V., Kotelnikov, A.I. and Levchenko, L.A. (1986) Methods of physical labels – a combined approach to the study of microstructure and dynamics in biological systems. *J. Biochem. Biophys. Methods* 12: 1–28.

Marsh, D. (1985) ESR spin label studies of lipid-protein interactions. *In*: A. Watts and J.J.H.H.M. De Pont (eds): *Progress in Protein-Lipid Interactions*. Vol. 1, Elsevier, Amsterdam, pp 143–172.

Marsh, D. (1987) Selectivity of lipid-protein interactions. *J. Bioenerg. Biomemb.* 19: 677–689.

Marsh, D. (1989) Experimental methods in spin label spectral analysis. *In*: L.J. Berliner and J. Reuben (eds): *Biological Magnetic Resonance. Spin Labeling Theory and Applications*. Vol. 8, Plenum, New York, pp 255–303.

Marsh, D. and Horváth, L.I. (1989) Spin label studies of the structure and dynamics of lipids and proteins in membranes. *In*: A.J. Hoff (ed.): *Advanced EPR. Applications in Biology and Biochemistry,* Elsevier, Amsterdam, pp 707–752.

Marsh, D. (1992a) Influence of nuclear relaxation on the measurement of exchange frequencies in CW saturation EPR studies. *J. Magn. Reson.* 99: 332–337.

Marsh, D. (1992b) Exchange and dipolar spin-spin interactions and rotational diffusion in saturation transfer EPR spectroscopy. *Appl. Magn. Reson.* 3: 53–65.

Marsh, D. and Horváth, L.I. (1992a) A simple analytical treatment of the sensitivity of saturation transfer EPR spectra to slow rotational diffusion. *J. Magn. Reson.* 99: 323–331.

Marsh, D. and Horváth, L.I. (1992b) Influence of Heisenberg spin exchange on conventional and phase-quadrature EPR lineshapes and intensities under saturation. *J. Magn. Reson.* 97: 13–26.

Marsh, D., (1993a) The nature of the lipid-protein interface and the influence of protein structure on protein-lipid interactions. *In*: A. Watts (ed.): *Protein-Lipid Interactions. New Comprehensive Biochemistry*. Vol. 25, Elsevier, Amsterdam, pp 41–66.

Marsh, D. (1993b) Progressive saturation and saturation transfer ESR for measuring exchange processes of spin labelled lipids and proteins in membranes. *Chem. Soc. Rev.*: 329–335.

Marsh, D. (1994) Spin labelling in biological systems. *In*: N.M. Atherton, M.J. Davies and B.C. Gilbert (eds): *Electron Spin Resonance*, Vol. 14. Royal Society of Chemistry, Cambridge, pp 166–202.

Marsh, D. (1995) Specificity of lipid-protein interactions. *In*: A.G. Lee (ed.): *Biomembranes*. Vol. 1, JAI Press, Greenwich, CT, pp 137–186.

Páli, T., Bartucci, R., Horváth, L.I. and Marsh, D. (1992) Distance measurements using paramagnetic ion-induced relaxation in the saturation transfer electron spin resonance of spin labeled biomolecules. Application to phospholipid bilayers and interdigitated gel phases. *Biophys. J.* 61: 1595–1602.

Páli, T., Horváth, L.I. and Marsh, D. (1993a) Continuous-wave saturation of two-component, inhomogeneously broadened, anisotropic EPR spectra. *J. Magn. Reson.* A101: 215–219.

Páli, T., Bartucci, R., Horváth, L.I. and Marsh, D. (1993b) Kinetics and dynamics of annealing during sub-gel phase formation in phospholipid bilayers. A saturation transfer electron spin resonance study. *Biophys. J.* 64: 1781–1788.

Pates, R.D., Watts, A., Uhl, R. and Marsh, D. (1985) Lipid-protein interactions in frog rod outer segment disc membranes. Characterization by spin labels. *Biochim. Biophys. Acta* 814: 389–397.

Peelen, S.J.C.J., Sanders, J.C., Hemminga, M.A. and Marsh, D. (1992) Stoichiometry, selectivity, and exchange dynamics of lipid-protein interaction with bacteriophage M13 coat protein studied by spin label electron spin resonance. Effects of protein secondary structure. *Biochemistry* 31: 2670–2677.

Ryba, N.J.P. and Marsh, D. (1992) Protein rotational diffusion and lipid/protein interactions in recombinants of bovine rhodopsin with saturated diacylphosphatidylcholines of different chain lengths studied by conventional and saturation transfer electron spin resonance. *Biochemistry* 31: 7511–7518.

Snel, M.M.E. and Marsh, D. (1993) Accessibility of spin labeled phospholipids in anionic and zwitterionic bilayer membranes to paramagnetic relaxation agents. Continuous wave power saturation EPR studies. *Biochim. Biophys. Acta* 1150: 155–161.

Snel, M.M.E., De Kruijff, B. and Marsh, D. (1993) Location of spin labelled apocytochrome *c* and cytochrome *c* relative to spin labelled lipids from spin-spin interactions in ESR saturation studies. *Biophys. J.* 64: A16.

Snel, M.M.E. and Marsh, D. (1994) Membrane location of apocytochrome *c* and cytochrome *c* determined from lipid-protein spin exchange interactions by continuous wave saturation electron spin resonance. *Biophys. J.* 67: 737–745.

Takumi, T., Ohkubo, H. and Nakanishi, S. (1988) Cloning of a membrane protein that induces a slow voltage-gated potassium current. *Science* 242: 1042–1045.

Thomas, D.D., Dalton, L.R. and Hyde, J.S. (1976) Rotational diffusion studied by passage saturation transfer electron paramagnetic resonance. *J. Chem. Phys.* 65: 3006–3024.

Todd, A.P., Cong, J., Levinthal, F., Levinthal, C. and Hubbell, W.L. (1989) Site-directed mutagenesis of Colicin E1 provides specific attachment sites for spin labels whose spectra are sensitive to local conformation. *Proteins* 6: 294–305.

Watts, A., Volotovski, I.D. and Marsh, D. (1979) Rhodopsin-lipid associations in bovine rod outer segment membranes. Identification of immobilized lipid by spin labels. *Biochemistry* 18: 5006–5013.

Bioradicals Detected by ESR Spectroscopy
H. Ohya-Nishiguchi & L. Packer (eds)
© 1995 Birkhäuser Verlag Basel/Switzerland

Electron spin resonance imaging of rat brain

M. Hiramatsu, M. Komatsu, K. Oikawa, H. Noda, R. Niwa, R. Konaka, A. Mori[1] and H. Kamada

Institute for Life Support Technology, Yamagata Technopolis Foundation, Numagi, Yamagata 990 and [1]Department of Neuroscience, Institute of Molecular and Cellular Medicine, Okayama University Medical School, Okayama 700, Japan

Introduction

Free radical mechanisms may be involved in several neurological disorders, including Parkinson's disease, Alzheimer's disease, brain ischemia, Down's syndrome, and aging (Sinet and Ceballos-Picot, 1992; Javoy-Agid, 1992; Mori et al., 1992). Thus, a diagnostic method for locating regions of pathological change related to free radicals in the brain may be of use. Ishida et al. (1992) obtained electron spin resonance – computed tomography (ESR-CT) images of the cephalic region of rats. However, they were unable to record ESR-CT images of the brain. In the present study we obtained rat brain images using an ESR-CT system with spin labels for image enhancement.

ESR-CT imaging agents

3-Carbamoyl-2,2,5,5-tetramethyl-1-pyrrolinyloxy (carbamoyl-PROXYL) and [2-(14-carboxy-tetramethyl)-2-ethyl-4,4-dimethyl-3-oxazolinyloxy] (16-doxyl-stearic acid, 16-DS), 2,2,6,6-tetramethylpiperidine-1-oxyl-4-maleimide (maleimide-TEMPO) and 3-carbamoyl-2,2,5,5-tetramethyl-3-pyrrolin-1-yloxy (CTPO) were from SIGMA Chemical Co. (St. Louis, MO). 5-Hydroxytryptamine acetamide-2,2,6,6-tetramethylpiperidinyloxy (serotonin-TEMPO), 2-chloro-10-[3-methyl-amino-3-(2,2,6,6-tetramethylpiperidinyloxy) aminopropyl] phenothiazine (phenothiazine-TEMPO) and N-(2,2,6,6-tetramethyl-1-oxy-4-piperidinyl)-(7-chloro-1,3-dihydro-2-oxo-5-phenyl-2H-1,4-benzodiazepin-1-yl) acetamide (nordiazepam-TEMPO, neurospin I) were synthesized in our laboratory. Chemical structures are shown in Figure 1.

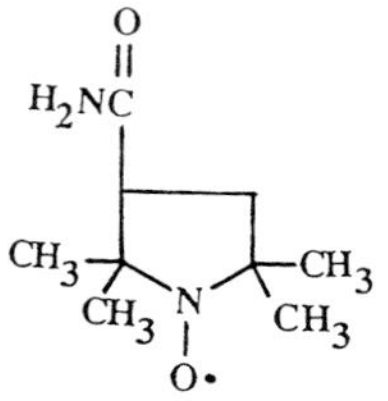

3-carbamoyl-2,2,5,5-tetramethyl
-1-pyrrolinyloxy
(carbamoyl-PROXYL)

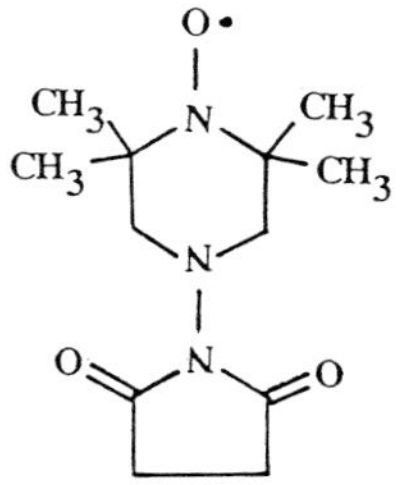

2,2,6,6-tetramethylpiperidine
-1-oxyl-4-maleimide (TEMPO)

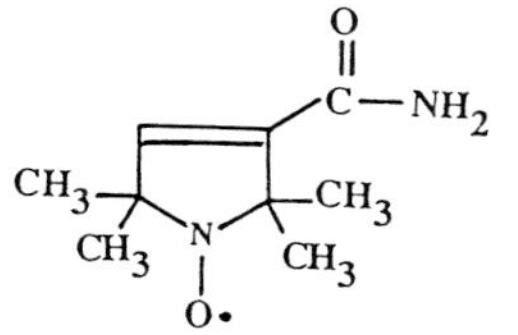

3-carbamoyl-2,2,5,5-tetramethyl
-3-pyrrolin-1-yloxy (CTPO)

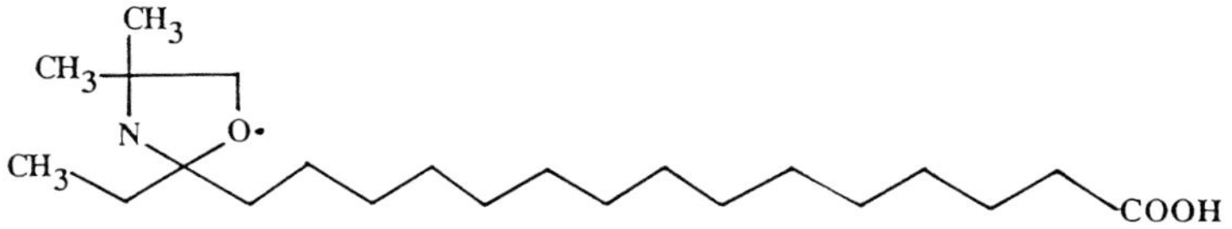

[2-(14-carboxytetramethyl)-2-ethyl-4,4-dimethyl-3-oxazolinyloxy]
(16-doxyl-stearic acid, 16-DS)

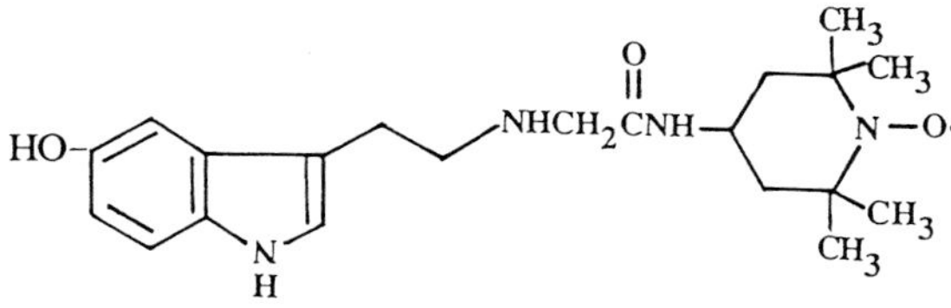

5-Hydroxytryptamine acetamide-2,2,6,6-tetramethylpiperidinyloxy
(serotonin-TEMPO)

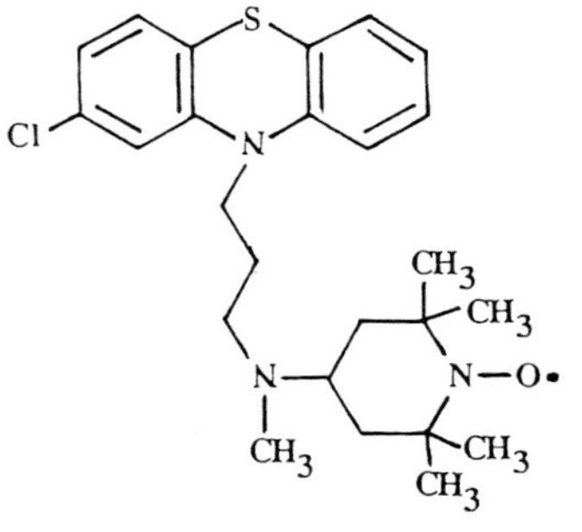

2-chloro-10-[3-methylamino-3-(2,2,6,6
-tetramethylpiperidinyloxy) aminopropyl]
phenothiazine (phenothiazine-TEMPO)

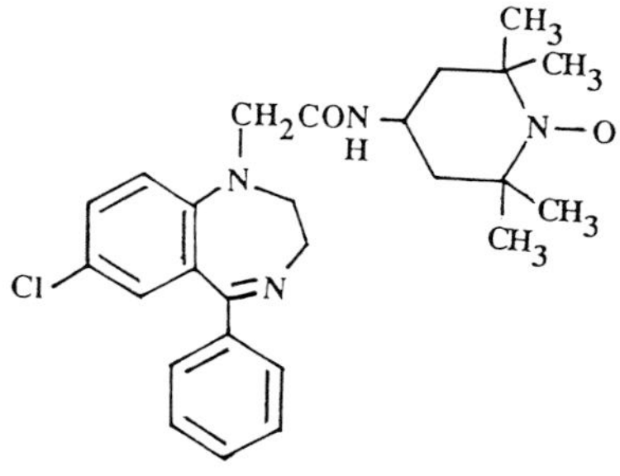

N-(2,2,6,6-tetramethyl-1-oxy-4-piperidinyl)
-(7-chloro-1,3-dihydro-2-oxo-5-phenyl-2H
-1, 4-benzodiazepin-1-yl) acetamide
(nordiazepam-TEMPO, neurospin I)

Figure 1. Chemical structures of spin labels as image enhancing reagents.

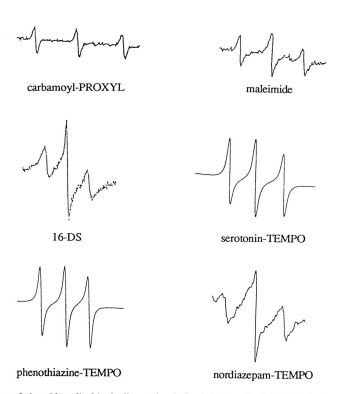

carbamoyl-PROXYL   maleimide

16-DS   serotonin-TEMPO

phenothiazine-TEMPO   nordiazepam-TEMPO

Figure 2. Spectra of nitroxide radical in the live rat head after injection of spin labels into the carotid artery.

EST-CT imaging method

Male Wistar rats, 200 g, were anesthetized with an intraperitoneal injection of 50 mg/kg pento-barbital. A polyethylene tube (0.25 x 0.75 mm), filled with saline was inserted into the cervical portion of the left carotid artery. Following arterial cannulation, the animal's head was inserted into a loop gap resonator centered on the external ear canal. Then the imaging agent (0.1 ml of 0.4 M solution) was infused over 30 s through the intra-carotid canula.

The L-band ESR imaging system consists of an L-band ESR spectrometer (JEOL, Tokyo), a pair of field gradient coils (Yonezawa Electric Wire Co., Ltd., Yonezawa, Yamagata) and a compu-ter (5450, Concurrent Computer Corporations, Massachusetts, USA). ESR spectra were recorded at 700 mHz by the L-band ESR spectrometer with an electrical shield in the loop of the gap reso-nator. The loop gap resonator was 41 mm in diameter and 10 mm in axial length. Conditions for ESR-CT were as follows; rapid response; magnetic field modulation, 0.2 mT; sweep time, 3 s; magnetic gradient, 1 mT/cm; angle, 0 ~ 160° at 20° intervals.

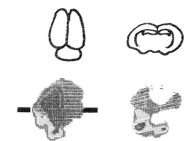

Figure 3. ESR-CT images of nitroxide radical in the rat head after injection of carbamoyl-PROXYL into the left carotid artery.

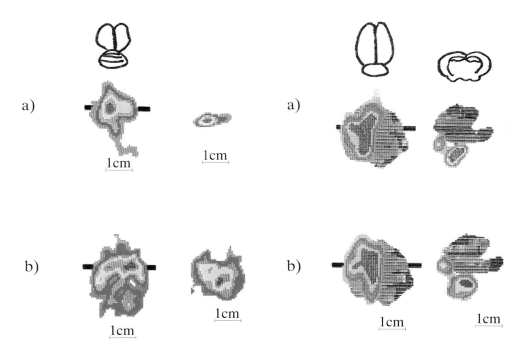

a)

1cm

1cm

b)

1cm

1cm

Figure 4. ESR-CT images of nitroxide radical in the brain after injection of neurospin I into the left cortex (a) and in the head after intraperitoneal injection of carbamoyl-PROXYL (b) in Mongolian gerbil.

a)

b)

1cm

1cm

Figure 5. ESR-CT images of nitroxide radical in the rat head after injection of 5-hydroxytryptamine-TEMPO (a) and phenothiazine-TEMPO (b).

ESR-CT imaging of rat brain

Endogenous free radicals in the brain could not be detected by the ESR-CT system when carbamoyl-PROXYL was used as the imaging agent. The spectra from the head of a live rat after injection of various spin labels are shown in Figure 2. When 0.1 ml of carbamoyl-PROXYL was perfused into the left carotid artery a spectrum of nitroxide radicals of carbamoyl-PROXYL was obtained. The half-life was approximately 7.8 min.

The position of the brain in the calvarium was determined by overlapping the image of the nitroxide radicals from the living rat with the nitroxide radical patterns from the brain surface from a dead rat. Although ESR-CT images of nitroxide radicals in the rat head were obtained after perfusion of carbamoyl-PROXYL into the left carotid artery, no brain image was detected (Fig. 3). ESR-CT images of nitroxide radicals from the cephalic region of the live male Mongolian gerbil (60 g) were obtained after intraperitoneal injection of carbamoyl-PROXYL (Fig. 4). Although carbamoyl-PROXYL in DMSO should pass the blood brain barrier, the ESR-CT images of nitroxide radicals were not observed in brain areas. These data suggest that carbamoyl-PROXYL may not penetrate the blood brain barrier of rats.

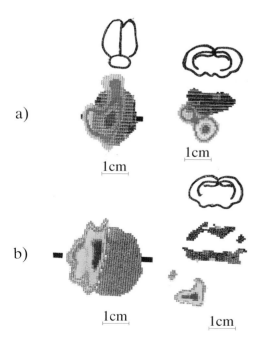

Figure 6. ESR-CT images of nitroxide radical in the rat head after injection of maleimide (a) and CTPO (b).

5-Hydroxytryptamine, an inhibitory neurotransmitter, and phenothiazine, a psychotropic agent, were synthesized as TEMPO derivatives, forming 5-hydroxytryptamine-TEMPO and phenothiazine-TEMPO. Perfusion of TEMPO-derivatives into the left carotid artery failed to show SR-CT brain imaging patterns (Fig. 5). The half-life of nitroxide radicals after perfusion of each TEMPO-derivative into the left carotid artery was 15 and 23 min, respectively. Further, maleimide, an agent used for the study of sulfur-containing protein also failed to yield ESR-CT images of nitroxide radicals (Fig. 6). The half-life of maleimide in the rat head was 5 min. The half-life of CTPO in rat head is 4 min following intravenous injection into a tail vein (Ishida et al., 1989). ESR-CT images of cranial structures can be obtained using ESR-CT with rapid scan. However, brain ESR-CT images were not obtained (Fig. 6).

Nordiazepam binds to the benzodiazepine receptors of inhibitory neurons. We synthesized a derivative of nordiazepam with a TEMPO label (neurospin I). The half-life in the cranium of the rat was 5 min following perfusion into the carotid artery. ESR-CT images of neurospin I can be obtained from the brain of Mongolian gerbils after direct injection into the cerebral cortex (Fig. 4). The spin label 16-DS has been used for study of membrane fluidity. Nitroxide radicals in the core of the lipid bilayer were obtained by X-band spectrometry with the nitroxide radical

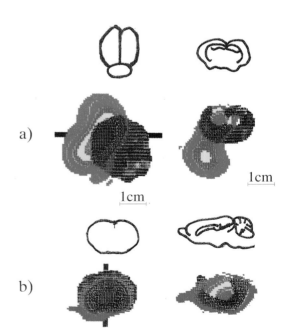

Figure 7. ESR-CT images of nitroxide radical in the coronal section (a) and in the horizontal section (b) of rat brain after injection of 16-DS into the left carotid artery.

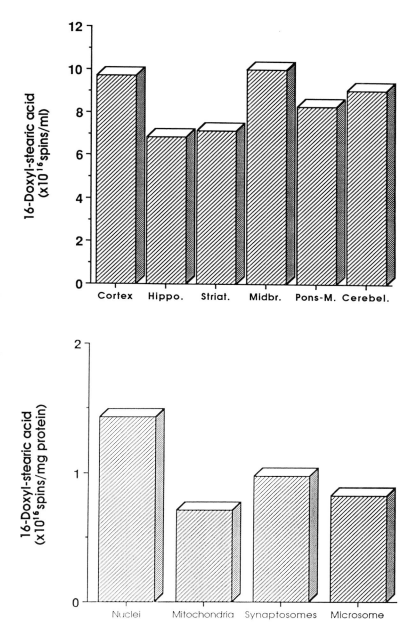

Figure 8. Uptake of 16-DS into various areas and brain fractions after injection of 16-DS into the left carotid artery. Each value represents the mean of 2 animals.

located on position 16 of the carbon chain (from the carboxylic group) of the fatty acid. The
signal height of nitroxide radicals in the rat cranium was unchanged 20 min after perfusion of 16-
DS into the left carotid artery. The ESR-CT images from 16-DS is shown in Figure 7.

An x-band ESR spectrometer was used to measure uptake of 16-DS in homogenates of brain
areas after perfusion into the left carotid artery. Following perfusion of 16-DS into the carotid
artery, rats were reperfused with physiological saline and were then decapitated. Six regions of
brain were dissected and homogenized in 0.9% saline. Uptake of 16-DS was found in the cortex,
hippocampus, striatum, midbrain, pons-medulla oblongata and cerebellum (Fig. 8). Further,
uptake of 16-DS was also found in the nuclear, mitochondrial, synaptosomal and microsomal
fractions from rat brain (Fig. 8). These data support the ESR-CT observation that nitroxide
radicals are present in rat brain following perfusion of 16DS into the carotid artery.

Free radical change in iron-induced focal epileptogenic regions of rat brain

The decay of nitroxide radicals in rat brain was studied in a standard model of post-traumatic
epilepsy. Rats were anesthetized with ether and placed in a stereotaxic apparatus. A burr hole was
made in the left calvarium 1 mm posterior and 1 mm lateral to the bregma. A freshly prepared

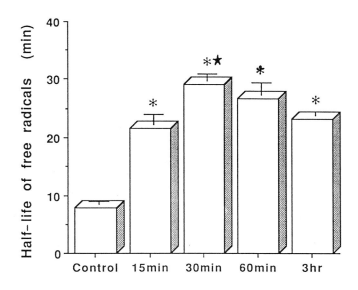

Figure 9. Half-life of nitroxide radicals of iron-injected rat after injection of carbamoyl-PROXYL into the left
carotid artery. Each value represents the mean ± SEM of 3–5 animals. *p < 0.01 *vs.* control, ★p < 0.01 *vs.* 15 min.

solution containing 100 mM ferric chloride (5 µl) was injected over a period of 5 min into the left cortex at a depth of 2.5 mm below the exposed dura as previously reported (Willmore et al., 1978). The decay of the signal height of nitroxide radicals in the rat calvarium was measured after intraperitoneal injection of carbamoyl-PROXYL. Spin label injection times were 15, 30, 60 min and 3 hours after the iron solution injection. The half-life of nitroxide radicals was increased 15 min after injection of iron solution and continued to increase for 30 min after the injection. The intensity of the ESR signals began to decrease by 60 min and three hours after injection. However, three hours after injection the half-life was still elevated when compared to controls (Fig. 9). The increased half-life of nitroxide radicals in the rat brain following iron injection into

Control

60 min after iron injection

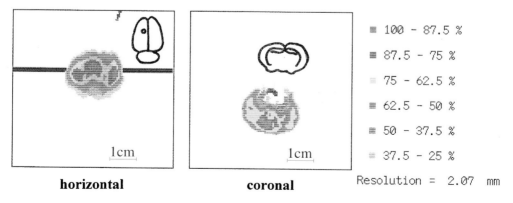

horizontal   **coronal**

Figure 10. ESR-CT images of nitroxide radical in iron-injected rat head after intraperitoneal injection of carbamoyl-PROXYL.

the cortex appeared to be due to decreased antioxidant activity in the region of cortical injection. Reactive oxygen species are generated in rat brain after injection of iron solution into the cortex. Free radicals are scavenged by antioxidant substances such as ascorbic acid, glutathione, and superoxide dismutase. Since antioxidant mechanisms are decreased, it is possible that nitroxide radicals could not be scavenged because of lowered antioxidant activity.

The ESR-CT image of nitroxide radical formation in rat brain was obtained 60 min after injection of the iron solution into the left cerebral cortex. Carbamoyl-PROXYL was perfused into the left carotid artery. No change in the EST-CT was found (Fig. 10). In the future an ESR-CT system with higher sensitivity and resolution will be used in these experiments. A strategy to develop imaging agents with higher penetration of the blood brain barrier should be helpful in detecting changes in free radicals in pathological foci in the brain.

Acknowledgement
We thank Dr. L.J. Willmore of The University of Texas, Health Science Center, Houston for reading our manuscript and providing us with valuable comments.

References

Ishida, S., Kumashiro, H., Tsuchihashi, N., Ogata, T., Ono, M., Kamata, H. and Yoshida, N. (1989) *In vivo* analysis of nitroxide radicals injected into small animals by L-band ESR technique. *Phys. Med. Biol.* 34: 1317–1323.

Ishida, S., Matsumoto, S., Yokoyama, H., Mori, N., Kumashiro, H., Tsuchihashi, N., Ogata, T., Yamada, M., Ono, M., Kitajima, T., Kamada, H. and Yoshida, E. (1992) An ESR-CT imaging of the head of a living rat receiving an administration of a nitroxide radical. *Magnetic Resonance Imaging* 10: 109–14.

Javoy-Agid, F. (1992) Dopaminergic cell death in Parkinson's disease and Down's syndrome. *In:* L. Packer, L. Prilipko and Y. Christen (eds): *Free Radicals in the Brain.* Springer-Verlag, Berlin, pp 99–108.

Mori, A., Hiramatsu, M. and Yokoi, I. (1992) Posttraumatic epilepsy, free radicals and antioxidant therapy. *In:* L. Packer, L. Prilipko and Y. Christen (eds): *Free Radicals in the Brain.* Springer-Verlag, Berlin, pp 109–122.

Sinet, P.-M. and Ceballos-Picot, I. (1992) Role of free radicals in Alzheimer's disease and Down's syndrome. *In:* L. Packer, L. Prilipko and Y. Christen (eds): *Free Radicals in the Brain.* Springer-Verlag, Berlin, pp 91–98.

Willmore, L.J., Sypert, G.W. and Munson, J.B. (1978) Chronic focal epileptiform discharges induced by injection of iron into rat and cat cortex. *Science* 200: 1501–1503.

Bioradicals Detected by ESR Spectroscopy
H. Ohya-Nishiguchi & L. Packer (eds)
© 1995 Birkhäuser Verlag Basel/Switzerland

ESEEM spectroscopy – probing active site structures of metalloproteins

J. Peisach

Department of Molecular Pharmacology, Albert Einstein College of Medicine, 1300 Morris Park Avenue, Bronx, NY 10461, USA

Summary. Electron spin echo envelope modulation (ESEEM) spectroscopy is a pulsed EPR method useful for the elucidation of structure in the vicinity of paramagnetic metal centers and free radicals, based on measurement of weak hyperfine interactions. In the two examples provided, Cu(II) and Co(II) coordinated to imidazole, we show how the measurement of nuclear quadrupole and weak nuclear hyperfine interaction provides a basis for understanding how local changes in a protein structure are transmitted to metal binding sites.

Introduction

Electron spin echo envelope modulation (ESEEM) spectroscopy based on pulsed EPR methodo-logy, has been shown to be a valuable addition to the arsenal of spectroscopic methods used for the elucidation of structure at or near paramagnetic centers in macromolecules (Mims and Peisach, 1979a, 1981, 1989). Spin echoes are generated subsequent to the application of short, high powered microwave pulses to paramagnetic samples (Fig. 1). The timing between pulses, τ, is incremented and the intensity of the echo is measured as a function of time. The resulting electron spin echo decay envelope, normally modulated by nuclear precession frequencies, is characteristic of nuclei coupled to the electron spin. Both two-pulse and three-pulse methods have commonly been used to examine biological materials although the greatest advances in structural studies of proteins have thus far been derived from three-pulse investigations. (For an application of a four-pulse method, see McCracken and Friedenberg, 1994.) Using a standard approach of Fourier transformation, one obtains a frequency, or ESEEM, spectrum containing the same type of information as from an ENDOR experiment (Shimizu et al., 1979). This spectrum (Fig. 2, as an example) is particularly useful in the identification and characterization of weakly coupled nuclei and, unlike ENDOR, provides under favorable circumstances both their number and distance.

For a totally unknown structure, one can sometimes identify the atom magnetically coupled to the electron spin of a paramagnetic metal center. In some cases, that atom can be shown to be a component of a ligand coordinated to the metal ion. For an $I = 1/2$ nucleus, such as ^1H or ^{15}N, or an $I = 1$ nucleus with a small nuclear quadrupole moment, such as ^2H, nuclear identification is

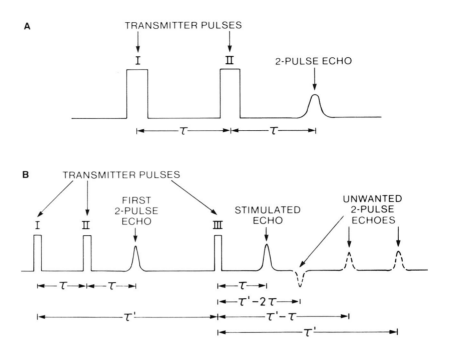

Figure 1. (A) Two-pulse and (B) three-pulse electron spin echo sequences commonly used for ESEEM investigations. In (A), the echo envelope is generated by gradually increasing the spacing τ between the microwave transmitter pulses and measuring the intensity in successive pulse sequences as a function of τ. In (B), τ is set to a fixed value and τ' is varied. The stimulated echo intensity measured at τ after the third pulse is used to generate the echo envelope. Unwanted two-pulse echoes are artefacts of the experiment.

often made from spectral frequency and magnetic field dependence of spectral components, as these relate to the nuclear Zeeman interaction.

In the case of a coupled ^{14}N nucleus where there is a nuclear quadrupole interaction and it is large as compared to the Zeeman interaction, the relative independence of ESEEM spectral components with magnetic field at constant g (thus requiring spectroscopic measurements at two microwave frequencies) not only provides a useful method for nuclear identification, but also provides a method for structural identification based on the determination of nuclear quadrupole interaction, largely through spectral simulation (Mims and Peisach, 1978, 1979a; Peisach et al., 1979; Cammack et al., 1988; Cornelius et al., 1990). In a simpler way, this can be done by experimentation at a microwave frequency near where A_{iso}, the isotropic hyperfine coupling, and the Zeeman interactions nearly cancel so that the spectrum is dominated by the quadrupole interaction (Mims and Peisach, 1978; Jiang et al., 1990), the so-called condition of "exact cancellation" (Flanagan and Singel, 1987).

Structural information can be obtained both from ESEEM frequencies and also from the depth of modulation in the electron spin echo decay envelope. For example, in the case of weakly coupled nuclei, such as ^{2}H belonging to a ligand not directly coupled to a metal ion (Mims et al., 1977, 1990), or ^{23}Na or ^{133}Cs bound to a paramagnetic metalloprotein near the metal center (Tipton et al., 1989), the depth of modulation pattern is directly related to number and distance based on an n/r^6 relationship, where r is the distance from the nucleus in question to a paramagnetic center and n is the number of such nuclei.

The question of whether the association of an exogenous ligand with a macromolecule leads to alteration of metal ligation can be addressed when it is possible to modify chemically a ligand with ^{2}H, ^{13}C, ^{15}N (Tipton et al., 1989), ^{95}Mo (Doi et al., 1988) or any nucleus with an altered nuclear spin. Information from experiments of this type can be used to determine whether ligand binding is *via* metal coordination or just by proximity. The demonstration of a contact interaction can be taken as a clear demonstration of bonding (Peisach et al., 1984) or other structural pathway for electron-nuclear interaction.

Where spectra are complicated, especially where lines arising from ^{14}N interfere, it is often useful to apply a method of division to segregate spectral components for materials where isotopic substitution of a particular ligand is under investigation (Mims and Peisach, 1979a; Zweier et al.,

$$Nd^{3+} - Na - ATP$$

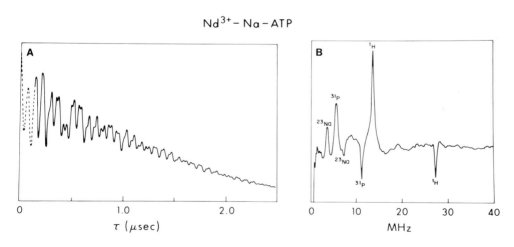

Figure 2. (A) Two-pulse ESEEM decay envelope and (B) Cosine Fourier transform for Nd^{3+}-ATP prepared with Na_3–ATP. The continuous line in (A) shows the experimental data and the broken line, an extension of the envelope to time $\tau = O$. The two-pulse ESEEM spectrum in (B) contains frequency components arising from magnetic coupling with ^{1}H, ^{23}Na, and ^{31}P. The presentation in (B) shows features arising from the Zeeman interaction (peaks) and twice this value (trough) for coupled nuclei at 3000 gauss where the contact interaction is several times smaller than the Zeeman energies. This spectrum is taken as a clear indication of bimetallic complex formation with ATP. (After Shimizu et al., 1979).

1979a; Magliozzo et al., 1987; McCracken et al., 1987; Lee et al., 1992, 1993, 1994; Magliozzo and Peisach, 1993; Jiang et al., 1993). As the modulation obtained in an ESEEM experiment is a product function representing interactions from all nuclei, dividing data for a sample containing a single isotopic substitution in a ligand with data for a sample with a different isotope (e.g., ^{13}C for ^{12}C, ^2H for ^1H, ^{15}N for ^{14}N) will, subsequent to Fourier transformation, leave the spectrum for the unusual isotope, with negative spectral contributions from the isotope that has been substituted (Mims and Peisach, 1981). As an example (Fig. 3), ESEEM data for a protein exchanged against D_2O divided by data for the same sample in H_2O provides spectral information, using the division procedure, concerning deuterium interaction, without interference from ^{14}N in proximity to the paramagnet (Mims and Peisach, 1979a; Lee et al., 1992, 1993, 1994).

For a protein sample in D_2O, under some circumstances the splitting of a deuterium line and the scaling of frequency with magnetic field can be taken as a direct indication of the presence of a deuterium on a metal ligand, possibly from water (Fig. 3) (Peisach et al., 1984). Where the nucle-

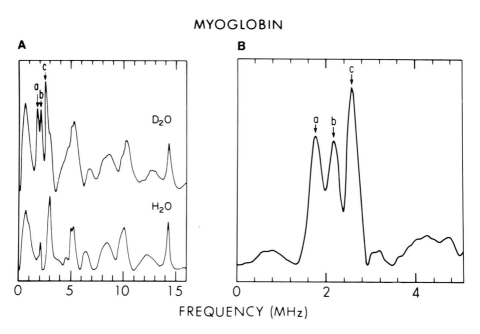

Figure 3. (A) Cosine Fourier transform of echo envelopes for metmyoglobin in D_2O and in H_2O obtained by the three-pulse procedure. (B) A spectrum obtained by Fourier transformation of the ratio of time wave forms for D_2O and H_2O samples. Features marked a, b, c are seen in the D_2O spectrum and arise from ^2H. The ratio procedure largely abolishes spectral contributions common to both samples. Peaks a and c are assigned to ^2H on a water molecule bound to heme Fe(III). Peak b is attributed to non-coordinated ^2H_2O and to exchangeable deuterons on the protein. (After Peisach et al., 1984).

ar Zeeman interaction is much greater than the nuclear quadrupolar interaction, such as for ^{2}H, the relative contributions of nuclear Zeeman, quadrupole, and superhyperfine (shf) interaction is a useful indicator of whether interacting nuclei are close to or far from a paramagnetic center, thereby affecting the nature of the second harmonic in a two-pulse experiment and the second harmonic frequency line. This finding is the basis for experiments which are used to demonstrate

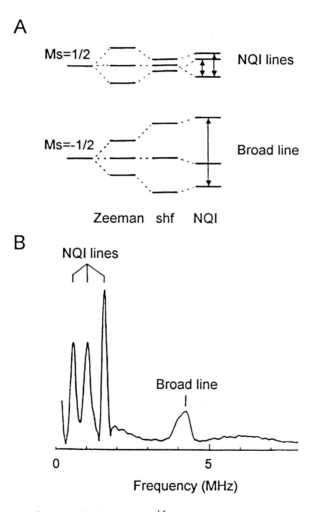

Figure 4. (A) Electron spin energy level scheme for ^{14}N and (B) Fourier transform of ESEEM data obtained at near cancellation of nuclear Zeeman and nuclear hyperfine (shf) terms of the spin Hamiltonian for ^{14}N. The three NQI lines in (B) arise from the nuclear quadrupole interaction where the Zeeman and shf terms cancel. For the lower submanifold, they add and give rise to the broad, high frequency line.

differences in ESEEM modulations for close coordinated D_2O and for ambient D_2O (McCracken et al., 1987; McCracken and Friedenberg, 1994).

[14]N interactions

With ^{14}N, the modulation depth may be large, and one observes lines in the spectrum arising from the nuclear quadrupolar interaction. This can be understood from an examination of the

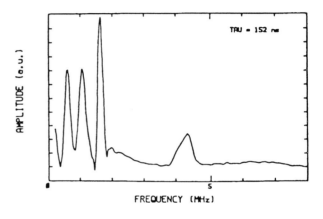

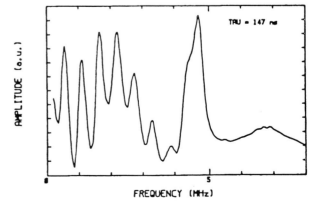

Figure 5. ESEEM spectra of (A) Cu(II) diethylene triamine 2-methylimidazole and (B) Cu(II) tetrakis 2-methylimidazole. In (A) the modulations arise from the remote ^{14}N of the 2-methylimidazole with the other ^{14}N in the sample making no spectral contribution. In (B), combination lines appear as additional peaks in the spectrum.

electron spin energy scheme for ^{14}N interacting with an $S = /2$ paramagnet, such as Cu(II) (Fig. 4) (Jiang et al., 1990). The spin Hamiltonian for ^{14}N consists of three terms, the nuclear Zeeman interaction, the electron nuclear superhyperfine interaction (shf) and the nuclear quadrupole interaction. For one of the $M_S = \pm 1/2$ submanifolds, under conditions where the electron nuclear superhyperfine interaction is approximately twice the nuclear Zeeman interaction, these two terms almost cancel, leaving the nuclear quadrupole interaction to dominate the energy level splitting. Such a condition, achieved with the remote or amino nitrogen of Cu(II)-coordinated C- or N-substituted imidazole, gives rise to three sharp, low frequency lines in the ESEEM spectrum, where the lower two add to give the third.

In the second submanifold, where the Zeeman and shf terms add, one obtains a broad line, whose magnitude is approximately equal to four times the nuclear Zeeman frequency, and encompasses a $\Delta M_I = 2$ transition. The two $M_I = 1$ transitions in this submanifold are highly orientation dependent and are not resolved in powder samples, but can be resolved in single crystal experiments (Colaneri et al., 1990; Colaneri and Peisach, 1992, 1993). Where more than a single ^{14}N nucleus having similar coupling interacts with the electron spin, one can observe combination frequencies in the Fourier transform spectrum. These harmonics and their intensities relative to the fundamental lines can be taken as evidence for the presence of two or more such ^{14}N nuclei, and further, can be used as a quantitative marker of absolute number (Fig. 5) (McCracken, et al., 1988). Structural alteration of the metal binding site in mutagenized protein where there is a reduction in number of a particular nucleus is easily recognized either from the alteration in relative intensity of combination lines, or their total abolition where the number has been reduced to unity (Balasubramanian et al., 1994).

The remote ^{14}N of Cu(II)-coordinated imidazole

ESEEM arises from weak magnetic coupling between a paramagnet and neighboring nuclei. For Cu(II) complexed to nitrogenous ligands, such as in a diethylene triamine complex, the coupling to directly coordinated equatorially bound ^{14}N is too large to give rise to modulations (Mims and Peisach, 1978). For Cu(II)-diethylenetriamine-imidazole complexes, coupling to the remote, amino ^{14}N of the coordinated imidazole does give rise to ESEEM. The spectrum obtained is characteristic of a large number of natural and synthetic Cu(II)-proteins and is in fact the basis for the first assignment of imidazole as a metal ligand for numerous Cu(II)-proteins (Peisach, 1993, and references therein). However, the characteristic low frequency 3-line features (Fig. 4) are reduced to two (Fig. 6), due to the degeneracy of the first two peaks (Mims and Peisach, 1978; Jiang et al., 1990). The broad peak at 4 MHz is the $\Delta M = 2$ transition alluded to above.

For comparison we show the spectrum of stellacyanin (Mims and Peisach, 1979b) (Fig. 6) a blue copper protein having structural homology with the electron transfer proteins azurin and plastocyanin. The major features of the model spectrum are duplicated. Additional peaks are combination lines arising from a second imidazole coordinated to copper, albeit with a different coupling.

In a number of instances, e.g., galactose oxidase, phenylalanine hydroxylase and the copper-containing amine oxidase (Jiang et al., 1990 and references therein) the low frequency lines of the spectrum attributable to the nuclear quadrupole frequencies are so far removed from those observed with a Cu(II)-diethylenetriamine imidazole model or even for stellacyanin, as to suggest

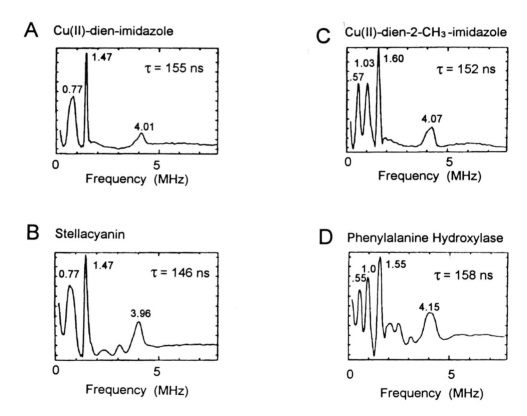

Figure 6. ESEEM spectra of (A) the model compound Cu(II) diethylenetriamine imidazole, (B) stellacyanin (C), the model Cu(II) diethylenetriamine 2-methylimidazole, and (D) *Chromobacterium violaceum* phenylalanine hydroxylase. The sharp, low frequency lines in the spectra arise from the zero field quadrupolar frequencies of the remote, protonated ^{14}N of Cu(II)-coordinated imidazole. The combination lines in (C) and (D) arise from multiple ^{14}N interactions.

that the nuclear quadrupole interaction at the remote ^{14}N of a imidazole ligand to Cu(II) is altered (Fig. 6). This alteration is believed to arise from local changes in the electric field gradient of ^{14}N, thereby effecting the sp^2 orbital occupancy. In model studies, it was shown that such changes came about from altering the chemical structure of imidazole (N- or 2-alkylation), for example, but structural changes of this type are not found in proteins. An alternative explanation for the situation in protein is that the environment of ^{14}N may be altered by local effects, such as H-bonding. This suggestion is verified in model studies (Jiang et al., 1990) demonstrating that in the presence of a high concentration of formate, a good H-bonding anion, for example, the nuclear quadrupole frequencies of a Cu(II)-substituted imidazole complex is altered (Fig. 7) although the

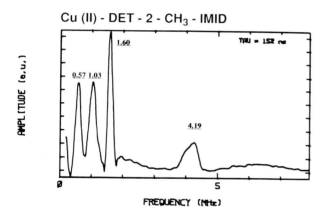

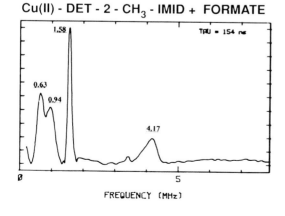

Figure 7. (upper) ESEEM spectrum of Cu(II)-dien-2 methylimidazole (lower) and in the presence of formate. Note the effect of a good hydrogen bond-forming reagent, formate, on the nuclear quadrupole line, effectively increasing the electric field gradient at the remote ^{14}N of the coordinated 2-methylimidazole.

continuous wave EPR spectrum, an indicator of equatorial coordination of Cu(II), is unchanged. In contrast, samples in D_2O show the opposite effect of formate – deuterium bonds are weaker than hydrogen bonds. Thus, nuclear quadrupole measurements in Cu(II) proteins can provide information concerning H-bonding of the remote N—H of a Cu(II)-coordinated histidine imidazole side chain. This effect on N—H bond polarization is believed to alter the properties of the coordinated metal and provides a pathway for protein-metal interaction.

Co(II) – ^{14}N interactions

EPR-active metal substitution for diamagnetic metal or EPR silent metals as probes of natural metal cofactors has been an extremely successful approach to the elucidation of active site structures in metalloproteins. Notable examples can be cited for Mn(II) replacing Mg(II); VO^{2+}, Co(II) and Cu(II) for Zn(II); and Co(II) for Fe(II). By and large, one of the most useful substitutions is Co(II) protoporphyrin IX for the naturally occurring Fe compound, heme, in myoglobin (Mb) (Hoffman and Petering, 1970). The metal ion in the native and metal-substituted form, is covalently attached *via* a proximal imidazole. When O_2 is bound to native, Fe-containing myoglobin, neutron diffraction studies demonstrate that the reversibly bound ligand is H-bonded to another imidazole, the one distal to the heme iron (Phillips and Schoenborn, 1981).

In the functional state, the Fe-containing protein has either even spin (deoxy) or zero spin (oxy) and is thus EPR silent. With Co(II) substitution for Fe(II), EPR active oxy and deoxy forms are produced, both of which can be used to probe the local environment of Co(II). For oxy CoMb, the continuous wave EPR spectrum is essentially that of superoxide anion, with resolved hyperfine interactions arising from the $I = 7/2$ Co nuclear spin (Hoffman and Petering, 1970). The majority of electron spin density resides on the bound superoxide. The major features of the ESEEM spectrum of the oxygenated Co(II)-containing-protein (Fig. 8) are attributed to the directly coordinated ^{14}N of the proximal imidazole (Lee et al., 1992). Spectral simulation allows us to determine the isotropic hyperfine interaction as 2.5 MHz. From a study of this protein in D_2O, one can also obtain a hyperfine coupling of 0.6 MHz for the hydrogen bonded deuteron from the distal imidazole which interacts with bound O_2.

In an analogous protein, *Glycera* hemoglobin (Hb) (Lee et al., 1993), the distal imidazole is substituted with a leucine and hydrogen bonding of the type found in myoglobin cannot take place. Indeed, the ESEEM spectrum of the D_2O-exchanged protein shows no deuterium hyperfine coupling of the type found in O_2CoMb. The nuclear hyperfine interaction to the axial ^{14}N in this case, however, is increased to 3.6 MHz (Fig. 8). It is noteworthy that CoMb has a higher affinity for O_2 than does Co *Glycera* Hb. We have attributed the differences in oxygen affinity, in part, to the presence or absence of hydrogen bonding, and further have related the spectroscopic differences to the functional difference.

If one considers a molecular orbital picture for the O_2Co unit in both proteins, we note that the unpaired spin resides in an orbital consisting of an $O_{2\pi}$ molecular orbital mixed with some $Co_{d\pi(x2, y2)}$ atomic orbitals (Tovrog et al., 1976). Hyperfine interaction with the Co and with the axial ^{14}N arises from polarization of the unpaired spin largely in the oxygen atom not coordinated to Co. With H-bonding to bound O_2, this interaction is reduced and less spin density

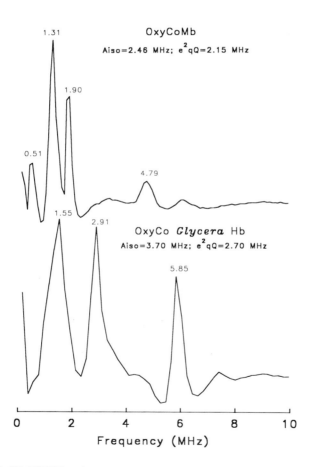

Figure 8. (upper) (A) ESEEM and computer simulated spectra of O_2Co myoglobin, studied at $g = 2.038$; microwave frequency, 9.0 GHz, (lower) (A) ESEEM and (B) computer simulated spectra of O_2Co *Glycera* hemoglobin studied at $g = 2.038$; microwave frequency, 10.1 GHz. (After Lee et al., 1992).

is found in Co (A∥ = 25 MHz in O_2CoMb and 28 MHz in O_2 Co(II) *Glycera* Hb) and concurrently, in the axially bound ^{14}N. A consequence of this is that H-bonding to O_2 promotes more superoxide character, or a more ionic, more stable structure.

The promotion of an ionic structure need not necessarily be promulgated by hydrogen bonding, but can, to some degree, take place by other alterations in the local environment of bound oxygen. A case in point is a series of human myoglobin mutants having distal substitutions for the native imidazole, ranging from amino acids with hydrophilic sidechains to those with hydrophobic sidechains (Lee et al., 1994 and references therein). In the heme-containing proteins, oxygen association appears to be depressed in those mutants with increasing distal hydrophobicity.

For the Co(II)-substituted proteins, measurements of the nuclear hyperfine coupling to the proximal imidazole ^{14}N demonstrate an increase with increasing hydrophobicity, suggestive of a less ionic CoO_2 structure, much in line with what was seen in the comparison of cobalt-substituted myoglobin and *Glycera* hemoglobin.

Acknowledgement
This work was supported by U.S. Public Health Service grants RR-02583 and GM-40168 from the National Institutes of Health.

References

Balasubramanian, S., Carr, R.T., Bender, C.J., Peisach, J. and Benkovic, S.J. (1994) Identification of histidine ligands to copper in *Chromobacterium violaceum* phenylalanine hydroxylase. *Biochemistry* 33: 8532–8537.

Cammack, R. Chapman, A., McCracken, J., Cornelius, J.B., Peisach, J. and Weiner, J.H. (1988) Electron spin-echo spectroscopic studies of *Escherichia coli* fumarate reductase. *Biochem. Biophys. Acta* 956: 307–312.

Colaneri, M.J., Potenza, J.A., Schugar, H.J. and Peisach, J. (1990) Single crystal electron spin-echo envelope modulation study of Cu(II)-doped zinc bis-1,2-dimethylimidazole dichloride. *J. Am. Chem. Soc.* 112: 9451–9458.

Colaneri, M.J. and Peisach, J. (1992) An electron spin echo envelope modulation study of Cu(II)-doped single crystals of L-histidine hydrochloride monohydrate. *J. Am. Chem. Soc.* 114: 5335–5341.

Colaneri, M.J. and Peisach, J. (1993) Enhanced resolution of EPR single crystal spectral parameters using field-swept electron spin-echo spectroscopy. *J. Mag. Reson.* 102: 360–363.

Cornelius, J.B., McCracken, J., Clarkson, R.B., Belford, R.L. and Peisach, J. (1990) ESEEM angle selection studies of axial pyridine coordination to copper(II) benzoylacetonate. *J. Phys. Chem.* 94: 6977–6982.

Doi, K., McCracken, J., Peisach, J. and Aisen, P. (1988) The binding of molybdate to uteroferrin: hyperfine interactions of the binuclear center with ^{95}Mo, ^1H, and ^2H. *J. Biol. Chem.* 263: 5757–5763.

Flanagan, K.L. and Singel, D.J. (1987) Analysis of ^{14}N ESEEM patterns of randomly oriented solids. *J. Chem. Phys.* 87: 5606–5616.

Hoffman, B.M. and Petering, D.H. (1970) Coboglobins: Oxygen-carrying cobalt-reconstituted hemoglobin and myoglobin. *Proc. Natl. Acad. Sci. USA* 67: 637–643.

Jiang, F., McCracken, J. and Peisach, J. (1990) Nuclear quadrupole interactions in Cu(II)-diethylenetriamine-substituted imidazole complexes and in Cu(II) proteins. *J. Am. Chem. Soc.* 112: 9035–9044.

Jiang, F., Karlin, K.D. and Peisach, J. (1993) An electron spin echo envelope modulation (ESEEM) study of electron-nuclear hyperfine and nuclear quadrupole interactions of d_{z2} ground state copper(II) complexes with substituted imidazoles. *Inorganic Chem.* 32: 2576–2582.

Lee, H.C., Ikeda-Saito, M., Yonetani, T., Magliozzo, R.S. and Peisach, J. (1992) Hydrogen bonding to the bound dioxygen in Oxy cobaltous myoglobin reduces the superhyperfine coupling to the proximal histidine. *Biochemistry* 31: 7274–7281.

Lee, H.C., Wittenberg, J.B. and Peisach, J. (1993) The role of hydrogen bonding to bound oxygen in soybean leghemoglobin. *Biochemistry* 32: 11500–11506.

Lee, H.C., Peisach, J., Dou, Y. and Ikeda-Saito, M. (1994) Electron-nuclear coupling to the proximal histidine in oxy cobalt-substituted distal histidine mutants of human myoglobin. *Biochemistry* 33: 7609–7618.

Magliozzo, R.S., McCracken, J. and Peisach, J. (1987) Electron-nuclear coupling in nitrosyl heme proteins and in nitrosyl ferrous and oxy cobaltous tetraphenylporphyrin complexes. *Biochemistry* 26: 7923–7931.

Magliozzo, R.S. and Peisach, J. (1993) Evaluation of nitrogen nuclear hyperfine and quadrupole coupling parameters for the proximal imidazole in myoglobin-azide, -cyanide, and -mercaptoethanol complexes by electron spin echo envelope modulation spectroscopy. *Biochemistry* 32: 8446–8456.

McCracken, J., Peisach, J. and Dooley, D.M. (1987) Cu(II) coordination chemistry of amine oxidases: Pulsed EPR studies of histidine imidazole, water, and exogenous ligand coordination. *J. Am. Chem. Soc.* 109: 4064–4072.

McCracken, J., Pember, S., Benkovic, S.J., Villafranca, J.J., Miller, R.J. and Peisach, J. (1988) Electron spin echo studies of the Cu(II) site in phenylalanine hydroxylase from *Chromobacterium violaceum*. *J. Am. Chem. Soc.* 110: 1069–1074.

McCracken, J. and Friedenberg, S. (1994) Electron spin echo envelope modulation studies of water bound to tetracyanonickelate(II). *J. Phys. Chem.* 98: 467–473.

Mims, W.B., Peisach, J. and Davis, J.L. (1977) Nuclear modulation of the electron spin echo envelope in glassy materials. *J. Chem. Phys.* 66: 5536–5550.

Mims, W.B. and Peisach, J. (1978) The nuclear modulation effect in electron spin echoes for complexes of Cu^{2+} and imidazole with ^{14}N and ^{15}N. *J. Chem. Phys.* 69: 4921–4930.

Mims, W.B. and Peisach, J. (1979a) Pulsed EPR studies of metallo-proteins. *In:* R.G. Shulman (ed.): *Biological Applications of Magnetic Resonance*. Academic Press, New York, pp 221–269.

Mims, W.B. and Peisach, J. (1979b) Measurement of ^{14}N superhyperfine frequencies in stellacyanin by an electron spin echo method. *J. Biol. Chem.* 254: 4321–4323.

Mims, W.B. and Peisach, J. (1981) Electron spin echo spectroscopy and the study of metalloproteins. *In:* L.J. Berliner and J. Reuben (eds): *Biological Magnetic Resonance*. Vol. 3, Plenum Press, New York, pp 213–263.

Mims, W.B. and Peisach, J. (1989) ESEEM and LEFE of metalloproteins and model compounds. *In:* A. Hoff (ed.): *Advanced EPR in Biology and Biochemistry*. Elsevier Publishers, Amsterdam, pp 1–57.

Mims, W.B., Davis, J.L. and Peisach, J. (1990) The exchange of hydrogen ions and of water molecules near the active site of cytochrome *c. J. Magn. Reson.* 86: 273–292.

Peisach, J., Mims, W.B. and Davis, J.L. (1979) Studies of the electron-nuclear coupling between Fe(III) and ^{14}N in cytochrome P-450 and in a series of low spin heme compounds. *J. Biol. Chem.* 254: 12379–12389.

Peisach, J., Mims, W.B. and Davis, J.L. (1984) Water coordination by heme iron in metmyoglobin. *J. Biol. Chem.* 259: 2704–2706.

Peisach, J. (1993) Pulsed EPR studies of copper proteins. *In:* K.D. Karlin and Z. Tyeklar (eds): *Bioinorganic Chemistry of Copper*, Chapman and Hall, New York, pp 21–33.

Phillips, S.E.V. and Schoenborn, B.P. (1981) Neutron diffraction reveals oxygen-histidine hydrogen bond in oxymyoglobin. *Nature* 292: 81–82.

Shimizu, T., Mims, W.B., Peisach, J. and Davis, J.L. (1979) Analysis of the electron spin echo decay envelope for Nd^{3+}: ATP complexes. *J. Chem. Phys.* 70: 2249–2254.

Tipton, P.A., McCracken, J., Cornelius, J. and Peisach, J. (1989) Electron spin echo envelope modulation studies of pyruvate kinase active-site complexes. *Biochemistry* 28: 5720–5728.

Tovrog, B.S., Kitko, D.J. and Drago, R.S. (1976) Nature of the bound O_2 in a series of cobalt dioxygen adducts. *J. Am. Chem. Soc.* 98: 5144–5153.

Zweier, J., Aisen, P., Peisach, J. and Mims, W.B. (1979) Pulsed EPR studies of copper complexes of transferrin. *J. Biol. Chem.* 254: 3512–3515.

Bioradicals Detected by ESR Spectroscopy
H. Ohya-Nishiguchi & L. Packer (eds)
© 1995 Birkhäuser Verlag Basel/Switzerland

Nitric oxide, a versatile biological ligand for hemeproteins

T. Yoshimura

Institute for Life Support Technology, Yamagata Technopolis Foundation, Kurumanomae-683, Numagi, Yamagata 990, Japan

Summary. In this article, the coordinating behavior of nitric oxide (NO) as a biological ligand for hemeproteins and enzymes is presented by reference to the spectral and stereochemical properties of nitrosylhemeproteins and their model nitrosylheme. The structural change in the heme coordination environments on NO binding to hemeproteins is described primarily in relation to the mechanisms of activation and inhibition of the enzymes by NO.

Introduction

Gaseous NO is known as an atmospheric pollutant and a potential health hazard. The finding by Hermann (1865) that NO can combine with hemoglobin (Hb) as do oxygen and carbon monoxide was followed by many studies on the affinity of Hb for NO. Although Gibson and Roughton (1957) found the affinity of Hb for NO *in vitro* to be about 1000 times higher than that for CO, only small traces of NO-Hb were detected in the blood samples of animals exposed to NO (Oda et al., 1975). It is likely that the difficulty in detection of NO-Hb is attributable to the quick conversion of NO-Hb to O_2-Hb in the presence of O_2 and met-Hb reductase (Kosaka et al., 1989). NO is now widely known as an endogenous molecule with critical physiological roles, which is biosynthesized from L-arginine (see the reviews Calver et al., 1993; Butler and Williams, 1993). Since NO is produced in endothelial cells, macrophages, neutrophils, and platelets by constitutive or inducible NO synthases, it is not too much to say that the Hb molecules in erythrocytes are always exposed to endogenous NO. Therefore, the quick conversion of NO-Hb to O_2-Hb mentioned above may be involved in the metabolic pathways of endogenous as well as exogenous NO.

The principal physiological functions of endogenous NO are associated with activation of heme-containing guanylate cyclase which catalyzes the biosynthesis of cyclic GMP from GTP. The activation of the enzyme is assumed to be due to NO coordination to heme iron and to subsequent conformational changes of the enzyme (Ignarro et al., 1984). On the other hand, NO inhibits peroxidases and oxygenases through its coordination to heme iron in place of oxygen (Henry et al., 1991; Stadler et al., 1994 and references therein). However, the detailed activation, deactivation and inhibition mechanisms remain to be clarified. Thus, diverse information about the interaction of NO with metalloproteins, especially hemeproteins, is required for understanding the physiological roles of endogenous NO.

NO free radical can coordinate as a nitrosyl ligand to the sixth-coordination site of heme iron in hemeproteins with a high affinity in both the reduced and oxidized states. Thus, NO has so far been employed as a useful electronic probe for elucidating the structure of heme iron and its environments. For understanding the properties of nitrosylhemeproteins, studies of their model, nitrosyl(porphyrinato)iron complexes, under various conditions have yielded useful information. In this article, we will attempt to illustrate the spectral and stereochemical properties of model nitrosylheme and to demonstrate the structural change of the heme coordination environments on NO binding to hemeproteins, to elucidate the mechanism of activation and inhibition by NO.

Abbreviations

The abbreviations used in this chapter: Hb, hemoglobin; Mb, myoglobin; cyt, cytochrome; HRP, horseradish peroxidase; IDO, indoleamine dioxygenase; CPO, chloroperoxidase; GC, guanylate cyclase; P, dianion of porphyrin; TPP, dianion of tetraphenylporphyrin; PPIX, dianion of protoporphyrin IX; PPIXDME, dianion of protoporphyrin IX dimethyl ester; DPIXDME, dianion of deuteroporphyrin IX dimethyl ester; OEP, octaethylporphyrin; *N*-MeIm, 1-methylimidazole; Py, pyridine; *n*BuNH$_2$, *n*-butylamine; Pip, piperidine; IHP, inositol hexaphosphate; SDS, sodium dodecyl sulfate; cyclic GMP, guanosine 3',5'-cyclic monophosphate; GTP, guanosine 5'-triphosphate; *A, Achromobacter*; *Rb, Rhodobacter*; EPR, electron paramagnetic resonance; IR, infrared; RR, resonance Raman; MCD, magnetic circular dichroism.

Various properties of nitrosylheme

A nitrosyl ligand is generally classified as a strong field ligand and shows considerable electronic and structural *trans* effects (Richter-Addo and Legzdins, 1992). The versatility of the NO ligand arises from the presence of an unpaired electron in the antibonding π orbital. Four coordination modes are known: (1) NO$^+$ (transfer of the odd electron to the metal ion), (2) NO, (3) NO$^-$ (transfer of an electron from the metal ion), and (4) bridging ligand.

The iron atom in a nitrosylheme with planar porphyrinato and axial nitrosyl ligands is in low-spin state. There are two types of coordination stereochemistry in the nitrosylheme, which differ in the coordination mode of the axial ligand (Fig. 1). The five- and six-coordinated nitrosylhemes

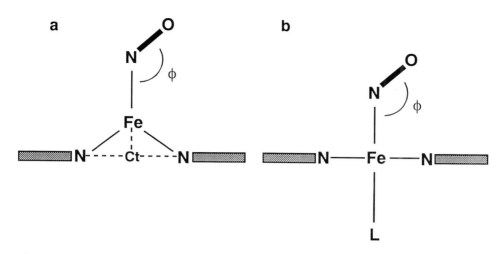

Figure 1. Two types of nitrosylheme: (a) 5- and (b) 6-coordinated complexes; Ct, the center of porphyrinato core; L, the axial ligand *trans* to NO; φ, 140–150° in the ferrous complexes and about 175° in the ferric complexes.

have notably different molecular structures and spectral properties such as electronic absorption, EPR, IR, resonance Raman, and MCD spectra. The coordination structure of nitrosylheme in nitrosylhemeproteins, closely related to the activation and inhibition of heme enzymes by NO binding, can be differentiated by the spectral properties.

In what follows, the presentation of various spectral properties for ferrous nitrosylhemes focuses on the differences between five- and six-coordinate complexes.

Since an unpaired electron on the NO group is delocalized towards an iron d orbital, it is unreasonable to assign a formal oxidation state to iron and NO (such as NO^-, NO, and NO^+) in the nitrosylheme. According to the formalism of Enemark and Feltham (1974), the electronic configuration of the Fe-NO unit in ferrous- and ferric nitrosylheme can be classified, respectively, as the $\{FeNO\}^7$ and $\{FeNO\}^6$, where the superscript 7 or 6 indicates the total number of electrons associated with metal-d and NO-π^* orbitals. The low-spin complexes with $\{FeNO\}^7$ and $\{FeNO\}^6$ configuration are in paramagnetic (EPR positive) and in diamagnetic (EPR negative) states, respectively.

Ferrous nitrosylheme with $\{FeNO\}^7$ group

The molecular structures of several five- and six-coordinated nitrosyl(porphyrinato)iron(II) (Fe(P)(NO)) complexes have been reported (see the reviews Scheidt, 1977; Scheidt and

Gouterman, 1983). The structures of representative complexes among them are described as follows. The five-coordinated Fe(TPP)(NO) complex has an Fe-N-O angle (ϕ) of 149.2°, an Fe-N_{NO} bond length of 1.717 Å, and an Fe – Ct distance of 0.21 Å (Scheidt and Frisse, 1975). The six-coordinated Fe(TPP)(NO)(N-MeIm) complex has an Fe-N-O angle (ϕ) of 140°, an Fe-N_{NO} bond length of 1.743 Å, an Fe-N_{Im} bond length of 2.180 Å, and an Fe – Ct distance of 0.07 Å (Scheidt and Piciulo, 1976). The bent structure of Fe-N-O unit is expected from $\{FeNO\}^7$ configuration (Enemark and Feltham, 1974). The shorter Fe-N_{NO} bond length and the larger Fe-N-O angle in the five-coordinated complex than in the six-coordinated one suggest the greater contribution of π bonding to the Fe-N_{NO} bond in the former complex. It is noted that the iron atom in the six-coordinated complex is almost in the porphyrinato plane, while in the five-coordinated complex it is displaced out-of-plane towards the NO ligand. The structural differences between five- and six-coordinated nitrosylheme can result in a difference in the interaction of d_π(Fe) with π^*(NO) and with π^*(porphyrin). The bending of the Fe-N-O unit can cause the interaction of d_{z2}(Fe) with π^*(NO). Zerner et al. (1966) reported theoretically that the displacement of iron

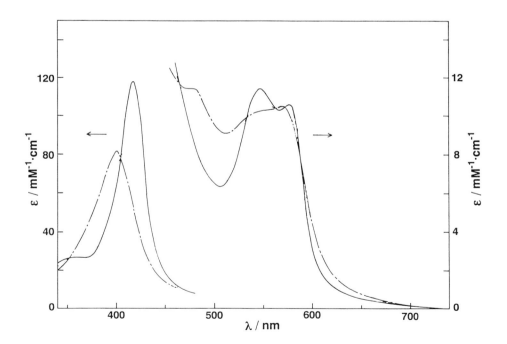

Figure 2. Electronic absorption spectra of five-coordinated Fe(PPIXDME)(NO) (chain line) and six-coordinated Fe(PPIXDME)(NO)(N-MeIm) (solid line) complexes in benzene at room temperature (Yoshimura and Ozaki, 1984).

from the porphyrinato plane enhances the mixing of $d_{z2}(Fe)$ into $a_{2u}(porphyrin)$ and lowers the a_{2u} in energy.

Figure 2 illustrates the electronic absorption spectra of five-coordinated Fe(PPIXDME)(NO) and six-coordinated Fe(PPIXDME)(NO)(N-MeIm) complexes at room temperature, in which Fe(PPIXDME) complexes have electronic spectra similar to those of naturally occurring proto-heme, Fe(PPIX), complexes (Yoshimura and Ozaki, 1984). The electronic spectral bands of both five- and six-coordinates were shifted to shorter wavelengths with an increase of solvent polarity. The band positions of six-coordinates with N-donor ligands were essentially insensitive to changes in an axial ligand *trans* to NO group. The spectral differences between five- and six-coordinates were observed in the Soret band rather than in visible α and β bands (Tab. 1, Fig. 2). The Soret band (415–420 nm) of six-coordinates is found at longer wavelengths with a greater intensity than that (395–400 nm) of five-coordinates. Further, the shoulder absorption at about 480 nm is characteristic of the spectrum of five-coordinates. The electronic spectra can be interpreted in relation to the molecular structure of nitrosylhemes.

Both the B or Soret and Q or visible bands in the electronic absorption spectra of porphyrins result theoretically from the mixing of the transitions $a_{1u} \rightarrow e_g^*$ and $a_{2u} \rightarrow e_g^*$ of porphyrin $\pi \rightarrow \pi^*$ by a configuration interaction (Gouterman, 1978). The a_{2u} state is lowered in energy with an increase in iron displacements from the porphyrinato plane, and the resulting increase in

Table 1. Electronic absorption spectral data for NO-ferrous hemeproteins and their model complexes at room temperature (sh; shoulder absorption)

	λ_{max} (nm) [ε(mM^{-1}·cm^{-1})]				
	Soret (γ)			β	α
Fe(PPIXDME)(NO) in benzene[a]	400.5 (81.6)		480sh(11)	550sh(10)	569.5(10.5)
Fe(PPIXDME)(NO)(N-MeIm) in benzene[a]		418.5(118)		546.5(11.5)	576.5(10.6)
Fe(PPIXDME)(NO)(Py) in pyridine[a]		416 (104)		548 (11)	570.5(10.9)
Fe(PPIXDME)(NO)(nBuNH$_2$) in benzene[a]		417 (128)		544 (11.9)	576 (12.1)
NO-Hb A (pH 6.5)[b]		416.7(129)		544 (13)	573 (13)
NO-Hb A + IHP (pH 6.5)[b]		415.2(99)		542 (13)	571 (12)
NO-Mb (sperm whale) (pH 7.4)[c]		420 (127)		548 (11.3)	579 (10.1)
NO-HRP (pH 7.0)[d]		421 (110)		542 (11.5)	570 (10.5)
NO-IDO (pH 7.0)[e]		418.5(127)		544 (12.2)	574 (12.5)
NO-cyt c (horse heart) (pH 5.3)[f]		412 (146)		538.5(11.6)	563.5(11.4)
NO-cyt c' (A. xylosoxidans) (pH 7.2)[g]	396.5 (78.9)	415sh	485 (9.8)	541 (10.4)	565sh(10)
NO-cyt c'(Rb. capsulatus B100) (pH 7.2)[h]	395sh(61)	417 (89)	480sh(7.8)	540 (10.8)	570sh(8.9)
NO-GC (pH 7.8)[i]	398 (79)		485sh	537 (12)	572 (12)

[a]Yoshimura and Ozaki, 1984. [b]Perutz et al., 1976. [c]O'Keefe et al., 1978. [d]Yonetani et al., 1972. [e]Sono and Dawson, 1984. [f]Yoshimura and Suzuki, 1988. [g]Yoshimura et al., 1986. [h]Yoshimura et al., 1987. [i]Stone and Marletta, 1994.

energy separation between a_{2u} and $e_g{}^*$ causes the shift to shorter wavelengths of absorption maxima. Since the iron in five-coordinates has much larger displacements from the porphyrinato plane than in six-coordinates, the absorption bands in the former can be situated at shorter wavelengths than in the latter.

In the NO derivatives of Hb, Mb, and cyt c peroxidase, the axial ligand *trans* to the NO group is an imidazolyl group of histidine residue. The electronic spectra of their NO derivatives (Yonetani et al., 1972) resembled those of six-coordinated Fe(PPIXDME)(NO) complexes with N-donor ligands (Tab. 1, Fig. 2). It has been shown that, upon the addition of IHP to NO-Hb, the quaternary structure of NO-Hb is switched from the oxy (R) to the deoxy (T) form, the heme iron to imidazole bond is stretched or cleaved, and consequently, the heme group of α subunits changes from six- to five-coordinate. Such quaternary structural changes have been confirmed by electronic spectral changes of NO-Hb in the absence and presence of IHP (Perutz et al., 1976). The cyt c' contains a five-coordinated heme with histidyl imidazole as an axial ligand (Meyer and Kamen, 1982). The observation that ferrous cyt c' reacts with NO to form a mixture of five- and six-coordinate nitrosylhemes (Tab. 1) demonstrates that the iron-to-histidine bond in ferrous cyt c' is labile on NO coordination. The lability of the bond may be attributed to the absence of an H-bonding group near the histidyl ligand and the presence of steric hindrance around vacant heme coordination sites (Weber, 1982; Yoshimura et al., 1986). The finding that the ratios of five-/six-coordinates in the NO reaction differ depending upon the bacterial source of cyt c' suggests that the heme environments also differ for each cyt c' (Yoshimura et al., 1986; Yoshimura et al., 1995) The NO adduct of cyt P-450 exhibited the Soret band at much longer wavelength (438 nm) than that of hemeproteins with a histidine as a heme fifth ligand (Ebel et al., 1975). This wavelength of the Soret band has been obtained from the system consisting of heme, methylmercaptide, and NO, suggesting the coordination mode of NO-Fe(II)-S$^-$ in the NO adduct of P-450 (Stern and Peisach, 1976).

MCD spectroscopy has been widely used in porphyrin and hemeprotein studies to probe changes in oxidation state, spin state, and axial ligation. Figure 3 illustrates the MCD spectra of five-coordinated Fe(PPIXDME)(NO) and six-coordinated Fe(PPIXDME)(NO)(N-MeIm) complexes at room temperature (Suzuki et al., 1987). The comparison of MCD spectra (Fig. 3) with electronic absorption spectra (Fig. 2) demonstrates that MCD bands are observable in the region of electronic absorption bands. The MCD bands of five-coordinates are observed as a negative extremum at 398 nm in the Soret band region, which is indicative of a Faraday B or C term; and two positive extrema at 514 and 558 nm and two negative extrema at 532 and 583 nm in the α and β band region, which can be composed to A term. The MCD bands of six-coordinates are observed as a derivative-shaped band with a positive extremum at 408 nm and a negative one at 421 nm in the Soret band region; and two positive extrema at 520 and 560 nm and a negative extrema at 578 nm in the α and β band region, all of which are probably due to A term. Significant differences in the MCD spectral line shape between five- and six-coordinates are observed, especially in the Soret band region. Thus, the Soret MCD band is more sensitive to the coordination structure of nitrosylheme than the MCD bands in the α and β band region (Suzuki et al., 1987; Yoshimura, 1990). The MCD spectra for NO derivatives of Hb, Mb, HRP, and IDO were quite similar to that for six-coordinated nitrosylhemes (Yamamoto et al., 1982; Sono and Dawson, 1984), while that for NO-cyt *c'* (*Achromobacter xylosoxidans* NCIB 11015) was essentially identical to that for five-coordinates (Suzuki et al., 1987).

The EPR spectra of five-coordinated Fe(P)(NO) complexes in non-donor organic solvents at room temperature exhibited a well-resolved triplet for [14]NO complexes or a doublet for [15]NO

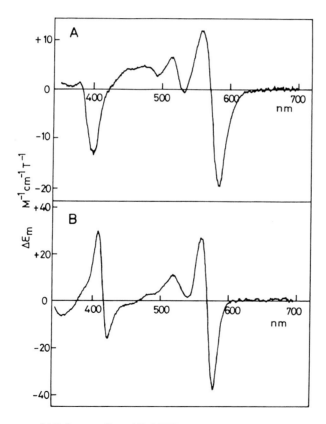

Figure 3. MCD spectra of (A) five-coordinated Fe(PPIXDME)(NO) complex in benzene and (B) six-coordinated Fe(PPIXDME)(NO)(*N*-MeIm) complex in dimethylacetamide at room temperature (Suzuki et al., 1987).

complexes. Five-coordinated Fe(P)(NO) complexes in non-donor solvents at low temperatures exhibited characteristic EPR spectra (Fig. 4) with an intense triplet in the g_3 (or g_z) absorption (Wayland and Olson, 1974; Yoshimura, 1978, 1991). The EPR parameters of g values and hyper fine coupling constants changed with the electron withdrawing power of porphyrin peripheral substituents (Yoshimura, 1991). EPR spectral results of five-coordinated Fe(P)(NO) complexes at low temperature have been utilized in elucidating the stereochemistry of the heme environment for nitrosylhemeproteins such as the NO adducts of Hbs in the presence of IHP (Rein et al., 1972) or SDS (Kon, 1968), those of mutant Hb M (Nagai et al., 1979), those of denatured cyt P-450 (or P-420) (O'Keefe et al., 1978), and those of cyt c' (Yoshimura et al., 1986), all of which contain five-coordinated nitrosylheme.

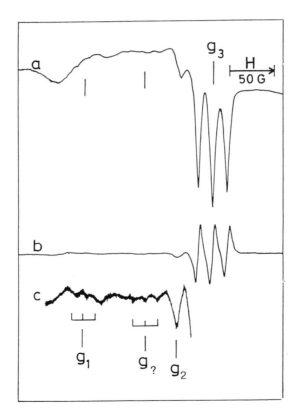

Figure 4. EPR spectra of five-coordinated Fe(PPIXDME)(NO) complex in frozen toluene at 77 K (Yoshimura, 1986b): (a) the first- and (b,c) second-derivative display, where (c) is the expansion of the ordinate of (b) ($g_1 = 2.105$, $a_1 = 13$ G; $g_2 = 2.036$; $g_3 = 2.0104$, $a_3 = 16.5$ G; $g_? = 2.059$, $a_? = 14$ G). For the assignments of absorptions, see the reference (Yoshimura, 1991).

The EPR spectra of six-coordinated Fe(P)(NO) complexes with imidazole derivatives at room temperature were slightly asymmetric singlets, suggesting an overlapping of two-components (Yoshimura et al., 1979; Yoshimura, 1991). The EPR spectra of six-coordinates with N-donor ligands at low temperature (Fig. 5) exhibited a line shape characteristic of randomly oriented systems with a rhombic symmetry and its central g_2 (or g_z) absorption had mostly nine hyperfine lines or a triplet of triplets. These hyperfine lines arise from the delocalization of an unpaired electron of NO toward the *trans* axial ligand through the d_{z2}(Fe) orbital and so the nine lines originate from the hyperfine interaction with two axial ^{14}N nuclei (Wayland and Olson, 1974; Yoshimura et al., 1979; Yoshimura, 1991). The EPR parameters of g values and hyperfine coupling constants varied systematically with a basicity of axial ligand *trans* to the NO group (Yoshimura, 1980, 1982a, 1986a) and with the electron withdrawing power of porphyrin peripheral

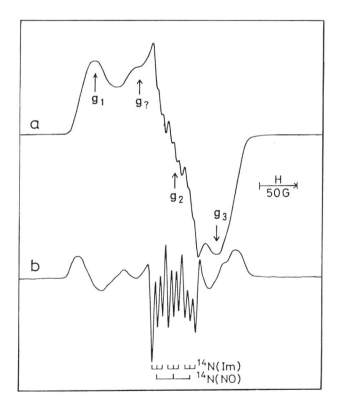

Figure 5. EPR spectra of six-coordinated Fe(PPIXDME)(NO)(*N*-MeIm) complex in frozen chloroform at 77 K (Yoshimura et al., 1979): (a) the first- and (b) second-derivative display ($g_1 = 2.074$; $g_2 = 2.004$, $a_2(^{14}N_{NO}) = 21.5$ G, $a_2(^{14}N_{Im}) = 6.9$ G; $g_3 = 1.971$; $g_? = 2.031$). For the assignments of absorptions, see the reference (Yoshimura, 1991).

substituents (Yoshimura, 1991, 1993a). These results could be explained by the fact that the electron densities on d_{z2}, $d\pi$, and π^* (NO) orbitals are altered by a change in electron donor or acceptor ability of *cis* or *trans* substitution. The nitrosyl derivatives of many hemeproteins at low temperatures exhibited EPR spectra similar to those of six-coordinates with N-donor ligands. These results have been utilized in identifying the fifth heme ligand, of which the most probable candidate is an imidazolyl group of the histidine residue (Yonetani et al., 1972; see the reviews Dickinson and Symmons, 1983; Palmer, 1983). On the other hand, the g_2 (or g_z) absorption in six-coordinated Fe(P)(NO) complexes with O- or S-donor ligands exhibited only three hyperfine lines, because neither oxygen nor sulfur has nuclear spin (Yoshimura, 1982b, 1983a). Similar g_2 (or g_z) absorption with three hyperfine lines has been observed in the spectra for NO adducts of cyt P-450 (O'Keefe et al., 1978), CPO (Chiang et al., 1975), and catalase (Yonetani et al., 1972). These results are consistent with the fact that the P-450 and CPO have a cystein thiolate-sulfur as a heme fifth ligand and the catalase has a tyrosine phenolic-oxygen.

The NO stretching bands in the IR spectra for five-coordinated Fe(PPIXDME)(NO) complex were located at higher frequencies than those for the six-coordinates (Tab. 2). Accompanying the coordination of N-donor ligand to an axial position *trans* to the NO group, the σ electrons are donated from the ligand to the iron, which results in an increase in electron density of d_{z2}(Fe) and π^*(NO). Consequently, the increased antibonding character of the NO bond can induce the observed decrease in NO stretching frequencies (Yoshimura, 1983b). It has been demonstrated that NO-Hb in the absence of IHP exhibited one NO stretching band for the six-coordinates, while in the presence of IHP it had two NO stretching bands for the five- and six-coordinates

Table 2. Infrared absorption spectral data for NO-hemoglobin and the model complexes at room temperature

		NO stretching frequencies (cm^{-1})	
Fe(PPIXDME)(NO)	in carbon tetrachloride[a,b]		1684
	in benzene[a,b]		1672
	in chloroform[a,b]		1675
	in benzonitrile[a,b]		1664
	in acetonitrile [a,b]		1658
Fe(PPIXDME)(NO)(*N*-MeIm) in *N*-MeIm[b]		1618	
Fe(PPIXDME)(NO)(Py) in pyridine[b]		1631	
Fe(PPIXDME)(NO)(*n*BuNH$_2$) in *n*BuNH$_2$[b]		1639	
Fe(PPIXDME)(NO)(Pip) in Pip[b]		1643	
NO-Hb A[c]		1615	
NO-Hb A + IHP[c]		1615	1668

[a]The order of solvent polarity is carbon tetrachloride < benzene < chloroform < benzonitrile < acetonitrile.
[b]Yoshimura, 1983b. [c]Maxwell and Caughey, 1976.

(Tab. 2). This suggests that the binding of IHP to the protein results in cleavage of the iron-to-proximal-histidine bonds in two of four subunits (Maxwell and Caughey, 1976). The NO stretching frequencies varied linearly with solvent polarity of the medium, basicity of axial ligand *trans* to NO group in the six-coordinates, and electron withdrawing power of porphyrin peripheral substituents in both five- and six-coordinates, consistent with EPR spectral results (Yoshimura, 1983c, 1991).

Resonance Raman (RR) spectra have been extensively measured both for Fe(P)(NO) complexes and for nitrosylhemeproteins (Lipscomb et al., 1993; Spiro, 1983 and references therein). Upon the addition of IHP, the RR spectrum of NO-Hb is expected to display an additional Fe-N_{NO} stretching band originating from five-coordinated nitrosylheme together with the band from six-coordinates as just mentioned in the IR description. However, the Fe-NO stretching band for five-coordinated nitrosylheme in NO-Hb in the presence of IHP has not been detected yet (Tsubaki and Yu, 1982). On the other hand, on adding IHP, the RR band of HO-Hb at 1633 cm^{-1} assigned to a depolarized porphyrin ring mode was weakened and a new band at 1643 cm^{-1} appeared (Szabo and Barron, 1975). The bands of the depolarized ring mode in the five- and six-coordinated Fe(PPIXDME)(NO) complexes have been observed at 1643 cm^{-1} and 1637 cm^{-1}, respectively (Stong et al., 1980). The Fe-N_{NO} stretching (or bending) frequencies of six-coordinated nitrosyl heme have been measured in nitrosylhemeproteins (at around 550 cm^{-1}) such as NO-Hb, -Mb, and -HRP (Tsubaki and Yu, 1982; Benko and Yu, 1983), and in Fe(OEP)(NO)(N-MeIm) (at 524 cm^{-1}) (Lipscomb et al., 1993). The five-coordinated Fe(TPP)(NO) complex displayed the N-O and Fe-N_{NO} stretching band at 1681 and 525 cm^{-1}, respectively (Choi et al., 1991).

The reactions of Fe(P)L with X and that of Fe(P)X with L in solution, where L = N-donor ligands and X = NO or CO, take place according to the following equations:

$$\text{Fe(P)L} + \text{X} \underset{}{\overset{K_1}{\rightleftharpoons}} \text{Fe(P)XL} \qquad [1]$$

$$\text{Fe(P)X} + \text{L} \underset{}{\overset{K_2}{\rightleftharpoons}} \text{Fe(P)XL} \qquad [2]$$

Equilibrium constants and rate constants of Equations [1] and [2] for hemoglobin, myoglobin, and their model complexes are tabulated in Table 3. The affinity for NO is much higher than that for CO. The ratio of the equilibrium constants, $K_1(NO)/K_1(CO)$, for Fe(PPIX) complexes is comparable to that for Hb, but is much less than that for Mb. The rate constant for NO coordination to the Fe(PPIX) complex is about ten times higher than that for Hb or Mb. The amino acid residues surrounding the NO binding site of Hb or Mb appear to restrict the access of NO (Rose and Hoffman, 1983).

For various N-donor ligands, the equilibrium constants (K_2) have been estimated from the dependence of EPR and IR spectra on the base concentration (Yoshimura, 1980, 1983b,c). The equilibrium constants increased linearly with increase in basicity of the ligand and were sensitive to both the polarity and the hydrogen bonding ability of the solvent. As shown in Table 3, the affinity of Fe(PPIXDME)(NO) for 1-methylimidazole (11 M^{-1}) is markedly lower than that of Fe(DPIXDME)(CO) for imidazole (4.3 × 10^7 M^{-1}), indicating that the *trans* effect of a NO ligand is much more significant than that of a CO ligand or that the vacant axial site of a five-coordinated nitrosylheme is much less accessible to ligands than that of a five-coordinated carbonylheme.

Ferric nitrosylheme with {FeNO}^{6} group

The molecular structures of two nitrosyl(porphyrinato)iron(III) complexes have been reported (Scheidt et al., 1984). The five-coordinated [Fe(OEP)(NO)]ClO$_4$ complex has an Fe-N-O angle (ϕ) of 176.9°, an Fe-N$_{NO}$ bond length of 1.644 Å, and an Fe – Ct distance of 0.32 Å. In the solid state, the cations [Fe(OEP)(NO)]$^+$ interact in pairs to form π-π dimers. The two planar cores are parallel with an interplanar separation of 3.36 Å. The six-coordinated [Fe(TPP)(NO)(H$_2$O)]ClO$_4$ complex has Fe-N-O angle (ϕ) of 174.4°, Fe-N$_{NO}$ bond length of 1.652 Å, Fe-O$_{H_2O}$ bond length of 2.001 Å, and Fe – Ct distance of ~ 0 Å. An Fe(III)-NO unit with {FeNO}^{6} configuration is

Table 3. Equilibrium constants and rate constants of Equation 1 and 2 for hemoglobin, myoglobin, and their model complexes[a]

	X (NO or CO)	L	K_1 (M^{-1})	K_2 (M^{-1})	k_{on} (M^{-1}·s^{-1})	k_{off} (s^{-1})
Fe(PPIX)[b]	NO	1-methylimidazole	5.8 × 10^{11}		1.8 × 10^8	2.9 × 10^{-4}
	CO	1-methylimidazole	7.8 × 10^8		1.8 × 10^6	2.3 × 10^{-3}
Fe(PPIXDME)[c]	NO	1-methylimidazole		11		
		pyridine		1.6		
Fe(DPIXDME)[d]	CO	imidazole	4.8 × 10^8	4.3 × 10^7		
		4-cyanopyridine	5.6 × 10^7	5.3 × 10^6		
Hb[e]	NO		2.5 × 10^{10}		1.5 × 10^7	6 × 10^{-5}
Hb[b,e]	CO		2.3 × 10^7			
Mb[b]	NO		1.4 × 10^{11}		1.7 × 10^7	1.2 × 10^{-4}
	CO		2.4 × 10^7		5.0 × 10^5	2.1 × 10^{-2}

[a]Equation 1: Fe(P)L + X $\xrightleftharpoons{K_1}$ Fe(P)XL; Equation 2: Fe(P)X + L $\xrightleftharpoons{K_2}$ Fe(P)XL. [b]in Tris·HCl buffer (pH 9) for Fe(PPIX) and in aqueous buffer (pH 6–7) for Hb and Mb (Rose and Hoffman, 1983). [c]in benzene (Yoshimura, 1983c). [d]in benzene (Rougee and Brault, 1975). [e]in phosphate buffer (pH 6.8) (Gibson and Roughton, 1957; Antonini and Brunori, 1971).

formally isoelectronic with an Fe(II)-CO unit and the porphyrin complexes with these units can have a linear Fe-N-O or Fe-C-O linkage, which maximizes $d_\pi(Fe) \rightarrow p_\pi(NO)$ back-bonding (Wayland and Olson, 1974).

The reaction of NO with Fe(III) porphyrin complexes results in the formation of nitrosyl iron(II) porphyrin complexes. In the reaction, NO first reduces the iron(III) to iron(II) and then coordinates to the iron(II) as a nitrosyl ligand, which is a typical case of reductive nitrosylation. Therefore, the distinct electronic absorption spectra of nitrosyl(porphyrinato)iron(III) complexes have not so far been reported in solution. However, the spectra of Fe(TPP)(Cl)(NO) with {FeNO}[6] group have been measured in frozen toluene glass and nujol mull state and assigned to the low-spin Fe(II) complex, Fe(II)(TPP)(Cl$^-$)(NO$^+$) (Wayland and Olson, 1974). The NO stretching frequencies in the IR spectra of [Fe(OEP)(NO)]ClO$_4$, [Fe(TPP)(NO)(H$_2$O)]ClO$_4$, and Fe(TPP)(Cl)(NO) complexes have been reported to be 1862, 1937, and 1880 cm^{-1}, respectively, indicating extensive π back-bonding.

The reaction of NO with ferric hemeproteins proceeds in various patterns. The NO reaction with ferric Hb and Mb results in the formation of relatively stable nitrosyl ferric analogs, but the products are gradually reduced in the presence of NO to form corresponding nitrosyl ferrous analogs (Antonini and Brunori, 1971). The NO derivatives of ferric peroxidases and oxygenases are not reduced by NO (Yonetani et al., 1972; Sono and Dawson, 1984). The NO reaction with ferric cyt c' from photosynthetic and denitrifying bacteria results in the formation of a nitrosyl ferric and a nitrosyl ferrous analog, respectively (Yoshimura et al., 1986, 1995). The difference of heme iron(III) in reducibility by NO appears to arise from the difference in the midpoint redox potential of these hemeproteins. The feature of spectral line shape and band positions for nitrosyl ferric peroxidases and oxygenases is characteristic of a low-spin ferrous hemeprotein, demonstrating the transfer of electrons from NO to the heme iron(III) (Yonetani et al., 1972).

In the RR spectrum of an Fe(III)(OEP)(NO)(Py) complex, the Fe-N$_{NO}$ stretching mode was detected at 602 cm^{-1}, but the bending mode was undetectable because of the linearity of the Fe-N-O unit (Lipscomb et al., 1993). On the other hand, in NO derivatives of ferric Mb and HRP, the Fe-N$_{NO}$ stretching mode was detected at 595 and 604 cm^{-1}, and further, the bending mode was detected at 573 and 574 cm$^{-1,}$ respectively (Benko and Yu, 1983). The observation of the bending mode is indicative of the Fe-N-O distortion in the nitrosyl ferric hemeproteins.

Guanylate cyclase: Activation by NO

Guanylate (guanylyl) cyclase (GC) catalyzes the biosynthesis of cyclic GMP from GTP (Waldman and Murad, 1987). The cytosolic (soluble) form of GC is activated by NO and is a heterodimer which contains a protoheme. In the electronic spectral study of GC, the Soret band at 433 nm for reduced GC shifts to 399 nm on the NO reaction (Gerzer et al., 1981). As described above, the appearance of a 399 nm Soret band in the reaction of NO with ferrous hemeproteins indicates the formation of a five-coordinated nitrosylheme. Recently, the GC has been carefully isolated and purified, and excellent electronic and resonance Raman spectra have been presented (Stone and Marletta, 1994; Yu et al., 1994). Both the ferric and ferrous GC have a five-coordinated high-spin spectrum, which is similar to that of deoxy ferrous Mb and ferric cyt c', respectively. In analogy with the deoxy ferrous Mb and ferric cyt c', the most probable candidate for the heme fifth ligand of GC is a histidine residue (Stone and Marletta, 1994). The RR spectral results for NO adduct of ferrous GC have supported the formation of a five-coordinated nitrosylheme (Yu et al., 1994). The heme of GC is five-coordinated and the ferrous heme binds NO to form a five-coordinated nitrosylheme (Stone and Marletta, 1994), similar behavior to cyt c' from denitrifying bacteria described above (Yoshimura et al., 1986).

It has been demonstrated that the iron of high-spin five-coordinated ferrous hemeproteins and their model complexes is situated $0.4-0.5$ Å out of heme plane toward the fifth heme ligand (Fermi et al., 1984; Scheidt and Gouterman, 1983), while that of ferrous nitrosylheme models is situated about 0.2 Å toward the nitrosyl ligand (Scheidt and Gouterman, 1983). On NO coordination to heme iron, therefore, the iron of high-spin five-coordinated ferrous GC can be shifted by $0.6-0.7$ Å from the proximal to the distal side. This displacement of iron may cause a conformational change in the enzyme.

Two mechanisms for the activation on NO binding to GC have been proposed: (1) the cleavage of the iron-to-histidine bond accompanied by the NO coordination causes subsequent conformational changes in the enzyme, resulting in the activation (Ignarro, 1992); (2) the conformational changes liberate proximal histidyl imidazole to act as a nucleophilic catalyst and to catalyze the conversion reaction of GTP to cyclic GMP (Traylor et al., 1993). The substrate-binding site is not clarified in these mechanisms. The binding of substrate to the five-coordinated nitrosyl heme seems to be stereochemically impossible. Assuming that the heme site in GC is not the substrate-binding site, the NO ligand, which induces conformational changes on coordinating to heme iron, may be an allosteric effector.

On the other hand, the deactivation mechanisms have not been proposed yet. The dissociation of coordinated NO is required to deactivate GC. The iron-to-NO bond in Fe(P)(NO) complexes is weakened by (1) an increase in basicity of the *trans* axial ligand or a strengthening of iron-to-ligand bond *trans* to NO group, and (2) an increase in electron-withdrawing power of porphyrin peripheral substituents or a decrease in electron density on porphyrin pyrrole nitrogen (Yoshimura, 1986a, 1991; Yoshimura et al., 1993a). On GTP binding to NO-GC, if an H-bond is induced and strengthened between histidine ligand and the other amino acid residues, the basicity

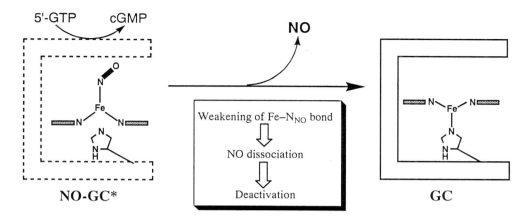

Figure 6. Possible deactivation mechanism of guanylate cyclase. The activated NO-GC* and resting GC are represented by dotted and solid patterns, respectively.

of histidyl imidazole can increase, and if hydrophobic interaction modes between heme side-chain and amino acid residues are changed, the electron density on the pyrrole nitrogen may decrease (it has been pointed out [Ignarro et al., 1984] that such interactions are essential for GC activation). These changes can result in a restoration of the iron-to-histidine bond and a dissociation of coordinated NO and consequently, the enzyme can be deactivated (Fig. 6).

Nitrosylheme detected in whole cells of denitrifying bacteria

Biological denitrification is a process which converts nitrate or nitrite to nitrous oxide or dinitrogen as follows:

$$NO_3^- \rightarrow NO_2^- \rightarrow NO \rightarrow N_2O \rightarrow N_2$$

Many bacteria, denitrifiers, are capable of performing this reaction (Zumft, 1992). Recently, nitrite reduction and NO reduction have been suggested to be interdependent (Zumft et al., 1994). Each reduction step is linked to the generation of ATP and all the reductases are known to be metallo-enzymes and to be associated with metal-containing cofactors. It has been demonstrated that NO formed as an intermediate in the process is not released into the environment under physiological conditions. Since NO has a high affinity for metal ions as described above, the interaction of NO with metalloproteins in the denitrification process in bacterial cells is noted.

The denitrifying bacterium, *Achromobacter xylosoxidans* (*Alcaligenes xylosoxidans* subsp. *xylosoxidans*) NCIB 11015 was cultivated in meat extract-peptone medium and in manganese (Mn)-free synthetic medium under denitrifying or non-denitrifying conditions (Yoshimura et al., 1993b). The characteristic three-line signal has been detected in the EPR spectrum at 77 K for whole cells of the bacterium under denitrifying conditions, but not under non-denitrifying conditions. The three-line signal was more distinctly detected in the cells cultured in Mn-free medium. This signal was quite similar to that of NO-adducts of cyt *c'* from *A. xylosoxidans* NCIB 11015 and that of five-coordinated nitrosyl(porphyrinato)iron(II) complexes (Fig. 4). Therefore, this three-line signal could be assigned to NO-cyt *c'* containing a five-coordinated nitrosylheme, which the NO formed during the biological denitrification process in the bacterium as a nitrosyl ligand to heme iron. The finding that a fraction of cyt *c'* in bacterial cells can exist in the form of NO-cyt *c'* indicates that the cyt *c'* may capture endogenous and exogenous NO to form NO-cyt *c'*.

Acknowledgements
I would like to acknowledge the contributions of my coworkers whose names appear in referenced papers.

References

Antonini, E. and Brunori, M. (1971) *Hemoglobin and Myoglobin in their Reactions with Ligands*, North-Holland, Amsterdam.

Benko, B. and Yu, N.-T. (1983) Resonance Raman studies of nitric oxide binding to ferric and ferrous hemoproteins: Detection of Fe(II)-NO stretching, Fe(III)-N-O bending, and Fe(II)-N-O bending vibration. *Proc. Natl. Acad. Sci. USA* 80: 7042–7046.

Butler, R.B. and Williams, D.L.H. (1993) The physiological role of nitric oxide. *Chem. Soc. Rev.* 22: 233–241.

Calver, A., Collier, J. and Vallance, P. (1993) Nitric oxide and cardiovascular control. *Exp. Physiol.* 78: 303–326.

Chiang, R., Makino, R., Spomer, W.E. and Hager, L.P. (1975) Chloroperoxidase: P-450 type absorption in the absence of sulfhydryl groups. *Biochemistry* 14: 4166–4171.

Choi, I.-K., Liu, Y., Feng, D., Paeng, K.-J. and Ryan, M.D. (1991) Electrochemical and spectroscopic studies of iron porphyrin nitrosyls and their reduction products. *Inorg. Chem.* 30: 1832–1839.

Dickinson, L.C. and Symmons, M.C.R. (1983) Electron spin resonance of Haemoglobin and myoglobin. *Chem. Soc. Rev.* 12: 387–414.

Ebel, R.E., O'Keefe, D.H. and Peterson, J.A. (1975) Nitric oxide complexes of cytochrome P-450. *FEBS Lett.* 55: 198–201.

Enemark, J.H. and Feltham, R.D. (1974) Principles of structure, bonding, and reactivity for metal nitrosyl complexes. *Coordin. Chem. Rev.* 13: 339–406.

Fermi, G., Perutz, M.F. and Schaanan, B. (1984) The Crystal Structure of Human Deoxyhaemoglobin at 1.74 Å Resolution. *J. Mol. Biol.* 175: 159–174.

Gerzer, R., Bohme, E., Hofman, F. and Schultz, G. (1981) Soluble guanylate cyclase purified from bovine lung contains heme and copper. *FEBS Lett.* 132: 71–74.

Gibson, Q.H. and Roughton, F.J.W. (1957) The kinetics and equilibria of the relations of nitric oxide with sheep hemoglobin. *J. Physiol.* 136: 507–526.

Gouterman, M. (1978) *In:* D. Dolphin (ed.): *The Porphyrins.* Vol. 3, Acad. Press, New York, pp 1–165.

Henry, Y., Ducrocq, C., Drapier, J.-C., Servant, D., Pellat, C. and Guissani, A. (1991) Nitric oxide, a biological effector. Electron paramagnetic resonance detection of nitrosyliron-protein complexes in whole cells. *Eur. Biophys. J.* 20: 1–15.

Hermann, L. (1865) Über die Wirkungen des Stickstoffoxydgasses auf das Blut. *Arch. Anat. Physiol.* 469–481.

Ignarro, L.J., Ballot, B. and Wood, K.S. (1984) Regulation of soluble guanylate cyclase activity by porphyrins and metalloporphyrins. *J. Biol. Chem.* 259: 6201–6207.

Ignarro, L.J. (1992) Haem dependent activation of cytosolic guanylate cyclase by nitric oxide: a widespread signal transduction mechanism. *Biochem. Soc. Trans.* 20: 465–469.

Kon, H. (1968) Paramagnetic resonance study of nitric oxide hemoglobin. *J. Biol. Chem.* 243: 4350–4357.

Kosaka, H., Uozumi, M. and Tyuma, I. (1989) The interaction between nitrogen oxides and hemoglobin and endothelium-derived relaxing factor. *Free Radical Biol. Med.* 7: 653–658.

Lipscomb, L.A., Lee, B.-S. and Yu, N.-T. (1993) Resonance Raman investigation of nitric oxide bonding in iron porphyrins: Detection of the Fe–NO stretching vibration. *Inorg. Chem.* 32: 281–286.

Maxwell, J.C. and Caughey, W.S. (1976) An infrared study of NO binding to heme B and hemoglobin A. Evidence for inositolhexaphosphate induced cleavage of proximal histidine to iron bonds. *Biochemistry* 15: 388–396.

Meyer, T.E. and Kamen, M.D. (1982) New perspectives on c-type cytochromes. *Adv. Protein Chem.* 35: 105–212.

Nagai, K., Hori, H., Morimoto, H., Hayashi, A. and Taketa, F. (1979) Influence of amino acid replacements in the heme pocket on the electron paramagnetic resonance spectra and absorption spectra of nitrosylhemoglobins M Iwate, M Boston, and M Milwaukee. *Biochemistry* 18: 1304–1308.

O'Keefe, D.H., Ebel, R.E. and Perterson, J.A. (1978) Studies of the oxygen binding site of cytochrome P-450. Nitric oxide as a spin label probe. *J. Biol. Chem.* 253: 3509–3516.

Oda, H., Kusumoto, S. and Nakajima, T. (1975) Nitrosyl-hemoglobin formation in the blood of animals exposed to nitric oxide. *Arch. Environ. Health.* 30: 453–456.

Palmer, G. (1983) *In:* A.B.P. Lever and H.B. Gray (eds): *Iron Porphyrins* Part II, Addison-Wesley, London, pp 43–88.

Perutz, M.F., Kilmartin, J.V., Nagai, K., Szabo, A. and Simmon, S.R. (1976) Influence of globin structures on the state of the heme. Ferrous low spin derivatives. *Biochemistry* 15: 378–387.

Rein, H., Ristau, O. and Scheler, W. (1972) On the influence of allosteric effectors on the electron paramagnetic spectrum of nitric oxide hemoglobin. *FEBS Lett.* 24: 24–26.

Richter-Addo, G.B. and Legzdins, P. (1992) *Metal Nitrosyls,* Oxford University Press, Oxford.

Rose, E.J. and Hoffman, B.M. (1983) Nitric oxide ferrohemes; Kinetics of formation and photodissociation quantum yields. *J. Am. Chem. Soc.* 105: 2866–2873.

Rougee, M. and Brault, D. (1975) Influence of trans weak or strong field ligands upon the affinity of deuteroheme for carbon monoxide monoimidazoleheme as a reference for unconstrained five-coodinate hemoproteins. *Biochemistry* 14: 4100–4106.

Scheidt, W.R. and Frisse, M.E. (1975) Nitrosylmetalloporphyrins. II. Synthesis and molecular stereochemistry of nitrosyl-$\alpha,\beta,\gamma,\delta$-tetraphenylporphynatoiron(II). *J. Am. Chem. Soc.* 97: 17–21.

Scheidt, W.R. and Piciulo, P.L. (1976) Nitrosylmetalloporphyrins. III. Synthesis and molecular stereochemistry of nitrosyl-$\alpha,\beta,\gamma,\delta$-tetraphenylporphinato(1-methylimidazole)iron(II). *J. Am. Chem. Soc.* 98: 1913–1919.

Scheidt, W.R. (1977) Trends in metalloporphyrin stereochemistry. *Acc. Chem. Res.* 10: 339–345.

Scheidt, W.R. and Gouterman, M. (1983) *In:* A.B.P. Lever and H.B. Gray (eds): *Iron Porphyrins.* Part I, Addison-Wesley, London, pp 89–139.

Scheidt, W.R., Lee, Y.J. and Hatano, K. (1984) Preparation and structural characterization of nitrosyl complexes of ferric porphyrinates. Molecular structure of aquonitrosyl(*meso*-tetraphenylporphinato)iron(III) perchlorate and nitrosyl(octaethylporphinato)iron(III) perchlorate. *J. Am. Chem. Soc.* 106: 3191–3198.

Sono, M. and Dawson, J.D. (1984) Extensive studies of the heme coordination structure of indoleamine 2,3-dioxygenase and of tryptophan binding with magnetic and natural circular dichroism and electron paramagnetic resonance spectroscopy. *Biochim. Biophys. Acta* 789: 170–187.

Spiro, T.G. (1983) *In:* A.B.P. Lever and H.B. Gray (eds): *Iron Porphyrins.* Part II, Addison-Wesley, London, pp 89–159.

Stadler, J., Trockfeld, J., Schmalix, W.A., Brill, T., Siewert, J.R., Greim, H. and Doehmer, J. (1994) Inhibition of cytochrome P4501A by nitric oxide. *Proc. Natl. Acad. Sci. USA* 91: 3559–3563.

Stern, J.O. and Peisach, J. (1976) A model compound for nitrosyl cytochrome P-450; Further evidence for mercaptide sulfur ligation to heme. *FEBS Lett.* 62: 364–368.

Stone, J.R. and Marletta, M.A. (1994) Soluble guanylate cyclase from bovine lung: Activation with nitric oxide and carbon monoxide and spectral characterization of the ferrous and ferric states. *Biochemistry* 33: 5636–5640.

Stong, J.D., Burke, J.M., Daly, P., Wright, P. and Spiro, T.G. (1980) Resonance raman spectra of nitrosyl heme proteins and of porphyrin analogues. *J. Am. Chem. Soc.* 102: 5815–5819.

Suzuki, S., Yoshimura, T., Nakahara, A., Iwasaki, H., Shidara, S. and Matsubara, T. (1987) Electronic and magnetic circular dichroism spectra of pentacoordinate nitrosylhemes in cytochrome c' from nonphotosynthetic bacteria and their model complexes. *Inorg. Chem.* 26: 1006–1008.

Szabo, A. and Barron, L.D. (1975) Resonance studies of nitric oxide hemoglobin. *J. Am. Chem. Soc.* 97: 660–662.

Traylor, T.G., Duprat, A.F. and Sharma, V.S. (1993) Nitric oxide-triggered heme-mediated hydrolysis: A possible model for biological reactions of NO. *J. Am. Chem. Soc.* 115: 810–811.

Tsubaki, M. and Yu, N.-T. (1982) Resonance Raman investigation of nitric oxide bonding in nitrosylhemoglobin A and -myoglobin: Detection of bound N-O stretching and Fe-NO stretching vibrations from the hexacoordinated NO-heme complex. *Biochemistry* 21: 1140–1144.

Waldman, S.A. and Murad, F. (1987) Cyclic GMP synthesis and function. *Phamacol. Rev.* 39: 163–196.

Wayland, B.B. and Olson, L.W. (1974) Spectroscopic studies and bonding model for nitric oxide complexes of iron porphyrins. *J. Am. Chem. Soc.* 96: 6037–6041.

Weber, P.C. (1982) Correlations between structural and spectroscopic properties of the high-spin heme protein cytochrome *c'*. *Biochemistry* 21: 5116–5119.

Yamamoto, T., Nozawa, T., Kaito, A. and Hatano, M. (1982) Experimental and calculated magnetic circular dichroism spectra of iron(II) low spin hemoglobin and myoglobin with CO, NO, O_2. *Bull. Chem. Soc. Jpn.* 55: 2021–2025.

Yonetani, T., Yamamoto, H., Erman, J.E., Leigh, J.S. and Reed, G.H. (1972) Electromagnetic properties of hemoproteins V. Optical and electron paramagnetic resonance characteristics of nitric oxide derivatives of metalloporphyrin-apohemoprotein complexes. *J. Biol. Chem.* 247: 2447–2455.

Yoshimura, T. (1978) The nitrogen oxide complex of the iron(II) protoporphyrin IX dimethyl ester. *Bull. Chem. Soc. Jpn.* 51: 1237–1238.

Yoshimura, T., Ozaki, T., Shintani, Y. and Watanabe, H. (1979) Electron paramagnetic resonance of nitrosylprotoheme dimethyl ester complexes with imidazole derivatives as model systems for nitrosylhemoproteins. *Arch. Biochem. Biophys.* 193: 301–313.

Yoshimura, T. (1980) Electron paramagnetic resonance study of the interaction of nitrosylprotoheme dimethyl ester with nitrogenous bases. *Inorg. Chim. Acta* 46: 69–76.

Yoshimura, T. (1982a) Electron paramagnetic resonance study of nitrosylprotoheme dimethyl ester with aliphatic nitrogenous bases: Characterization of the axial ligand *trans* to the nitrosyl group in nitrosylhemoproteins. *Arch. Biochem. Biophys.* 216: 625–630.

Yoshimura, T. (1982b) Electron paramagnetic resonance study of the interaction of nitrosyl(protoporphyrin IX dimethyl ester)iron(II) with sulfur- and oxygen-donor ligands. *Inorg. Chim. Acta* 57: 99–105.

Yoshimura, T. (1983a) Q-band electron paramagnetic resonance study of nitrosylprotoheme dimethyl ester complexes with N-, O-, and S-donor ligands as model systems for nitrosylhemoproteins. *J. Inorg. Biochem.* 18: 263–277.

Yoshimura, T. (1983b) Infrared and electron paramagnetic resonance study of nitrosyl(protoporphyrin IX dimethyl ester)iron(II) and its complexes with nitrogenous bases as model systems for nitrosylhemoproteins: Effect of solvent polarity. *Arch. Biochem. Biophys.* 220: 167–178.

Yoshimura, T. (1983c) Nitrosyl(protoporphyrin IX dimethyl ester)iron(II) complexes with nitrogenous bases. The basicity dependence of the NO stretching frequency. *Bull. Chem. Soc. Jpn.* 56: 2527–2528.

Yoshimura, T. and Ozaki, T. (1984) Electronic spectra of nitrosyl(protoporphyrin IX dimethyl ester)iron(II) and its complexes with nitrogenous bases as model systems for nitrosylhemoproteins. *Arch. Biochem. Biophys.* 229: 126–135.

Yoshimura, T. (1986a) Azolate complexes of nitrosyl(protoporhyrin IX dimethyl esterato)iron(II). *Inorg. Chem.* 25: 688–691.

Yoshimura, T. (1986b) EPR spectra of five-coordinate nitrosyl(protoporphyrin IX dimethyl ester)iron(II) in toluene under various conditions. *Inorg. Chim. Acta* 125: L27–L29.

Yoshimura, T., Suzuki, S., Nakahara, A., Iwasaki, H., Masuko, M. and Matsubara, T. (1986) Spectral properties of nitric oxide complexes of cytochrome *c'* from *Alcaligenes* sp. NCIB 11015. *Biochemistry* 25: 2436–2442.

Yoshimura, T., Suzuki, S., Iwasaki, H. and Takakuwa, S. (1987) Spectral properties of nitric oxide complex of cytochrome *c'* from *Rhodopseudomonas capsulata* B100. *Biochem. Biophys. Res. Commun.* 145: 868–875.

Yoshimura, T. and Suzuki, S. (1988) The pH dependence of the stereochemistry around the heme group in NO-cytochrome *c* (horse heart). *Inorg. Chim. Acta* 152: 241–249.

Yoshimura, T. (1990) Substituent effects on the electronic absorption and MCD spectra of five- and six-coordinate nitrosyl(tetraphenylporphyrinato)iron(II) complexes. *Bull. Chem. Soc. Jpn.* 63: 3689–3691.

Yoshimura, T. (1991) Five- and six-coordinated nitrosyl iron(II) complexes of tetrakis(*p*-substituted phenyl)porphyrins. Substituent effects on the EPR parameters and the NO stretching frequencies. *Bull. Chem. Soc. Jpn.* 64: 2819–2828.

Yoshimura, T., Kamada, H., Toi, H., Inaba, S. and Ogoshi, H. (1993a) Nitrosyl iron(II) complexes of porphyrins substituted with highly electron-withdrawing CF_3 groups: electronic absorption, MCD, and EPR spectral study. *Inorg. Chim. Acta* 208: 9–15.

Yoshimura, T., Shidara, S., Ozaki, T. and Kamada, H. (1993b) Five coordinated nitrosylhemoprotein in the whole cells of denitrifying bacterium, *Achromobacter xylosoxidans* NCIB 11015. *Arch. Microbiol.* 160: 498–500.

Yoshimura, T., Fujii, S., Kamada, H., Yamaguchi, K., Suzuki, S., Shidara, S. and Takakuwa, S. (1995) Spectroscopic characterization of nitrosylheme in nitric oxide complex of cytochrome *c'* from phothosynthetic bacteria. *Biochim. Biophys. Acta; in press*.

Yu, A.E., Hu, S., Spiro, T.G. and Burstyn, J.N. (1994) Resonance Raman spectroscopy of soluble guanylyl cyclase reveals displacement of distal and proximal heme ligands by NO. *J. Am. Chem. Soc.* 116: 4117–4118.

Zerner, M., Gouterman, M. and Kobayashi, H. (1966) Porphyrins VIII. Extended Huckel calculations on iron complexes. *Theoret. Chim. Acta.* 6: 363–400.

Zumft, W.G. (1992) *In:* A. Balows, H.G. Truper, M. Dworkin, W. Harder and K.-H. Schleifer (eds): *The Prokaryotes.* Vol. I, Springer-Verlag, Berlin, pp 554–582.

Zumft, W.G., Braun, C. and Cuypers, H. (1994) Nitric oxide reductase from *Pseudomonas stutzeri.* Primary structure and gene organization of a novel bacterial cytochrome *bc* complex. *Eur. J. Biochem.* 219: 481–490.

Bioradicals Detected by ESR Spectroscopy
H. Ohya-Nishiguchi & L. Packer (eds)
© 1995 Birkhäuser Verlag Basel/Switzerland

Vitamin E and antioxidant interactions in biological systems

L. Packer

Department of Molecular and Cell Biology, University of California, Berkeley, CA 94720-3200, USA

Introduction

Vitamin E is the collective name for a group of naturally-occurring tocopherols and tocotrienols found abundantly in plants, especially in plant oils (Sheppard et al., 1993). Both tocopherols and tocotrienols have a chromanol head group and a phytyl side chain. Differing methyl substitutions around the aromatic ring of the head group determine whether a tocopherol or a tocotrienol is designated alpha, beta, gamma, or delta (α, β, γ or δ) (Packer, 1994). The tocotrienols and tocopherols thus designated have identical structures in the chromanol nucleus but the hydrophobic tail, which anchors vitamin E molecules into membranes or in lipoproteins, differs. In tocotrienols, there are three unsaturated linkages in the tail, whereas in tocopherols it is fully saturated. It is not surprising that some of the biological actions of tocotrienols and tocopherols differ because of this structural difference.

Each natural form of vitamin E occurs in nature as a single stereoisomer. However, synthetic vitamin E contains eight stereoisomers arising from the 3 chiral centers (2, 4', 8'). Each of these isomers has a different biological activity. Only one form, *RRR*-α-tocopherol, is naturally occurring; it comprises about 12% of synthetic vitamin E (*all racemic* α-tocopherol). d-α-Tocopherol is about 36% more biologically active than the synthetic all racemic mixture. Biological activity has been traditionally defined as the amount of vitamin E needed to prevent resorption in the rat fetal resorption assay.

Antioxidant properties of vitamin E

The function of vitamin E was uncertain for quite a while after its discovery in 1922 by Herbert Evans at the University of California, Berkeley (Evans and Bishop, 1922). Vitamin E was found to retard the oxidation of polyunsaturated fatty acids, and thus protect against rancidification of fats and oils (Mason, 1980). Hence, it became important in preservation of food. After its molecular structure was determined, it became evident that the phenolic hydroxyl group located at the C6 position on the aromatic ring was important to its antioxidant properties (Burton and Ingold, 1981).

The rancidification of fats and oils is a free radical-mediated process; such oxidation of polyunsaturated fatty acids of phospholipids in lipoproteins or in membranes can lead to lipid peroxidation. A polyunsaturated fatty acid will become a lipid radical after hydrogen abstraction by a strong oxidant (such as hydroxyl radical) or interaction with another free radical molecule, like a lipid radical (L·). The L· reacts rapidly with oxygen to form a peroxyl radical (LOO·). This initiates a chain reaction in which different lipid radicals react with each other to form lipid hydroperoxides (LOOH) or lipid alkoxy radicals (LO·) and the chain of propagation of free radical reactions continues (Halliwell and Gutteridge, 1989). This process destroys lipids and neighboring molecules, including proteins and nucleic acids, with which radicals react. It is generally recognized that lipid peroxidation is an important factor in the progression of many chronic and degenerative diseases of aging (Halliwell and Chirico, 1993).

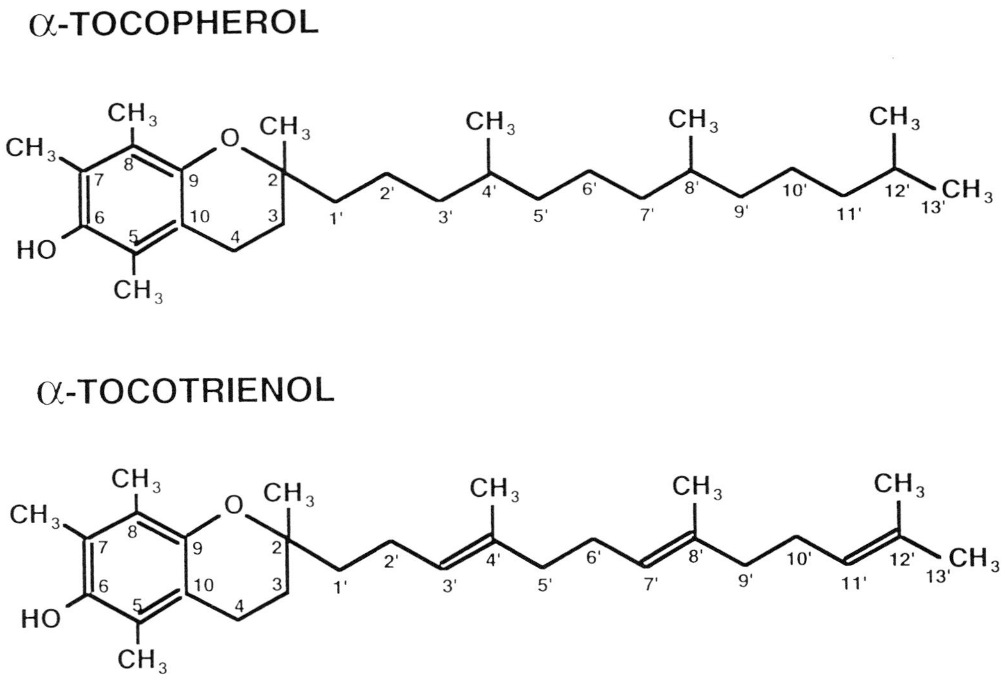

α-TOCOPHEROL

α-TOCOTRIENOL

Figure 1. Molecular structures of d-α-tocopherol and d-α-tocotrienol

Vitamin E can break this chain reaction during the propagation phase of these free radical reactions (Burton and Ingold, 1989). However, in the process, vitamin E itself becomes a free radical. It may be asked – what has been gained when vitamin E becomes a free radical? The answer is that when vitamin E becomes a free radical, the unpaired electron is delocalized around the aromatic ring. Hence, it is not as reactive a radical as the radical which it quenched. Because of the longer lived, persistent nature of vitamin E chromanoxyl radicals it is possible to regenerate the tocopherol or tocotrienol form by biochemical mechanisms.

To evaluate the antioxidant activity of different molecular forms of vitamin E, we have developed a number of *in vitro* and *in vivo* assays which assess their relative reactivity in radical quenching of peroxyl radicals formed by several means.

In vitro, we have used parinaric acid as a reporter group for evaluating the ability of vitamin E to prevent free radical reactions (Kuypers et al., 1987). Parinaric acid is a fluorescent lipid molecule whose fluorescence is destroyed when it reacts with free radicals. A constant stream of free radi-

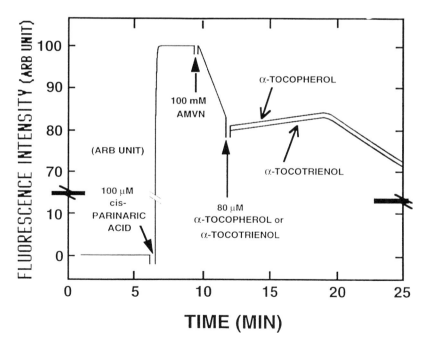

Figure 2. Comparison of the antioxidant activity of α-tocopherol and α-tocotrienol in hexane, using *cis*-parinaric acid as a reporter molecule. At the arrows, reactants were added to hexane, to the indicated final concentrations. When α-tocopherol or α-tocotrienol was added, the time delay in return of fluorescence decay corresponds to the time during which these substances were quenching peroxyl radicals. Because the concentrations of antioxidants and rate of radical production from AMVN under these conditions are known, the stoichiometry of radicals quenched per antioxidant molecule can be calculated (T = 40°C).

cals can be produced by 2, 2'-azobis (2,4-dimethylvaleronitrile) (AMVN), a lipophilic azo-initiator of peroxyl radicals which, at a given temperature, produces peroxyl radicals at a known rate. If parinaric acid in hexane solution is exposed to AMVN, one can observe, as shown in Figure 2, that the fluorescence of parinaric acid is rapidly lost (Suzuki et al., 1993).

Upon addition of either α-tocopherol or α-tocotrienol in a small but equal amount, a temporary delay in the fluorescence decay rate is observed (until the amount of vitamin E added is exhausted – whereupon loss of fluorescence commences once again). It is clear from this experiment that the antioxidant potency of quenching peroxyl radicals by d-α-tocopherol or d-α-tocotrienol in a pure chemical system is identical, and this is in agreement with other studies that show that the length or unsaturation of the hydrocarbon chain of forms of vitamin E does not affect their antioxidant potency in homogenous solution (Burton et al., 1985; Burlakova et al.,

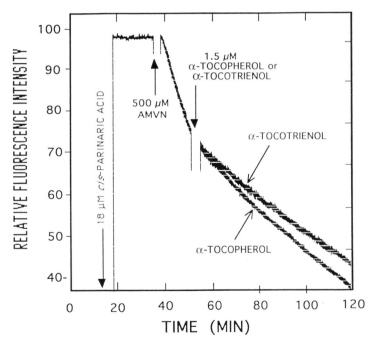

Figure 3. Comparison of the antioxidant activity of α-tocopherol and α-tocotrienol in dipalmitylphosphatidyl choline liposomes, using *cis*-parinaric acid as a reporter molecule. Generation of peroxyl radicals by AMVN was detected by monitoring decay of fluorescence of *cis*-parinaric acid (λ_{excit} 328 nm, λ_{emiss} 415 nm). The reaction mixture (2 ml) contained AMVN (500uM) and *cis*-parinaric acid (18 μM) in 20 mM Tris-HCl (pH 7.4). α-Tocopherol or α-tocotrienol (1.5 μM) dissolved in ethanol was incorporated into the membrane by further sonicating the DPPC liposomes, which were initially prepared by sonication with AMVN under nitrogen. (T = 40°C).

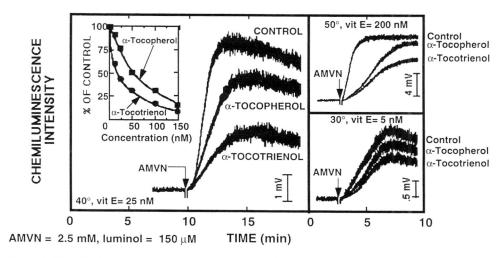

Figure 4. Chemiluminescence assays for the comparison of the antioxidant activity of α-tocopherol and α-tocotrienol in DPPC liposomes. Generation of peroxyl radicals by AMVN (2.5 mM) was detected by monitoring the chemiluminescence of luminol (150 μM) at various temperatures, as noted in the figure. At 40°C, [vitamin E] = 25 nM; at 50°C, [vitamin E] = 200 nM; at 30°C, [vitamin E] = 5 nM.

1980). It can be calculated that two peroxyl radicals are quenched per mole of either α-tocopherol or α-tocotrienol.

In a membrane, however, the phytyl side chains may well contribute to a difference in antioxidant activities (Niki et al., 1985; Burton and Ingold, 1986). Tocotrienols were shown to exert stronger antitumor action than tocopherols, which was dependent on their antioxidant properties (Kato et al., 1985; Sund'ram et al., 1989). Tocotrienols have been reported to possess higher protective activity against cardiotoxicity of the antitumor redox cycling drug adriamycin (Komiyama et al., 1989). It was also found that α-tocotrienol showed higher inhibitory effect on lipid peroxidation induced by adriamycin in rat liver microsomes than α-tocopherol (Kato et al., 1985). Can evidence be found relating the different molecular structures of vitamin E to activity?

To answer this, we have used two different systems. In the first system, cis-parinaric acid was incorporated in dipalmitoleylphosphatidyl choline (DPPC liposomes). AMVN initiated the loss of fluorescence; then either α-tocopherol or α-tocotrienol was added. Subsequently, the rate of fluorescence loss due to generation of peroxyl radicals was less for a given quantity of α-tocotrienol compared with an equal amount of α-tocopherol.

One cannot determine the stoichiometry of radical quenching in the membrane system owing to the absence of a clear discontinuity in the fluorescence loss (Suzuki et al., 1993). Therefore, in the second system, dipalmitoleylphosphatidyl choline (DPPC) or dioleylphosphatidyl choline

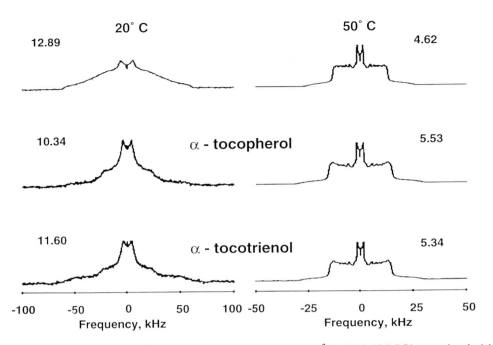

Figure 5. ^{2}H NMR spectra for [^{2}H$_{31}$] 16:0-16:0 PC (top two traces), [^{2}H$_{31}$]16:0-16:0 PC/α-tocopherol, 4:1 molar ratio (middle two traces), and [^{2}H$_{31}$] 16:0-16:0 PC/α-tocotrienol, 4:1 molar ratio (bottom two traces) at 20 and 50°C. The samples were multilamellar liposomes of approximately 100 mg of lipid in 50% by wt. 20 mM Tris (pH 7.5). First moments M_1 (10^4 s^{-1}) are stated next to the spectra.

(DOPC) liposomes containing various amounts of α-tocopherol or α-tocotrienol were used. However, in these studies, generation of peroxyl radicals, induced by AMVN, was detected by monitoring the chemiluminescence of luminol at various temperatures. It was found that both α-tocopherol and α-tocotrienol quenched the AMVN-induced luminol-enhanced chemiluminescence in DPPC liposomes with a half quenching concentration of 50 and 15 nanomolar for α-tocopherol and α-tocotrienol, respectively. Similar data were also obtained using DOPC liposomes. Such findings indicate that α-tocotrienol is a more efficient scavenger of peroxyl radicals than α-tocopherol in these model membrane systems.

It is clear from these experiments that at all concentrations of the vitamin E forms used, α-tocotrienol is always more effective in quenching chemiluminescence than α-tocopherol. The antioxidant activity of α-tocopherol and α-tocotrienol was compared previously by Serbinova et al. (1991), who used several systems for inducing lipid peroxidation (ascorbate + Fe^{+2}, or

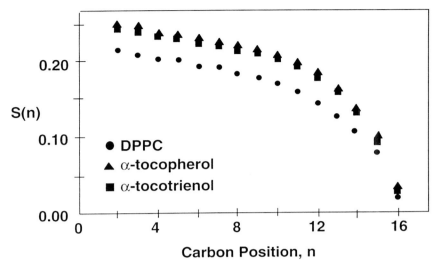

Figure 6. Order parameter profiles for [$^2H_{31}$] 16:0-16:0 PC, [$^2H_{31}$]16:0-16:0 PC/α-tocopherol (4:1), and [$^2H_{31}$] 16:0-16:0 PC/α-tocotrienol (4:1) membranes at 50°C.

NADPH + Fe^{+2}) in isolated microsomal suspensions, and observed that lipid peroxidation was inhibited to a much greater extent by α-tocotrienol compared with α-tocopherol.

To what can we attribute this greater antioxidant activity of tocotrienol? When high molar ratios of α-tocopherol or α-tocotrienol are incorporated into phosphatidylcholine (PC) multilamellar liposomes, proton NMR spectra are affected. Figure 5 shows that there is a marked change in the proton NMR spectra of PC liposomes below and above the phase transition.

Above the phase transition in the liquid crystalline phase (50°C), well defined sharp edges to the spectra are observed. Similar spectra are seen in PC liposomes made with α-tocotrienol or α-tocopherol. The vitamin E in these liposomes show reduced values in the proton NMR spectra due to this molecular ordering. It is clear that above the phase transition, vitamin E, whether it be α-tocopherol or α-tocotrienol, has a stabilizing effect on the structure. However, no significantly different effects between α-tocopherol or α-tocotrienol were observed in these studies.

Profiles of the order parameter for the three different types of liposomes were investigated. Figure 6 shows that molecular ordering decreased rapidly in all liposomes at carbon position 12, or greater, of the palmitate tails.

These proton NMR spectra, taken at the liquid crystalline phase, clearly reveal that when vitamin E is present there is a distinct difference in the order parameter as a function of carbon position only at some positions. This difference disappears at the very terminal end of the hydrophobic tail (C16) of the molecule where the length of the vitamin E tails are too short to affect the methyl end of the palmitate at C16.

ESR studies did not reveal differences between the two forms of vitamin E. Similar effects were shown in curves plotting the temperature dependence of the order parameter (S) and correlation times (τ) from conventional ESR spectra of 5- and 16- doxylstearic acid in DPPC liposomes.

Slow motions of the 5-doxylstearic acid (5-DSA) spin label were studied in DPPC liposomes, using saturation transfer ESR studies, which provide information about the reorientation dynamics. Typical saturation transfer spectra are shown in Figure 8. The positions of the central field [C/C'] and low field [L/L"] parameters are indicated in the control sample at 5°C.

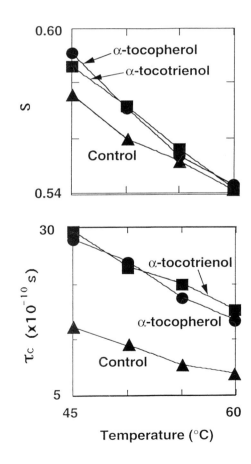

Figure 7. Temperature dependence of order parameter and correlation time from conventional ESR spectra of 5- and 16-doxylstearic acid in DPPC liposomes made in the absence or presence of either α-tocopherol or α-tocotrienol. [DPPC] = 68 mM; [5-DSA] = 0.68 mM; [16-DSA] 0.68 mM; [α-tocopherol] = 13.6 mM; [α-tocotrienol] = 13.6 mM.

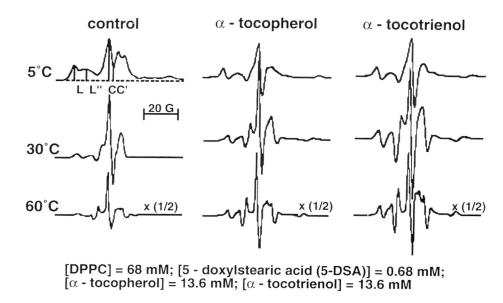

**[DPPC] = 68 mM; [5 - doxylstearic acid (5-DSA)] = 0.68 mM;
[α - tocopherol] = 13.6 mM; [α - tocotrienol] = 13.6 mM**

Figure 8. Saturation transfer ESR spectra of 5-doxylstearic acid, at 5°, 30° and 60°C made in DPPC liposomes in the absence and presence of either α-tocopherol or α-tocotrienol. [DPPC] = 68 mM; [5-DSA] = 0.68 mM; [α-tocopherol] = 13.6 mM; [α-tocotrienol] = 13.6 mM.

Temperature dependence of the central field [C/C'] and low field [L/L"] parameters were calculated for each of these spectra and are plotted in Figure 9.

Using the data shown in Figure 9, the anisotropic motion at various temperatures was computed by the method of Marsh (Marsh, 1980), and is shown in Figure 10.

Higher anisotropic motion was found at all temperatures in the presence of vitamin E. However, tocotrienol had a greater anisotropic effect than did tocopherol at all temperatures. Although introduction of α-tocopherol increased the anisotropy of motion in the gel phase, as also shown by Severcan and Cannistraro (1990), our data established that the motion anisotropy is greater with α-tocotrienol. Thus, a clear distinction between the effects of molecular organization of α-tocopherol and α-tocotrienol were identified at the molecular level in these membranes. α-Tocotrienol causes a higher degree of membrane disorganization, which allows it an increased mobility compared with α-tocopherol. Further, this increased mobility should allow α-tocotrienol a greater access to radicals.

From these results, we are led to speculate that the differences in membrane reorientation dynamics and ordering that have been identified correlate with the greater antioxidant activity of d-α-tocotrienol as compared with d-α-tocopherol.

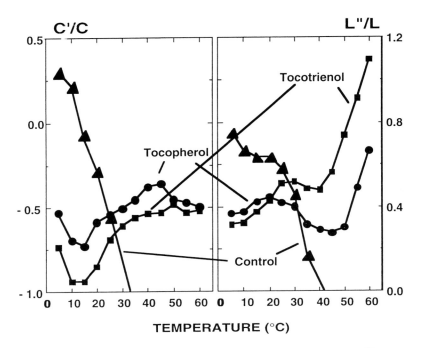

Figure 9. Temperature dependence of parameters C'/C and L"/L from saturation transfer ESR spectra of 5-doxyl-stearic acid. [DPPC] = 68 mM; [5-DSA] = 0.68 mM; [α-tocopherol] = 13.6 mM; [α-tocotrienol] = 13.6 mM.

Factors in evaluating antioxidant potency

A number of factors need to be considered when evaluating the potency of antioxidant substances such as tocopherols and tocotrienols. Some of the important factors are the location of the antioxidants in either the aqueous or the lipophilic domains or both, their mobility, their rate constants in radical reactions, their metal chelating activity, their effective concentrations, and their capacity for regeneration.

In the case of vitamin E, the various forms are exclusively in the lipophilic domains. Vitamin E is usually present at low molar ratios in natural membranes, perhaps one molecule per 2000–3000 phospholipids (Esterbauer et al., 1992). Hence, the mobility of vitamin E in the membrane may have great consequences for its potency in antioxidant action, and the effects of the different tails, as discussed above. Thus, in reactions with peroxyl radicals in pure hexane, rate constants appear equal; however, in membrane systems the rate constant for tocotrienols appear to be greater, due to their greater mobility compared to tocopherols (Suzuki et al., 1993).

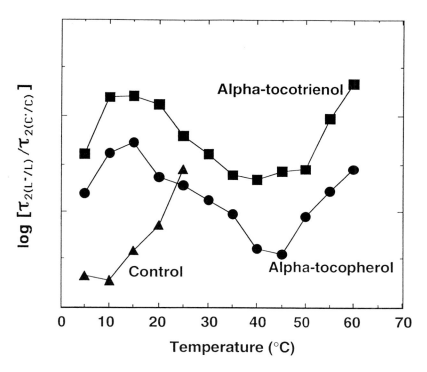

Figure 10. Temperature dependence of log of the ratio of correlation times derived from L"/L and C'/C DPPC liposomes in the presence of 5-doxylstearic acid and either α-tocopherol or α-tocotrienol. [DPPC] = 68 mM; [5-DSA] = 0.68 mM; [α-tocopherol] = 13.6 mM; [α-tocotrienol] = 13.6 mM.

Vitamin E does have some iron-binding activities (Stoyanovsky et al., 1989) which may, to an extent, affect its potency in quenching radicals depending on the presence of free reactive iron in the system.

About 24 hours after dietary consumption, d-α-tocopherol is preferentially enriched in plasma compared with other forms of vitamin E (Traber et al., 1992). This probably occurs as a result of specific vitamin E binding protein present in rat (Sato et al., 1991; Yoshida et al., 1992; Sato et al., 1993) and human liver (Kuhlenkamp et al., 1993). This protein discriminates between stereoisomers of d-α-tocopherol and negatively discriminates against all forms of vitamin E except d-α tocopherol (Sato et al., 1991). Patients who apparently lack this protein are unable to maintain normal plasma α-tocopherol concentrations and are unable to discriminate between forms of vitamin E (Traber et al., 1993). Apparently, in normal subjects d-α-tocopherol is preferentially incorporated into lipoproteins as they are synthesized by the liver, as was demonstrated using perfused monkey livers (Traber et al., 1990). In human supplementation studies we have found that there appears to be no rate-limiting effect in the absorption of d-α-tocopherol or d-α-tocotrienol from the intestinal tract into the chylomicron fraction, but its subsequent appearance in human lipoproteins is much diminished (unpublished results). This can be ascribed to the presence of the specific vitamin E tocopherol binding protein, which regulates vitamin E metabolism in hepatocytes (Traber, 1994).

Importantly, the rate of regeneration, or recycling, of the vitamin E radicals that form during its antioxidant action may affect both its efficiency of antioxidant action and its lifetime in biological systems. To study the recycling and regeneration of vitamin E during its antioxidant action, we have carried out studies using human low density lipoprotein suspensions (LDL) and membranes from animal tissues that had been enriched with vitamin E by feeding.

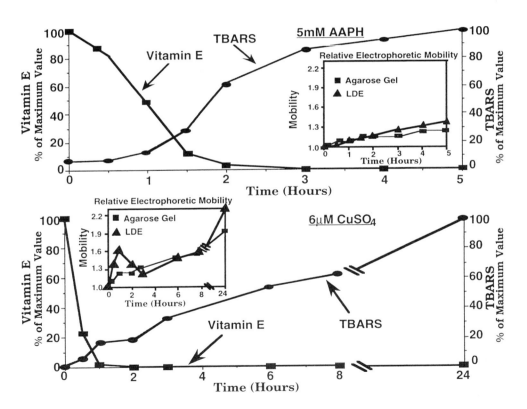

Figure 11. Kinetics of vitamin E disappearance and TBARS accumulation during AAPH and CuSO₄ induced oxidation of human LDL. LDL concentration = 1 mg protein (apo B)/ml, AAPH and CuSO4 were at the concentrations indicated. Insets: change in negative charge on LDL measured on agarose gels or by laser doppler electrophoresis (LDE).

LDL is the transport vehicle for vitamin E to the peripheral tissues (Traber and Kayden, 1984). In human lipoproteins which contain perhaps 2000–2500 molecules, the molar ratio of vitamin E to phospholipids is between 0.5 and 1 mole percent. In a typical human LDL preparation, on average there are about seven molecules vitamin E per LDL particle, or about 88% of all the lipophilic antioxidants in the particle (Esterbauer et al., 1992). The other antioxidants in LDL are the carotenoids; perhaps up to eight different carotenoids may be observed. On average, they constitute about ten percent of the antioxidants. Also, there are trace amounts of ubiquinol 10, usually about one percent of the total antioxidants.

LDL suspensions can be oxidatively stressed. A commonly used method is copper-induced oxidation, and we also use water soluble azo-initiators such as 2, 2'-azobis (2-amidinopropane) hydrochloride (AAPH) for generation of peroxyl radicals. This causes the loss of aqueous anti-oxidants, like Vitamin C and others. When these defenses are diminished, the vitamin E content of the LDL suspension begins to be drastically decreased. Figure 11 shows the kinetics of vita-min E disappearance and accumulation of thio-barbituric acid reactive substances (TBARS, markers of lipid peroxidation) during AAPH and copper sulfate induced oxidation of human LDL. As the levels of vitamin E are depleted, accumulation of lipid peroxidation products (TBARS) are observed. Changes in the physicochemical properties of the surface of the LDL preparation can be seen by agarose gel electrophoresis, or more sophisticated methods such as laser doppler electrophoresis (inset, Fig. 11), which shows that the surface of LDL become more net negatively charged as oxidation proceeds (Arrio et al., 1993).

As vitamin E is depleted during the oxidation of LDL, significant amounts of chromanoxyl radicals are present, which can be observed directly by ESR spectroscopy. This is because the molar ratio of vitamin E is high enough without supplementation in human LDL to observe such signals directly. Note this is not true for natural biological membranes, because the molar ratio of vitamin E is too low for detection by ESR. If, during this time, radical-radical reactions of vitamin E with itself or other lipid radicals occur, vitamin E is degraded and slowly lost from the system. If, however, regeneration of vitamin E is accomplished by adding ascorbic acid, the steady state signal of the vitamin E radicals is drastically lowered, such that the vitamin E radical signal cannot even be detected by ESR (Fig. 12).

We have compared the ability of d-α-tocopherol, d-α-tocotrienol and a short chain homologue of α-tocopherol, C6 chromanol (shorter phytyl tail), to be recycled after a given amount of vitamin C is added to the suspension (Kagan et al., 1992a). That is, in the presence of vitamin C the vitamin E radical signal is suppressed, but when the vitamin C is consumed, the vitamin E radical appears because the vitamin C is no longer present to regenerate tocopherol from the tocopheroxyl radical. We term this process "recycling". Figure 13 shows the kinetics of vitamin E radicals, with and without vitamin C, during the time course of oxidation of LDL.

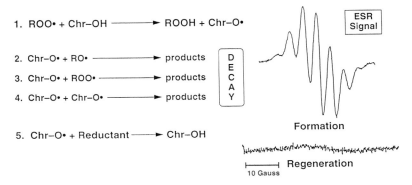

Figure 12. Some reactions of chromanoxyl radicals in the course of lipid peroxidation.

In such experiments, we have found that α-tocotrienol is more efficiently recycled than α-tocopherol (note the longer lag time before the reappearance of the ESR signal); the short chain homologue of vitamin E displays even greater capacity for recycling.

From these experiments, we have concluded that the greater antioxidant activity seen in isolated membranes and lipoproteins *in vitro* with α-tocotrienol compared to α-tocopherol is due to a number of different factors: its more uniform distribution in the membrane bilayer, which has

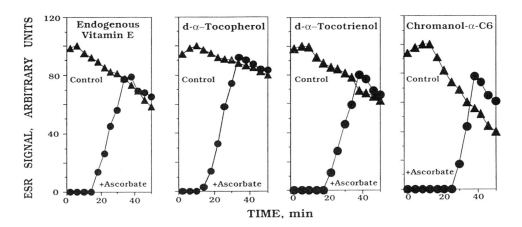

Figure 13. Time-course of chromanoxyl radical ESR signals generated by the lipoxygenase + linolenic acid oxidation system in LDL from endogenous vitamin E or exoogenously added α-tocopherol, α-tocotrienol, or chromanol-α-C6. Effect of ascorbate. In each case ascorbate (1.5 mM) was added before the recording of the first ESR spectrum. The concentrations of endogenous vitamin E were 6.2 nmol/mg protein in the sample to which no exogenous chromanols were added and 2.5 nmol/mg protein in the samples to which chromanols (80 nmol/mg protein) were added. All values are given as a percentage of the maximal magnitude obtained.

been observed in fluorescence self-quenching studies (Serbinova et al., 1991), the more effective collision of radicals with vitamin E as determined by its behavior in model membranes (described above), and the greater recycling activity of chromanoxyl radicals by Vitamin C, shown here for LDL (Kagan et al., 1992a) and in membranes (Serbinova et al., 1991). We have found that the greater recycling activity correlates with the increased inhibition of lipid peroxidation. The different molecular structures of tocotrienols and tocopherols described here appear to account *in vitro* for its more effective antioxidant activity in membranes.

The interaction between the vitamin E cycle and other redox based antioxidants

Vitamin E radical may be generated by a variety of oxidants, such as UVB irradiation, which may create a vitamin E radical directly; interaction of vitamin E with lipid radicals (lipoxygenase + linolenic acid or arachidonic acid); incubation with azo-generators of peroxyl radicals; interaction with neutrophils or macrophages which produce reactive oxygen species, and so forth. Regeneration of vitamin E may occur in the membrane by the action of reduced cytochrome c or reduced ubiquinols. $O_2^{-\cdot}$ (superoxide anion) may also react with vitamin E chromanoxyl radicals to rege-

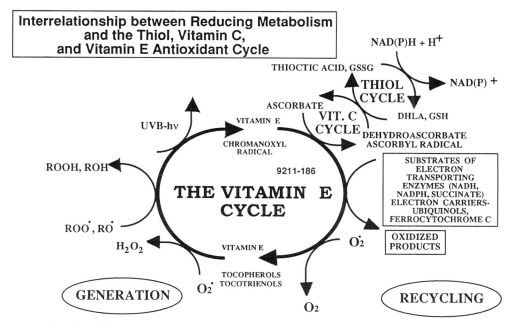

Figure 14. Pathways for generation of chromanoxyl radicals and their recycling to vitamin E.

nerate them. In the aqueous phase, the series of interacting redox-based antioxidant systems—from reducing metabolism to thiols to vitamin C—participate in vitamin E recycling.

Summarizing all these interactions, the diagram in Figure 14 reveals the various pathways whereby chromanoxyl radicals of vitamin E may be produced, then regenerated.

In the previous section, recycling of vitamin E by vitamin C was discussed. Besides vitamin C in the aqueous phase, which is capable of converting the one-electron oxidation product (chromanoxyl radical) to the reduced form of vitamin E (the native molecule), we have found a number of other pathways that are important in this redox cycling reaction. Another primary preventative antioxidant in biological systems is glutathione. Glutathione is often present in millimolar concentrations in tissues. Glutathione and other thiols appear to act to regenerate vitamin C from its semi-ascorbyl radical form, produced when it is acting to recycle vitamin E.

Reducing thiols, such as dihydrolipoic acid, are also capable of driving the reduction of vitamin C, which in turn drives the vitamin E cycle. Dihydrolipoic acid is more effective than glutathione at low concentrations, owing to its greater (two electron) negative redox potential—lipoic acid being −320 mV, 80 mV more negative than glutathione. There is a interacting series of redox based, antioxidant activities generated from NADH and NADPH, formed by metabolism to maintain thiols in their reduced state. The reduced thiols in turn keep vitamin C reduced, and vitamin C keeps the vitamin E in its reduced native state.

Besides interacting with antioxidants in the aqueous phase, vitamin E in the lipid domain interacts with electron donors present in the membrane. This has been observed in a number of different experiments (Maguire et al., 1989; Packer et al., 1989; Kagan et al., 1990; Kagan et al.,

Table 1. Content of α-tocopherol or α-chromanol-C6 in proteoliposomes containing complex II following exposure to a radical generating system for 30 min

Sample	% Residual Chromanol	
	α-tocopherol	α-chromanol-C6
DOPC liposomes		
+ chromanol (control)	100	100
+ (lipoxygenase + linoleate)	4.9	6.1
+ (lipoxygenase + linoleate) + succinate (20 mM)	5.0	6.0
+ (lipoxygenase + linoleate) + CoQ10 (or Q1) (5.0 mM)	14.6	5.5
+ (lipoxygenase + linoleate) + CoQ10 (or Q1) (5.0 mM) + succinate (20 mM)	94.5	97.6

1992b; Constantinescu et al., 1993). Ubiquinols, generated by the ubiquinone oxo-reductase system of complex II in the mitochondrial inner membrane, are capable of regenerating vitamin E chromanoxyl radicals under conditions where the presence of reduced ubiquinols correlate with protection of vitamin E against vitamin E loss (Tab. 1).

In experiments using liposomes containing vitamin E [d-α-tocopherol or the short chain homologue of α-tocopherol, 2,2,5,7,8-pentamethyl-6-hydroxychromane (PMC)], we have found that the redox state of cytochrome c can determine the steady state level of vitamin E chromanoxyl radicals. When vitamin E radicals are generated by lipoxygenase and arachidonic acid, the presence of *oxidized* cytochrome c has no effect on the vitamin E radical signal. However, *reduced* cytochrome c can abolish the radical (Fig. 15).

The steady state of reduction of the cytochrome c is directly related to the signal intensity of tocopheroxyl radicals seen in these model systems. The presence of reduced cytochrome c prevents loss of vitamin E from the system (as determined separately by HPLC analysis of the membranes).

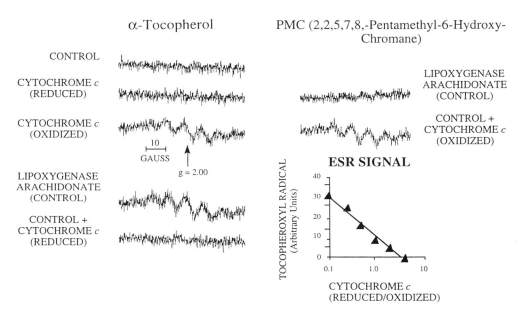

Figure 15. Effect of oxidized and reduced cytochrome c on chromanoxyl radicals in liposomes containing vitamin E or PMC. Spectra on the left are from liposomes with α-tocopherol (1.5 mg lipid/ml, 0.04 mg/ml α-tocopherol) in 50 mM MOPS, pH 7.5, which were oxidized by either oxidized cytochrome c (2.8 mM), or lipoxygenase (0.06 mg, 15 units/ml) and arachidonic acid (1 mM). Reduced cytochrome c was at 2 mM. Spectra on the right were from liposomes prepared without α-tocopherol but α-C-1chromanol was added in ethanol to 0.86 mM. The insert shows the tocopheroxyl radical signal height in the presence of 2.8 mM cytochrome c reduced and oxidized as indicated.

We have also observed in our laboratory that short chain homologues of vitamin E are capable of overcoming the block in electron transport between complex II through complex III of the respiratory chain (Maguire et al., 1992) (Fig. 16).

The antimycin inhibition of electron transport through complex II can be bypassed by short chain vitamin E homologues, allowing respiration to be partially restored-- again indicating a link in electron transfer between ubiquinols, vitamin E, and cytochrome c. It is possible that this kind of interesting and novel effect of vitamin E may be important in certain age-related mitochondrial dysfunctions where genetic lesions in complex III have arisen, such that electron transport is impaired through this complex. This may possibly be partially overcome by vitamin E therapy to restore energy functions of mitochondria in aging.

The relative importance of aqueous *versus* membrane pathways for driving the vitamin E cycle, protecting it from loss and bolstering its antioxidant activity in biological systems, is still something of a mystery. Good methods for separating the relative contributions of these different pathways to vitamin E recycling activity have not yet been developed.

Action of vitamin E as a biological response modifier

Besides its antioxidant action, vitamin E has been observed to have effects on transcellular signaling mechanisms and gene expression. It is not entirely clear whether the effects of vitamin E observed in these systems are due to its antioxidant activity or to other properties of vitamin E. Some examples of its action in this regard are given below.

Together with A. Pentland's laboratory (Pentland et al., 1992), we have shown that various structural homologues of vitamin E, with alterations of its sidechain, have effects on pathways involving arachidonic acid metabolism. Suppression of ionophore stimulated prostaglandin synthesis in human keratinocytes by the short chain homologue of vitamin E, PMC, the alpha C6 and the alpha C11 homologues were readily observed. These effects appear to be directly due to inhibition of phospholipase A2 activity, since the isolated enzyme was inhibited by PMC. The short chain homologues are more effective than the natural form of vitamin E, which we have attributed to the poorer cellular permeability of the natural form.

Another transcellular signaling mechanism, protein kinase C (PKC) has been studied in Angelo Azzi's laboratory (Chatelain et al., 1993). They demonstrated that various tocopherols and tocotrienols can downregulate protein kinase C activity in a variety of cell types, e.g., smooth muscle cells *in vitro*. In this regard, α-tocopherol exhibits a much greater inhibition that β-tocopherol (which is almost without action) whereas α-tocotrienol has only a small effect. These highly specific effects of different tocopherol and tocotrienol derivatives on PKC activity have been

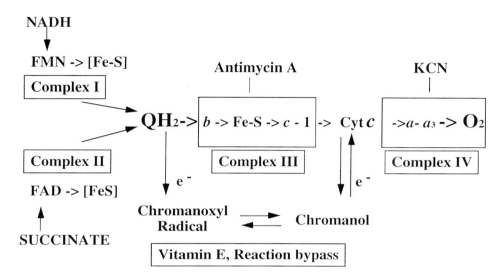

Figure 16. Vitamin E mediates electron transfer in mitochondria, bypassing antimycin A inhibition.

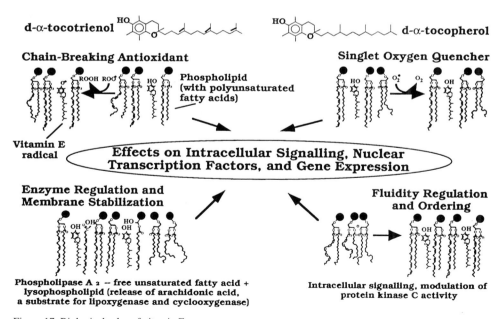

Figure 17. Biological roles of vitamin E.

shown to be linked with specific times in the cell growth cycle. Investigations in Azzi's laboratory suggest that this may not be due to the antioxidant properties of vitamin E.

Vitamin E and its homologues have also been shown to regulate the activation of oxidatively induced nuclear transcription factors, such as NFκB. When NFκB is activated by tumor necrosis factor or phorbol esters, its activation can be prevented by forms of vitamin E which are cell permeable. These include vitamin E acetate and vitamin E succinate and the short chain homologue of vitamin E, PMC. In addition, it has been observed (Suzuki and Packer, 1993) that vitamin E succinate has a specific effect, not shown by PMC or vitamin E acetate, in inhibiting the binding of NFkB to DNA. The DNA binding inhibition is not demonstrated by these other homologues. It is difficult to explain these effects purely by the antioxidant properties of vitamin E.

The actions of vitamin E as a biological response modifier are summarized in Figure 17. In this diagram, α-tocopherol and α-tocotrienol are depicted as active as chain-breaking antioxidants and as quenchers of singlet oxygen in the lipophilic domain, and also to exhibit effects on trans-cellular signaling, nuclear transcription and gene expression.

Thus vitamin E exhibits actions in biological systems from the membrane to the gene. All of these need to be factored together in evaluating its biological effects in aging, chronic and degenerative diseases, and in acute clinical conditions where vitamin E exerts actions in health and disease.

References

Arrio, B., Bonnefort-Rousselot, D., Catudioc, J. and Packer, L. (1993) Electrophoretic mobility changes of oxidized human low density lipoprotein measured by laser doppler electrophoresis. *Biochem. Mol. Biol. Int.* 30: 1101–1114.

Burkalova, Y.B., Kuchtina, T.E., Ol'khovskhaya, I.P., Sarycheva, I.K., Sinkiona, Y.B. and Khrapova, N.G. (1980) Study of the antiradical activity of tocopherol analogues and homologues by the method of chemiluminescence. *Biofizika* 24: 989–993.

Burton, G.W. and Ingold, K.U. (1981) Autoxidation of biological molecules. I. The antioxidant activity of vitamin E and related chain-breaking phenolic antioxidants *in vitro*. *J. Amer. Chem. Soc.* 103: 6472–6477.

Burton, G.W., Doba, T., Gabe, E.J., Hughes, L., Lee, F.L., Prasad, L. and Ingold, K.U. (1985) Autoxidation of biological molecules. 4. Maximizing the antioxidant activity of phenols. *J. Am. Chem. Soc.* 107: 7053–7065.

Burton, G.W. and Ingold, K.U. (1986) Vitamin E: application of the principles of physical organic chemistry to the exploration of its structure and function. *Acc. Chem. Res.* 19: 194–201.

Burton, G.W. and Ingold, K.U. (1989) Vitamin E as an *in vitro* and *in vivo* antioxidant. *Ann. N.Y. Acad. Sci.* 570: 7–22.

Chatelain, E., Boscoboinik, D.O., Bartoli, G.-M., Kagan, V.E., Gey, F., Packer, L. and Azzi, A. (1993) Inhibition of smooth muscle cell proliferation and protein kinase C activity by tocopherols and tocotrienols. *Biochim. Biophys. Acta* 1176: 83–89.

Constantinescu, A., Han, D. and Packer, L. (1993) Vitamin E recycling in human erythrocyte membranes. *J. Biol. Chem.* 15: 10906–10913.

Esterbauer, H., Gebicki, J., Puhl, H. and Jurgens, G. (1992) The role of lipid peroxidation and antioxidants in oxidative modification of LDL. *Free Rad. Biol. Med.* 13: 341–90.

Evans, H.M. and Bishop, K.S. (1922) On the existence of a hitherto unrecognized dietary factor essential for reproduction. *Science* 56: 650–651.

Halliwell, B. and Gutteridge, J.M.C. (1989) *Free Radicals in Biology and Medicine*, Clarendon Press, Oxford.

Halliwell, B. and Chirico, S. (1993) Lipid peroxidation: its mechanism, measurement, and significance. *Am. J. Clin. Nutr.* 57: 715S–725S.

Kagan, V.E., Serbinova, E.A. and Packer, L. (1990) Recycling and antioxidant activity of tocopherol homologs of differing hydrocarbon chain lengths in liver microsomes. *Arch. Biochem. Biophys.* 282: 221–225.

Kagan, V.E., Serbinova, E.A., Forte, T., Scita, G. and Packer, L. (1992a) Recycling of vitamin E in human low density lipoproteins. *J. Lipid Res.* 33: 385–397.

Kagan, V.E., Serbinova, E.A., Safadi, A., Catudioc, J.D. and Packer, L. (1992b) NADPH-dependent inhibition of lipid peroxidation in rat liver microsomes. *Biochem. Biophys. Res. Commun.* 186: 74–80.

Kato, A., Yamaoka, M., Tamaka, A., Komyama, K. and Umezawa, I. (1985) Physiological effects of tocotrienol. *Abura Kagaku* 34: 375–376.

Komiyama, K., Iizuka, K., Yamaoaka, M., Watanabe, H., Tsuchiya, N. and Umezawa, I. (1989) Studies on the biological activity of tocotrienols. *Chem. Pharm. Bull.* 37: 1369–1371.

Kuhlenkamp, J., Ronk, M., Yusin, M., Stolz, A. and Kaplowitz, N. (1993) Identification and purification of a human liver cytosolic tocopherol binding protein. *Prot. Exp. Purific.* 4: 382–389.

Kuypers, F.A., van den Berg, J.J., Schalkwijk, C., Roelofsen, B. and Op den Kamp, J.A. (1987) Parinaric acid as a sensitive fluorescent probe for the determination of lipid peroxidation. *Biochim. Biophys. Acta* 921: 266–274.

Maguire, J.J., Wilson, D.S. and Packer, L. (1989) Mitochondrial electron transport-linked tocopheroxyl radical reduction. *J. Biol. Chem.* 264: 21462–21465.

Maguire, J.J., Kagan, V.E. and Packer, L. (1992) Electron transport between cytochrome c and α tocopherol. *Biochem. Biophys. Res. Comm.* 188: 190–197.

Marsh, D. (1980) Molecular motion in phospholipid bilayers in the gel phase: long axis rotation. *Biochem.* 19: 1632–1637.

Mason, K.E. (1980) The first two decades of vitamin E history. *In*: L.J. Machlin (ed.):*Vitamin E: A Comprehensive Treatise.* New York, Marcel Dekker Inc., pp 1–6.

Niki, E., Kawakami, A., Saito, T., Yamamoto, Y., Tsuchiya, J. and Kamiya, Y. (1985) Effect of phytyl side chain of vitamin E on its antioxidant activity. *J. Biol. Chem.* 260: 2191–2196.

Packer, L., Maguire, J.J., Mehlhorn, R.J., Serbinova, E. and Kagan, V.E. (1989) Mitochondria and microsomal membranes have a free radical reductase activity that prevents chromanoxyl radical accumulation. *Biochem. Biophys. Res. Comm.* 159: 229–235.

Packer, L. (1994) Vitamin E is Nature's master antioxidant. *Sci. Am. Sci. Med.* 1: 54–63.

Pentland, A.P., Morrison, A.R., Jacobs, S.C., Hruza, L.L., Hebert, J.S. and Packer, L. (1992) Tocopherol analogs suppress arachidonic acid metabolism *via* phospholipase inhibition. *J. Biol. Chem.* 267: 15578–15584.

Sato, Y., Hagiwara, K., Arai, H. and Inoue, K. (1991) Purification and characterization of the α-tocopherol transfer protein from rat liver. *FEBS Lett.* 288: 41–45.

Sato, Y., Arai, H., Miyata, A., Tokita, S., Yamamoto, K., Tanabe, T. and Inoue, K. (1993) Primary structure of α-tocopherol transfer protein from rat liver. Homology with cellular retinaldehyde-binding protein. *J. Biol. Chem.* 268: 17705–10.

Serbinova, E., Kagan, V., Han, D. and Packer, L. (1991) Free radical recycling and intramembrane mobility in the antioxidant properties of α-tocopherol and α-tocotrienol. *Free Rad. Biol. Med.* 10: 263–275.

Severcan, F. and Cannistraro, S. (1990) A spin label ESR and saturation transfer ESR study of α-tocopherol containing model membranes. *Chem. Phys. Lipids* 53: 17–26.

Sheppard, A.J., Pennington, J.A.T. and Weihrauch, J.L. (1993) Analysis and distribution of vitamin E in vegetable oils and foods. *In*: L. Packer and J. Fuchs (eds): *Vitamin E in Health and Disease.* New York, NY, Marcel Dekker Inc., pp 9–31.

Stoyanovsky, D.A., Kagan, V.E. and Packer, L. (1989) Iron binding to α-tocopherol-containing phospholipid liposomes. *Biochem. Biophys. Res. Comm.* 160: 834–838.

Sund'ram, K., Khor, H.T., Ong, A.S. and Pathmanathan, R. (1989) Effects of dietary palm oil on mammary carcinogenesis in female rats induced by 7,12-dimethylbenz(α)anthracene. *Canc. Res.* 49: 1447–1451.

Suzuki, Y.J. and Packer, L. (1993) Inhibition of NF-KB by vitamin E derivatives. *Biochem. Biophys. Res. Comm.* a93: 277–283.

Suzuki, Y.J., Tsuchiya, M., Wassall, S.R., Choo, Y.M., Govil, G., Kagan, V.E. and Packer, L. (1993) Structural and dynamic membrane properties of α-tocopherol and α-tocotrienol: implication to the molecular mechanism of their antioxidant potency. *Biochemistry* 32: 10692–10699.

Traber, M.G. and Kayden, H.J. (1984) Vitamin E is delivered to cells *via* the high affinity receptor for low density lipoprotein. *Am. J. Clin. Nutr.* 40: 747–751.

Traber, M.G., Rudel, L.L., Burton, G.W., Hughes, L., Ingold, K.U. and Kayden, H.J. (1990) Nascent VLDL from liver perfusions of cynomolgus monkeys are preferentially enriched in *RRR-* compared with *SRR-α* tocopherol: studies using deuterated tocopherols. *J. Lipid Res.* 31: 687–694.

Traber, M.G., Burton, G.W., Hughes, L., Ingold, K.U., Hidaka, H., Malloy, M., Kane, J., Hyams, J. and Kayden, H.J. (1992) Discrimination between forms of vitamin E by humans with and without genetic abnormalities of lipoprotein metabolism. *J. Lipid Res.* 33: 1171–1182.

Traber, M.G., Sokol, R.J., Kohlschütter, A., Yokota, T., Muller, D.P.R., Dufour, R. and Kayden, H.J. (1993) Impaired discrimination between stereoisomers of α-tocopherol in patients with familial isolated vitamin E deficiency. *J. Lipid Res.* 34: 201–210.

Traber, M.G. (1994) Determinants of plasma vitamin E concentrations. *Free Rad. Biol. Med.* 16: 229–239.

Yoshida, H., Yusin, M., Ren, I., Kuhlenkamp, J., Hirano, T., Stolz, A. and Kaplowitz, N. (1992) Identification, purification and immunochemical characterization of a tocopherol-binding protein in rat liver cytosol. *J. Lipid Res.* 33: 343–350.

Bioradicals Detected by ESR Spectroscopy
H. Ohya-Nishiguchi & L. Packer (eds)
© 1995 Birkhäuser Verlag Basel/Switzerland

Antioxidant activity of vitamin E and vitamin C derivatives in membrane mimetic systems

Z.-L. Liu

National Laboratory of Applied Organic Chemistry, Lanzhou University, Lanzhou, Gansu 730000, China

Summary. This review outlines a decade of research on antioxidant activities of vitamin E, vitamin C and its lipophilic derivatives, ascorbyl-6-caprylate (VC-6), 6-laurate (VC-12) and 6-palmitate (VC-16) in micelles and artificial liposomes, aiming to shed light on the effects on the activity of the side-chain of the VCs and the micro-enviroment of the reaction medium. Nitroxides with various lipophilicities, i.e., 4-hydroxy-2,2,6,6-tetramethyl-piperidine-1-oxyl (TEMPO-OH), 4-methoxy-(TEMPO-OMe), 4-hexanoyloxy-(TEMPO-6), 4-lauroyloxy-(TEM-PO-12) and 4-palmitoyloxy-(TEMPO-16) were used as model compounds of alkyl peroxyl radicals to react with the antioxidants. By use of a stopped-flow technique ESR signals of both the nitroxide and the radical inter-mediate of the antioxidant were detected simultaneously and their kinetics followed. It was found that the critical factors which govern the reactivity involved the lipophilicity of both the oxidant and the antioxidant, the inter- and/or intra-micellar/liposomal diffusion rate, and the surface charge of the micelle or the liposome. In the case of liposomes, substrate-induced fusion may also contribute significantly. Up to four orders of magnitude of activity enhancement could be achieved by changing the lipophilicity of the antioxidant and conducting the reaction in an appropriate microenviroment. Preliminary biological assay showed that the information obtained from the chemical kinetic studies is also applicable to biological systems.

Introduction

Free radical initiated peroxidation of membrane lipids is believed responsible for the aetiology of a variety of chronic health problems such as aging, cancer, atherosclerosis and cataracts (Pryor, 1976–1986; McBrien and Slater, 1982; Halliwell and Gutteridge, 1985; Slater, 1987; Schwartz et al., 1993; Favier et al., 1994). Inhibition of the peroxidation by administration of antioxidants has proved beneficial in protection against amplification of these disease processes, and thus antioxidant therapy has developed rapidly in recent years (Miguel, 1988; Sies, 1992; Rice-Evans and Diplock, 1993; Bendich, 1993; Gey, 1993). α-Tocopherol (the most effective component of vitamin E) and L-ascorbic acid (vitamin C) are well-known chain-breaking antioxidants which have been extensively studied from the chemical, biochemical and medical points of view in the past decade (Burton and Ingold, 1986; Niki, 1987; Packer and Fuchs, 1992; Barclay, 1993). It has been recognized that the antioxidant activity in homogeneous solutions may not be the same as that in heterogeneous media, let alone *in vivo* (Castle and Perkins, 1986; Pryor et al., 1988; Barclay, 1993). One of the reasons is apparently that the microenvironment which the antioxidant experiences in biomembranes is significantly different from that in homogeneous solutions.

Indeed, we have found that changing the microenvironment may make vitamin E become a prooxidant (Liu et al., 1991a).

In order to bridge the gap between the chemical and biological antioxidant activities, one must at least understand the relationship between the activity and the microenvironment of the reaction medium. A general methodology is to carry out the reaction in membrane mimetic systems, i.e., micelles or artificial liposomes (Barclay, 1993). We have developed a method to determine reaction kinetics in micelles and liposomes by using nitroxides as a model compound of alkyl peroxy radicals. Nitroxides, although generally considered as stable and extensively used in spin labeling and spin trapping, can react as an oxidant with vitamin E and vitamin C in the similar way as peroxyl radicals (Liu et al., 1988; Takahashi et al., 1989). Nitroxides are isoelectronic with alkyl peroxyl radicals, and both have large dipole moments of ca. 3.0 and 2.6 Debye respectively (Barclay and Ingold, 1981). The advantage of using nitroxides is that they are very stable in comparison with peroxyl radicals, thus their reaction kinetics can easily be quantitatively monitored by ESR spectroscopy at ambient temperatures. Furthermore, nitroxides are spin probes that make it possible to get information about the micropolarity and microviscosity of the reaction medium by analyzing the ESR line shape. Therefore, nitroxides may serve as a good model oxidant for reactivity and mechanistic studies carried out in membrane mimetic systems.

This article outlines our decade of research on the antioxidant activity of vitamin E, vitamin C and its lipophilic derivatives in micelles and artificial liposomes, using this approach. The antioxi-

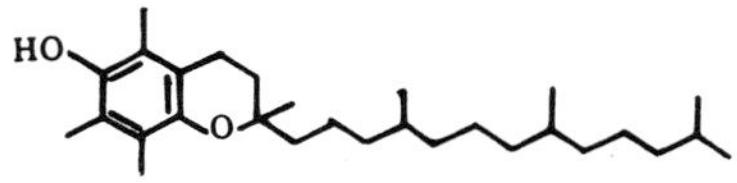

VE

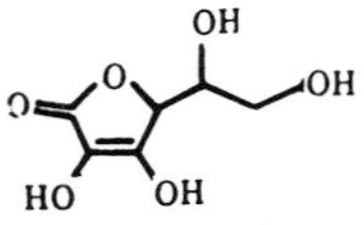

VC

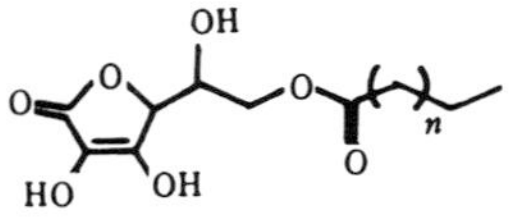

VC-8 (n = 6)
VC-12 (n = 10)
VC-16 (n = 14)

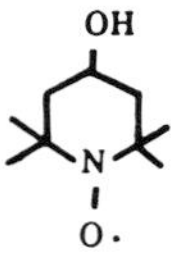

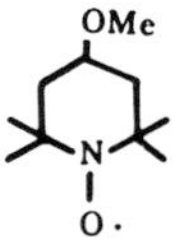

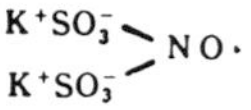

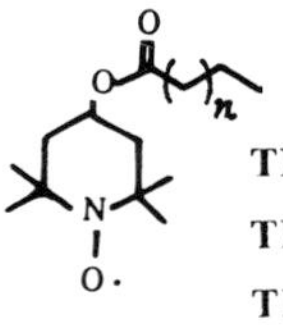

TEMPO-OH TEMPO-OMe Fremy's salt

TEMPO-6 (n = 4)
TEMPO-12 (n = 10)
TEMPO-16 (n = 14)

dants which have been investigated are α-tocopherol (VE), L-ascorbic acid (VC) and its lipophilic derivatives ascorbyl-6-caprylate (VC-8), 6-laurate (VC-12) and 6-palmitate (VC-16). Nitroxides with various lipophilicity, i.e., 4-hydroxy-2,2,6,6-tetramethylpiperidine-1-oxyl (TEMPO-OH), 4-methoxy-(TEMPO-OMe), 4-hexanoyloxy-(TEMPO-6), 4-lauroyloxy-(TEMPO-12) and 4-palmitoyloxy-TEMPO (TEMPO-16), and the ionic Fremy's salt, were used as oxidants. The membrane mimetic systems used include a cationic micelle cetyl trimethylammonium bromide (CTAB), an anionic micelle sodium dodecyl sulphate (SDS) and a non-ionic micelle polyoxyethylene diisobutylphenol (Triton X-100), as well as dipalmitoyl phosphatidylcholine liposome (DPPC).

Reactions of vitamin C in micelles

Ingold and Rassat (Ebel et al., 1985) and Okazaki and Kuwata (1985) have studied the reaction of vitamin C and nitroxides in β-cyclodextrin solution and proposed somewhat different mechanisms. The same reaction was studied independently at the same time in our laboratory. By using a stopped-flow ESR technique signals of both the nitroxide and the reaction intermediate, ascorbate radical (VC⁻·) were detected simultaneously and their kinetics monitored directly by ESR (see, for example, Fig. 1). A more detailed mechanism, which involves two consecutive single electron transfer steps and the disproportionation of the VC⁻·, was proposed (Liu et al., 1985).

For reactions conducted in micelles, the reactions occurring in the bulk water and in the micellar phase must be evaluated separately to take into account the formation of the nitroxide-micelle aggregate in the case of lipophilic nitroxides. Therefore, the mechanism in micelles may be described by the following equations (Liu et al., 1989):

$$M + NO\cdot \underset{}{\overset{K}{\rightleftarrows}} M + NO\cdot \qquad\qquad [1]$$

$$H_2A \rightleftarrows HA^- + H^+ \qquad pK = 4.8 \qquad [2]$$

$$HA^- + NO\cdot \overset{k_1^0}{\longrightarrow} HA\cdot + NO^- \qquad\qquad [3]$$

$$HA^- + M\text{–}NO\cdot \overset{k_1^m}{\longrightarrow} HA\cdot + M\text{–}NO^- \qquad\qquad [4]$$

$$HA\cdot \rightleftarrows A^{-\cdot} + H^+ \qquad pK = 4.25 \qquad [5]$$

$$A^{-\cdot} + NO\cdot \overset{k_2^0}{\longrightarrow} A + NO^- \qquad\qquad [6]$$

$$A^{-\cdot} + M\text{–}NO\cdot \overset{k_2^m}{\longrightarrow} A + M\text{–}NO^- \qquad\qquad [7]$$

$$2\,A^{-\cdot} \overset{k_3}{\longrightarrow} A + A^{2-} \qquad\qquad [8]$$

$$NO^- + H_2O \longrightarrow NOH + OH^- \qquad\qquad [9]$$

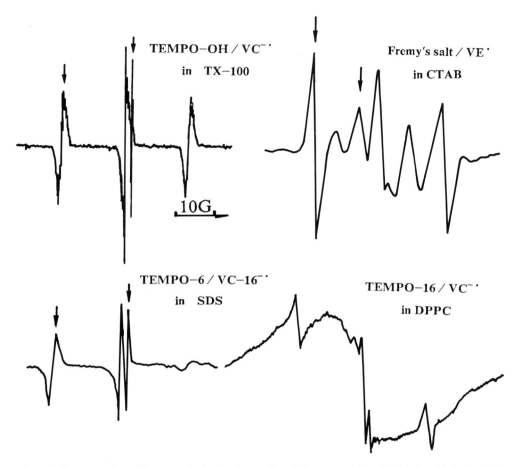

Figure 1. Representative ESR spectra obtained under fast flow during the reaction of TEMPOs with VE and VCs. Arrows indicate the position for field locking to determine the decay traces after stopping the flow.

where M, M—NO· and H_2A designate the micelle, the nitroxide-micelle aggregate and ascorbic acid, respectively. k^0 and k^m are rate constants for reactions conducted in the bulk water and in the micellar phase, respectively. Computer simulation of the kinetic results evaluated the rate constants and equilibrium constants for every elementary reaction (Tabs 1 and 2, Liu et al., 1989). Figure 2 depicts the relationship between the apparent pseudo first order decay rate of the nitroxide and the surfactant concentration. It is seen that the reactions related to the participation of the lipophilic nitroxides were accelerated by the cationic micelle CTAB and retarded by the

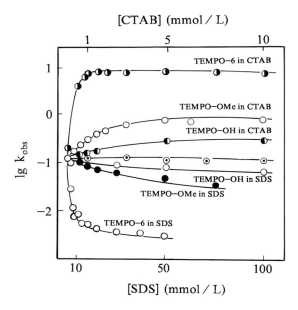

Figure 2. Correlation between the pseudo first-order decay of TEMPOs and the surfactant concentration in the reaction with VC in micelles.

anionic micelle SDS, while not appreciably influenced by the non-ionic micelle Triton X-100 (results not shown). And the more lipophilic the nitroxide, the more pronounced the rate variation. As high as 3600-fold of rate enhancement was observed for the reaction of TEMPO-6 with VC conducted in CTAB over that in SDS. Because VC existed predominantly in its anion form in neutral solutions, this result clearly indicates that the electrostatic interaction between VC$^-$ and the head group of the micelles is the rate-determining factor for the hydrophilic antioxidant. Furthermore, the abrupt change of the rate around the critical micellar concentration (CMC) indicates that the rate variation must also be related to the formation of the nitroxide-micelle aggregate, i.e., related to the lipophilicity of the oxidant. The binding constants of the nitroxides with the micelles obtained kinetically coincide well with those determined by the diffusion coefficient method (Liu et al., 1992a).

Reactions of lipophilic vitamin C and vitamin E in micelles

As indicated by Ingold (Doba et al., 1985) and Niki (1987), VC was ineffective as an antioxidant for lipid peroxidation initiated in the oil phase of micelles and liposomes, although it worked very

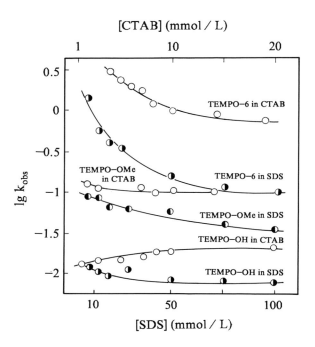

Figure 3. Correlation between the pseudo first-order decay of TEMPOs and the surfactant concentration in the reaction with VC-16 in micelles.

well if the lipid peroxidation was initiated in the water phase. On the other hand, Pryor et al. (1985) reported that the lipophilic ascorbyl palmitate (VC-16) was much more active as peroxidation inhibitor for linoleic acid than the lipophobic VC in SDS in micelles, despite the fact that they showed similar activity in homogeneous solutions (Mukai et al., 1987). Therefore, it is worth seeing the detailed behaviour of the lipophilic VCs in micelles and liposomes.

Taking into account the aggregation of both the lipophilic TEMPOs and the lipophilic VCs with micelles, the reaction mechanism becomes very complex. The two substrates may both come from the bulk water, one from bulk water and the other from the micelle, from different micelles, or from the same micelle. Therefore, there is no simple way to derive a comprehensive kinetic representation for the reactions. Nevertheless, it is possible to evaluate the effective rate constants for the three principal reactions as shown in Equations 3, 6 and 8 with the implication that the aggregation state of the two substrates is not expressed and several elementary processes should be involved in each of the reactions (Liu et al., 1991b). The result is listed in Table 2. It indicates that the rate is enhanced in both CTAB and in SDS micelles, in sharp distinction to the behaviour

Table 1. Rate constants and equilibrium constants for the reaction of TEMPOs and VC in SDS and CTAB micelles

	TEMPO-OH	TEMPO-OMe	TEMPO-6	micelle
k_1^0 (M^{-1}s^{-1})	3.0	2.5	2.0	
k_1^m (M^{-1}s^{-1})	1.4	0.49	0.067	SDS
	8.3	19.3	207	CTAB
k_2^0 (10^3 M^{-1}s^{-1})	9.0	8.5	10.0	
k_2^m (10^3 M^{-1}s^{-1})	7.8	3.4	0.34	SDS
	23.2	46.6	274	CTAB
k_3 (10^5 M^{-1}s^{-1})	4.7	5.0	5.6	
K (10^2 M^{-1})	9.1	30.9	1950	SDS
	14.8	31.6	1036	CTAB
[M–NO·] / [NO·]T	0.57	0.81	0.99	SDS
	0.16	0.25	0.60	CTAB

for the reactions of VC with TEMPOs where the rate is increased in CTAB but decreased in SDS (*vide supra*). This demonstrates that the electrostatic interation is no longer the operative factor for the activity in the case of lipophilic antioxidant. Instead, aggregation of the lipophilic VCs with the micelles plays a crucial role. It is impressive that a variation in rate of up to 2×10^4 could be achieved by making the VC lipophilic and put it into an appropriate microenvironment.

It is also interesting to compare the different surfactant concentration dependence of the rates for VC and VC-16 (Figs 2 and 3). In the case of VC the rate reaches a plateau above the CMC of the micelles, while in the case of VC-16 the rate is decreased with increase of the surfactant concentration, and at higher surfactant concentrations the rate is approximately reciprocal proportional to the concentration of the surfactant, an indication that inter-micellar diffusion may contribute to the rate-limiting step (Burkey and Griller, 1985). Castle and Perkins (1986) has reached the same conclusion in the study on the reactivity of VE and its derivatives in SDS micelles.

In order to compare the activity of the VCs with that of VE, Fremy's salt, which is a more powerful oxidant than the TEMPOs and able to oxidize VE, was used to react with these

Table 2. The effective rate constants for the reaction of TEMPO-6 with VC and VC-16 in micelles

	CTAB		SDS	
	VC–16	VC	VC–16	VC
k_1^ψ (M^{-1} s^{-1})	1.4×10^3	2×10^2	96	0.062
k_2^ψ (M^{-1} s^{-1})	3.7×10^4	2.2×10^5	2.8×10^3	3.5×10^2
k_3^ψ (M^{-1} s^{-1})	1.1×10^5	5.2×10^5	1.3×10^4	4.0×10^5

Table 3. Rate constants for reactions of VE and VCs with Fremy's salt

	f	k_{obs} (10^3 M^{-1} s^{-1})	k_{int} (10^3 M^{-1} s^{-1})	micelle
VE	1	7.9	7.9	CTAB
VC	0.048	4.3	89.6	CTAB
VC-8	0.33	35	106.1	CTAB
VC-12	0.71	53	74.6	CTAB
VC-16	0.89	56	62.9	CTAB
VE	1	0.0022	0.0022	SDS
VC		3.3		SDS
VC-8	0.16	2.7	16.9	SDS
VC-12	0.60	1.2	2.0	SDS
VC-16	0.90	0.86	1.0	SDS

antioxidants (Liu et al., 1990). It was found that the second order rate constant for the reaction of VE and Fremy's salt was 7.9×10^3 M^{-1}s^{-1} in CTAB, but dropped to 2.2 M^{-1}s^{-1} in SDS, obviously due to the electrostatic interaction between the anionic oxidant and the surface charg of the micelles in which VE is tightly anchored with its phenoxyl group pointing towards the surface of the micelle. On the other hand, the rates of VCs were significantly dependent upon their side-chain lengths (Tab. 3). It is worth pointing out that the activity sequence for the apparent rate, k_{obs}, is VC-16 > VC-12 > VC-8 >> VE > VC in CTAB micelle, whereas VC > VC-8 > VC-12 > VC-16 >> VE in SDS micelle. Similar activity sequences have also been observed in the inhibition of linoleic acid autoxidation using the oxygen uptake method (Liu et al., 1992c). This may be rationalized by the fact that in CTAB micelle the oxidant can adsorb on the surface of the micelle to react readily with the antioxidant which aggregates with the micelle, thus the stronger the aggregation (with longer side-chains), the faster the rate. Conversely, in SDS micelle the oxidant is found in the bulk water and far away from the micelle surface. It must wait for the antioxidant diffusing out from the micelle before it can react. Therefore, the shorter the side-chain, the easier the diffusion of the antioxidant, thus the faster the rate. It is worth pointing out, however, that if the intrinsic rate constant, k_{int}, which is defined by k_{obs}/f, is compared, the reactivity sequence becomes VC-8 > VC-12 > VC-16 in both CTAB and SDS micelles, which goes in the opposite direction to the sequence of the partition coefficients, f, of the antioxidants in the micelles. These results clearly demonstrate that the intermicellar diffusion rate is a crucial factor governing the activity of antioxidants with the same active group but bearing different lengths of side-chains in micelles.

Antioxidant synergism of vitamin E and lipophilic vitamin C derivatives

The antioxidant synergism of vitamin E and vitamin C has been well recognized and extensively studied (Packer et al., 1979; Burton and Ingold, 1986; Liu et al., 1988; Barclay, 1993). The key step of the synergism is the reaction between vitamin E radical and vitamin C:

$$VE\cdot + VC^- \longrightarrow VE + VC^{-\cdot} \qquad\qquad [10]$$

It has been reported previously (Mukai et al.,1989) that introducing a long hydrocarbon chain at the 5- or 6-position in VC did not significantly change its reactivity towards VE radical in homogeneous solutions. On the other hand, Scarpa et al. (1984) found that the rate constant for the reaction (10) was about one order of magnitude lower in soybean phosphatidylcholine liposomes than in homogeneous solutions. We have generated vitamin E radical by oxidizing vitamin E with Fremy' salt and studied its reaction with VC and the lipophilic VCs in CTAB micelles (Liu et al., 1992b). Because Vitamin E and its radical are believed to reside exclusively in the micellar phase, the water phase reaction can be neglected and the reaction mechanism may be simply represented as follows:

$$M + HA^- \underset{}{\overset{K_1}{\rightleftharpoons}} M\text{---}HA^- \qquad K_1 = k_1/k_{-1} \qquad [11]$$

$$M\text{---}VE\cdot + HA^- \overset{k_2}{\longrightarrow} M\text{---}VE + A^{-\cdot} \qquad\qquad [12]$$

The effective rate constant may be expressed as:

$$k_{eff} = k_2/(1 + K_1\,[M]) \qquad\qquad [13]$$

It was found that the effective rate constant was markedly influenced by the side-chain length of the VCs, being 9×10^5 $M^{-1}s^{-1}$ for VC and 3×10^3 $M^{-1}s^{-1}$ for VC-16. If we assume the entry rate, k_1, is diffusion limited, then the exit rate, k_{-1}, can be deduced from the equilibrium constant. The results are listed in Table 4. It is seen that the exit rate decreases with the increase of the side-chain length of the VCs, which parallels the activity sequence of k_{eff}. This enables us to depict a mechanism for the reaction in micelles. For a lipophilic VC which shows a strong preference for the micellar phase to react with VE· which resides in another micelle, it must diffuse out from the micelle in which it located and enter another micelle. It probably does so repeatedly until it encounters a VE· for the reaction to occur. It is also possible that the lipophilic VC must undergo intramicellar diffusion in one micelle to find a VE· in the right place, because k_2 is also side-chain length dependent.

Table 4. Rate constants for the reaction of VE· with VCs in CTAB micelles

	k_{eff} (M^{-1} s^{-1})	k_2 (M^{-1} s^{-1})	K_1 (M^{-1})	k_{-1} (s^{-1})
VC	9.0×10^5	1.1×10^6	9.2×10^2	1.1×10^6
VC-8	3.0×10^5	9.2×10^6	9.3×10^3	1.1×10^5
VC-12	7.0×10^4	7.9×10^5	4.6×10^4	2.0×10^4
VC-16	3.0×10^3	1.2×10^5	1.5×10^5	6.7×10^3

Reactions of vitamin C and its lipophilic derivatives in liposomes

The antioxidation of vitaines E and C, especially vitamin E, has been extensively studied in phospholipids (Lohmann and Winzenburg, 1983; Niki et al., 1985; Barclay, 1993). However, little is known about the behaviour of lipophilic VCs in liposomes. As a logical extension of our studies in micelles, reactions of nitroxides with lipophilic VCs were studied in DPPC liposomes and compared with those of hydrophilic VC (Wu et al., 1993). The reaction of VC with TEMPO-16 and TEMPO-12 in DPPC showed a biphasic kinetics, i.e., a relatively fast decay of the nitroxide followed by a very slow one (Fig. 4). In the case of TEMPO-16 about 20% of the initial amount of the nitroxide remained intact after a period of a few hours. Computer simulation of the kinetic

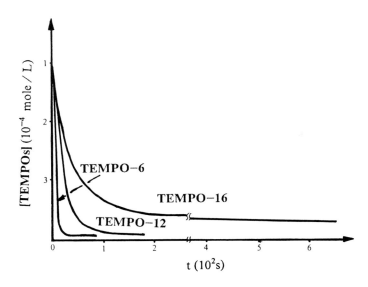

Figure 4. The pseudo first-order decay of TEMPOs during the reaction with VC in DPPC liposomes.

Table 5. Rate contants for the reaction between TEMPOs and VCs in DPPC liposomes

| | k_2 $(M^{-1} s^{-1})$ | | | | k_1 $(10^{-3} s^{-1})$ |
	VC	VC-8	VC-12	VC-16	VC
TEMPO-16	0.37	285	252	224	0.79
TEMPO-12	1.07	412	362	301	1.36
TEMPO-6	6.11	705	620	539	6.18

traces indicated that the initial fast decay was first order with respect to both the nitroxide and VC, being an overall second order reaction, while the latter slow decay was first order in the nitroxide but independent of the concentration of VC. Therefore, it is reasonable to propose that the initial stage of the reaction deals with the reaction of VC with the nitroxide residing in the external monolayer of DPPC liposomes, while the latter first-order reaction deals with the transverse motion, i.e., flip-flop (Kornberg and McConnel, 1971) of the nitroxide from the internal to the external monolayer of DPPC (Fig. 5). It is seen from Table 5 that both rates decrease with increasing side-chain length of the nitroxides as expected, revealing again the crucial importance of diffusion for the activity.

   The most striking observation was that, when the lipophilic VCs were incorporated into DPPC liposomes and allowed to react with the lipophilic TEMPOs which were separately incorporated into another portion of the liposomes, the decay of the nitroxide was second-order throughout the whole course of the reaction and the rate was dramatically enhanced, the shorter the side-chain, the faster the rate (Tab. 5). This two orders of magnitude rate enhancement in comparison with the hydrophilic VC reaction (*vide supra*) was really unexpected. It was anticipated that the rate should be slowed because the lipophilic VCs incorporated into the liposomes would have to diffuse out from the liposomes to carry out the reaction *via* an interliposomal exchange, which might be a very slow, unnoticeable process below the phase transition temperature (Fendler, 1982). Therefore, we propose a liposome fusion mechanism to rationalize this remarkable activity enhancement. It is well known that liposomes are dynamically stable systems below the phase transition temperature due to the tremendous hydration repulsion force against the get-together of two liposomes. However, fusion could be induced by some ionophores or fusogens, such as calcium ion which may form a calcium bridge between two liposomes to facilitate fusion (Ellens et al., 1985). In the present case incorporation of the lipophilic VCs into the DPPC liposomes would make the liposomal surface negatively charged.On the other hand, the nitroxide is polar and electrophilic and the high polarity of the liposomal surface would favour the polarization of the nitroxide, thus incorporation of the nitroxide into the liposome would bring some positive charge to the surface. Obviously, electrostatic interaction between the oppositely charged lipo-somes would serve as a driving force to induce the fusion. Furthermore, the rate is also side-chain length dependent with the shorter side-chain bringing about faster reaction. This implies that the intraliposomal diffusion may also contribute, because subsequent to the liposome fusion the two

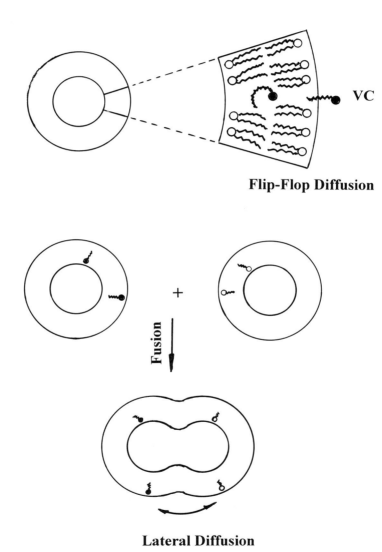

Figure 5. Schematic representation of substrate motions in DPPC liposomes.

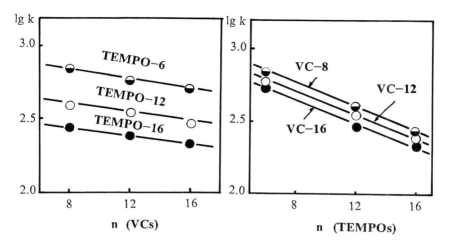

Figure 6. Correlation between the second-order rate constants for the reaction of TEMPOs with VCs and the number of carbon atoms in the side-chains.

substrates must subject to fast lateral intraliposomal diffusion to find each other for the reaction to take place. Interestingly, plotting the logarithm of the rate *versus* the number of carbon atoms

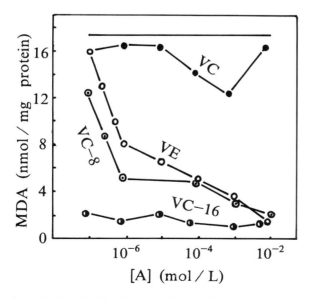

Figure 7. Inhibition of peroxidation of rat liver liposomes by antioxidants *in vitro*, initiated by Fe^{++}-cysteine.

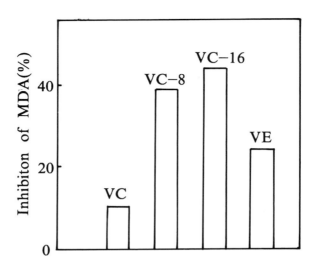

Figure 8. Inhibition of lipid peroxidation of rat liver by antioxidants *in vivo*, initiated by CCl$_4$.

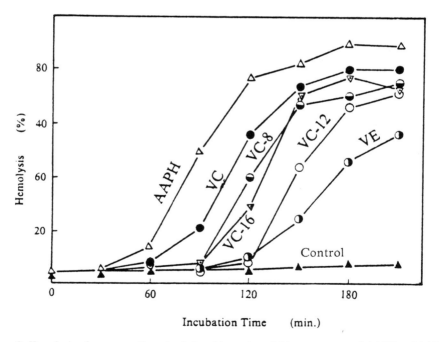

Figure 9. Hemolysis of mouse erythrocytes induced by water-soluble azo-compound AAPH and inhibited by antioxidants.

of the side-chain gave fairly good straight lines with the slopes constant for a series of substrates (Fig. 6). The significance of this interesting phenomenon remains to be uncovered.

Antioxidant activity of lipophilic VCs in biological membranes

Some preliminary biological assays have been performed to check if the criteria obtained from the membrane mimetic systems mentioned above also work in biological membranes. Representative results are illustrated in Figures 7, 8 and 9. It is seen that the antioxidant activity of VC-8 and VC-16 against lipid peroxidation is much better than that of VC, and even better than that of VE both *in vitro* and *in vivo* (Liu et al., 1991b). An informative observation is that in the antihemolysis experiment all three lipophilic VCs are more active than VC with VC-12 being the most active (Kuang et al., 1994). This demonstrates that the overall antioxidant activity of lipophilic VCs may be controlled by a subtle balance of the two effects of the side-chain mentioned above. A longer side-chain would make the antioxidant easier to insert into the membrane but more difficult to diffuse along the surface of the membrane to find a peroxyl radical that is expected to be trapped.

Conclusion

Despite an overwhelming accumulation of research data revealing the beneficial effects of antioxidants on health, research to uncover the mechanisms, the structure/activity relationship and the microenvironment/activity relationship of the antioxidant action is relatively scarce. This work and that of Ingold (Burton and Ingold, 1986), Niki (1987), Barclay (1993) and others has convincingly proved that application of the principles and techniques of physical organic chemistry to problems of biological significance may produce valuable information which may help us to gain an insight into the details of the antioxidant mechanisms in biological systems and help us in antioxidant drug design. To a larger extent, it is expected that the new interdiscipline, physical organic molecular biology, as coined by Fersht (1990), will develop further in future.

Acknowledgements
I am deeply grateful to Professor You-Cheng Liu for his scientific leadership which initiated and helped to accomplish this research. Thanks are also due to those colleagues, especially Professor Lung-Min Wu and Dr. Zheng-Xu Han, and students, whose names are in the references. Without their enthusiastic participation this work would never have been possible. Continuous financial support from the National Natural Science Foundation of China and the Education Commission of China is gratefully acknowledged.

References

Barclay, L.R.C. and Ingold, K.U. (1981) The autoxidation of a model membrane: a comparison of the autoxidation of egg lecithin phosphatidylcholine in water and in chlorobenzene. *J. Am. Chem. Soc.* 103: 6478–6485.

Barclay, L.R.C. (1993) Model Membranes: quantitative studies of peroxidation, antioxidation action, partitioning, and oxidative stress. *Can. J. Chem.* 71: 1–16.

Bendich, A. (1993) Physiological role of antioxidants in the immune system. *J. Dairy Sci.* 76: 2789–2794.

Burkey, T.J. and Griller, D. (1985) Micellar systems as devices for enhancing the lifetimes and concentration of free radicals. *J. Am. Chem. Soc.* 107: 246–249.

Burton, G.W. and Ingold, K.U. (1986) Vitamin E: Application of the principles of physical organic chemistry to the exploration of its structure and function. *Acc. Chem. Res.* 19: 194–201.

Castle, L. and Perkins, M.J. (1986) Inhibition kinetics of chain-breaking phenolic antioxidants in SDS micelles: evidence that intermicellar diffusion rates may be rate-limiting for hydrophobic inhibitors such as α-tocopherol. *J. Am. Chem. Soc.* 108: 6381–6382.

Doba, T., Burton, G.W. and Ingold, K.U. (1985) Antioxidant and co-antioxidant activity of vitamin C: The effect of vitamin C either alone or in the presence of vitamin E or a water soluble vitamin E analogue, upon the peroxidation of aqueous multilamellar phospholipid liposomes. *Biochim. Biophys. Acta* 835: 298–303.

Ebel, C., Ingold, K.U., Michon, J. and Rassat, A. (1985) Kinetics of reduction of a nitroxide radical by ascorbic acid in the presence of β-cyclodextrin: determination of the radical β-cyclodextrin association constant and rate constants for reaction of the free and complexed nitroxide radical. *Nouv. J. Chim.* 9: 479–485.

Ellens, H. Bentz, J. and Szoka, F.C. (1985) H^+- and Ca^{2+}-induced fusion and destabilization of liposomes. *Biochem.* 24: 3099–3106.

Favier, A., Sappey, C, Leclere, P., Faure, P. and Micoud, M. (1994) Antioxidant Status and lipid peroxidation in patients infected with HIV. *Chem. Biol. Interac.* 91: 165–180.

Fendler, J.H. (1982) *Membrane Mimetic Chemistry*, Wiley, New York, Chapter 6, pp 113–183.

Fersht, A.R. (1990) Protein structure and activity: physical organic molecular biology, 10th IUPAC Conference on Physical Organic Chemistry, Haifa, Israel, Abstracts, PL-2.

Gey, K.F. (1993) Prospects for the prevention of free radical diesease, regarding cancer and cardiovascular disease. *Br. Med. Bull., 49:* 679–699.

Halliwell, B. and Gutteridge, J.M.C. (1985) *Free Radicals in Biology and Medicine*, Clarendon Press, Oxford.

Kornberg, R.D. and McConnell, H.M. (1971) Inside-outside transitions of phospholipids in Vesicle membranes. *Biochem.* 10: 1111–1120.

Kuang, Z.H., Wang, P.F., Zheng, R.L., Liu, Z.L. and Liu, Y.C. (1994) Making vitamin C lipo-soluble enhances its protective effect against radical induced hemolysis of erythrocytes. *Chem. Phys. Lipids* 71: 95–97.

Liu, Y.C., Wu, M.L., Liu, Z.L. and Han, Z.X. (1985) A kinetic ESR study on the oxidation of ascorbic acid by a nitroxide. *Acta Chimica Sinica (Engl. Ed.)* 4: 342–348.

Liu, Y.C., Liu, Z.L. and Han, Z.X. (1988) Radical intermediates and antioxidant activity of ascorbic acid. *Rev. Chem. Intermed.* 10: 269–289.

Liu, Y.C., Han, Z.X., Wu, L.M., Chen, P. and Z.L. Liu (1989) Micellar effect on the reduction of nitroxides by vitamin C. *Science in China (B) Engl. Ed.* 32: 937–947.

Liu, Z.L., Han, Z.X., Chen, P. and Liu, Y.C. (1990) Stopped-flow ESR study on the reactivity of vitamin E, vitamin C and its lipophilic derivatives towards Fremy's salt in micellar systems. *Chem. Phys. Lipids* 56: 73–80.

Liu, Z.L., Wang, L.J. and Liu, Y.C. (1991a) Prooxidation of vitamin E on the autoxidation of linolic acid in sodium dodecyl sulfate micelles. *Science in China (B), Engl. Ed.* 34: 787–795.

Liu, Y.C., Liu, Z.L., Han, Z.X., Chen, P. and Wang, L.J. (1991b) Microenvironmental effects in the action of bioantioxidants. *Progr. Natur. Sci. (Engl. Ed.)* 1: 297–306.

Liu, Z.L., Han, Z.X., Chen, C, Wu, L.M. and Liu, Y.C. (1991c) An ESR study on the antioxidant efficiency of ascorbyl palmitate in micelles. *Chin. J. Chem.* 9: 144–151.

Liu, Y.C., Han, Z.X. and Liu, Z.L. (1992a) Determination of binding constants of lipophilic vitamin C derivatives and nitroxides in micelles. *Chem. J. Chin. Univ.* 13: 214–216.

Liu, Z.L., Wang, P.F. and Liu, Y.C. (1992b) Inhibition of autoxidation of linoleic acid by vitamin E, vitamin C and lipophilic vitamin C derivatives in micelles. *Science in China (B), Engl. Ed.* 35: 1307–1314.

Liu, Z.L., Han, Z.X., Yu, K.C., Zhang, Y.L. and Liu, Y.C. (1992c) Kinetic ESR studies on the reaction of vitamin E radical with vitamin C and its lipophilic derivatives in cetyl trimethylammonium bromide micelles. *J. Phys. Org. Chem.* 5: 33–38.

Lohmann, W. and Winzenburg, J. (1983) Structure of ascorbic acid and its biological function: V. Transport of ascorbate and isoascorbate across artificial membranes as studied by the spin label technique. *Z. Naturforsch.* 38C: 923–925.

McBrien, D.C.H. and Slater, T.F. (eds) (1982) *Free Radicals, Lipid Peroxidation and Cancer*, Academic Press, New York.

Miguel, J. (1988) *Handbook of Free Radicals and Antioxidants in Biomedicine*, Vol 3., CRC Press, Boca Raton, Fla.

Mukai, K., Fukuda, K., Ishizu, K. and Kitamura, Y. (1987) Stopped-flow investigaiton of the reaction between vitamin E radical and vitamin C in solution. *Biochem. Biophys. Res. Commun.* 146: 134–139.

Mukai, K., Nishimura, A., Nagano, K., Tanaka, K. and Niki, E. (1989) Kinetic study on the reaction of vitamin C derivatives with tocopheroxyl (vitamin E radical) and substituted phenoxyl radicals in solution. *Biochim. Biophys. Acta* 993: 168–173.

Niki, E., Kawakami, K., Yamamoto, Y. and Kamiya, Y. (1985) Synergistic inhibition of oxidation of phosphatidylcholine liposome in aqueous dispersion by vitamin E and vitamin C. *Bull. Chem. Soc. Jpn.* 58: 1971–1975.

Niki, E. (1987) Lipid antioxidants: How they may act in biological systems. *Br. J. Cancer, Suppl.* 55: 153–157.

Okazaki, M. and Kuwata, K. (1985) A stopped-flow ESR study on the reactivity of some nitroxide radicals with ascorbic acid in the presence of β-cyclodextrin. *J. Phys. Chem.* 89: 4437–4440.

Packer, J.E., Slater, T.F. and Willson, R.L. (1979) Direct observation of a free radical interaction between vitamin E and vitamin C. *Nature* 278: 737–738.

Packer, L. and Fuchs, J. (1992) *Vitamin E in Health and Disease*, Marcel Dekker, New York.

Pryor, W.A. (1976–1986) *Free Radicals in Biology*. Vols. 1–6, Academic Press, New York.

Pryor, W.A., Kaufman, M.T. and Church, D.F. (1985) Autoxidation of micelle-solubilized linoleic acid: relative inhibitory effectiveness of ascorbate and ascorbyl palmitate. *J. Org. Chem.* 50: 281–282.

Pryor, W.A., Strickland, T. and Church, D.F. (1988) Comparison of the effectiveness of several natural and synthetic antioxidants in aqueous sodium dodecyl sulfate micelle solution. *J. Am. Chem. Soc.* 110: 2224–2229.

Rice-Evans, C.A. and Diplock, A.T. (1993) Current status of antioxidant therapy. *Free Radical Biol. Med.* 15: 77–96.

Scarpa, M., Rigo, A., Maiorano, M., Ursini, F. and Gregolin, C. (1984) Formation of α-tocopherol radical and recycling of α-tocopherol by ascorbate during peroxidation of phosphatidylcholine liposomes. *Biochim. Biophys. Acta* 801: 215–219.

Sies, H., Stahl, W. and Sundquist, A.R. (1992) Antioxidant functions of vitamins: vitamin E and C, β-carotene, and other carotenoids. *Ann. N.Y. Acad. Sci.* 669: 7–20.

Slater, T.F. (1987) Free radicals and tissue injury: fact and fiction. *Br. J. Cancer, Suppl.* 55: 5–10.

Schwartz, J.L., Antoniades, D.Z. and Zhao, S. (1993) Molecular and biological reprogramming of oncogenesis through the activity of prooxidants and antioxidants. *Ann. N.Y. Acad. Sci.* 686: 262–278.

Takahashi, M., Tsuchiya, J. and Niki, E. (1989) Scavenging of radicals by vitamin E in the membranes as studied by spin labeling. *J. Am. Chem. Soc.* 111: 6350–6353.

Wu, L.M., Guo, F.L., Liu, Z.L. Zhang, Y.L., Jia, X.Q. and Liu, Y.C. (1993) Antioxidant activity of lipophilic vitamin c derivatives in dipalmitoyl phosphatidylcholine vesicles. *Res. Chem. Intermed.* 19: 657–663.

Bioradicals Detected by ESR Spectroscopy
H. Ohya-Nishiguchi & L. Packer (eds)
© 1995 Birkhäuser Verlag Basel/Switzerland

Action of antioxidants as studied by electron spin resonance

E. Niki

Research Center for Advanced Science and Technology, The University of Tokyo, 4-6-1 Komaba, Meguro, Tokyo 153, Japan

Summary. In this overview, the results of our study on the actions of radical-scavenging antioxidants by electron spin resonance spectroscopy are shown and discussed. The potency for hydrogen atom donation of antioxidants can easily be estimated from the rate of reaction with stable radicals such as galvinoxyl by following the decay of its ESR signal. In general, a potent antioxidant gives a stable radical whose ESR signal gives important information on its stability, location and reactivity toward other reductants. The spin probe is also a versatile tool for studying the action of antioxidant in the membranes and lipoproteins.

Introduction

As experimental, clinical and epidemiological evidence is accumulated which supports the causative role of free radical-mediated oxidation of biological membranes and tissues in the progress of various pathological events, cancer and aging, the importance of antioxidants, especially those from foods, has received renewed attention. Foods contain a variety of antioxidants such as vitamin C, vitamin E, carotenoids, phenolic compounds, flavonoids and metals. It is essential to elucidate the action of these antioxidants in order to understand their role fully and use them properly. Aerobic organisms are protected from oxidative damage by an array of defense systems (Tab. 1). The radical-scavenging antioxidants function as the second-line defense by scavenging active radicals to inhibit chain initiation and break chain propagation. In this reaction, a new radical is formed from the antioxidant and the fate of this antioxidant-derived radical is important in determining the total antioxidant potency. Therefore, ESR is a useful tool for investigation of these antioxidants.

Reaction of antioxidant with stable radical

A simple test for evaluating the activity of any compound as hydrogen atom-donating, radical-scavenging antioxidant is to measure the reactivity with a stable radical such as galvinoxyl or 1,1-diphenyl-2-picrylhydrazyl (DPPH). The reaction can be followed from the decay of the ESR

Table 1. Defense system against oxidative damage

1.	Preventive antioxidants: suppress the formation of free radicals	
	(a) Non-radical decomposition of hydroperoxides and hydrogen peroxide	
	catalase	decomposition of hydrogen peroxide $2H_2O_2 \rightarrow 2H_2O_2 + O_2$
	glutathione peroxidase (cellular)	decomposition of hydrogen peroxide and free fatty acid hydroperoxides $H_2O_2 + 2GSH \rightarrow 2H_2O + GSSG$ $LOOH + 2GS \rightarrow LOH + H_2O + GSSG$
	glutathione peroxidase (plasma)	decomposition of hydrogen peroxide and phospholipid hydroperoxides $PLOOH + 2GSH \rightarrow PLOH + H_2O + GSSG$
	phospholipid hydroperoxide glutathione peroxidase	decomposition of phospholipid hydroperoxides
	peroxidase	decomposition of hydrogen peroxide and lipid hydroperoxides $LOOH + AH_2 \rightarrow LOH + H_2O + A$ $H_2O_2 + AH_2 \rightarrow 2H_2O + A$
	glutathione-S-transferase	decomposition of lipid hydroperoxides
	(b) Sequestration of metal by chelation	
	transferrin, lactoferrin haptoglobin hemopexin ceruloplasmin, albumin	sequestration of iron sequestration of hemoglobin stabilization of heme sequestration of copper
	(c) Quenching of active oxygens	
	superoxide dismutase (SOD)	disproportionation of superoxide $2O_2^{-\cdot} + 2H^+ \rightarrow H_2O_2 + O_2$
	carotenoids, vitamin E	quenching of singlet oxygen
2.	Radical-scavenging antioxidants: scavenge radicals to inhibit chain initiation and break chain propagation	
	hydrophilic:	vitamin C, uric acid, bilirubin, albumin
	lipophilic:	vitamin E, ubiquinol, carotenoids, flavonoids
3.	Repair and *de novo* enzymes: repair the damage and reconstitute membranes	
	lipase, protease, DNA repair enzymes, transferase	
4.	Adaptation: generate appropriate antioxidant enzymes and transfer them to the right site at the right time and in the right concentration	

signal of the stable radical or absorption spectroscopy. This is a reversible reaction but a strong hydrogen-donating antioxidant reacts rapidly with the stable radical and consumes it completely. The relative antioxidant activity can be estimated from the reactivity of the compound with the stable radical.

α-Tocopherol, the most active form of vitamin E, reacts rapidly with galvinoxyl and DPPH to give α-tocopheroxyl radical. The ESR spectrum and hyperfine coupling constant of α-tocopheroxyl radical in different media (Fig. 1) give us information on the location and environment (Iwatsuki et al., 1994). The hyperfine signal of α-tocopherol can be observed in nonpolar solvents in the absence of oxygen and the hyperfine splitting constants were obtained as $a_H^{5CH_3} = 0.51$ mT, $a_H^{7\ CH_3} = 0.45$ mT, $a_H^{4CH_3} = 0.15$ mT, and $a_H^{8CH_3} = 0.10$ mT. In the presence of oxygen, the spectrum of only seven lines was obtained as shown in Figure 1. Their hyperfine splitting constants are the average of those of $a_H^{5CH_3}$ and $a_H^{7CH_3}$. They were larger in aprotic solvents than in protic solvents. The smaller coupling onstants in protic solvents may be

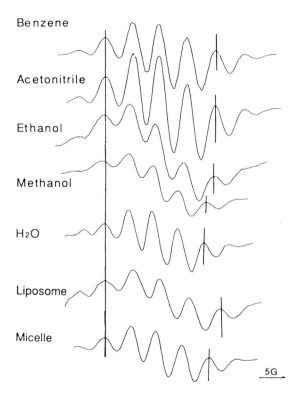

Figure 1. ESR spectrum of α-tocopheroxyl radical in various solvents and media. In benzene, acetonitrile, methanol and ethanol, α-tocopherol was treated with DPPH. In water, trolox was used instead of α-tocopherol and AAPH was used as a radical source. In liposomes, 5.0 mM α-tocopherol was incorporated into 59 mM 14:0 PC liposomal membrane and 200 mM AAPH was used as an initiator. In micelle, 10 mM α-tocopherol and 52 mM methyl linoleate were dispersed in 0.5 M SDS and 200 mM AAPH was used as an initiator. Samples dissolved in hexane or acetonitrile were taken into capillary tubes and measured under reduced pressure (approximately 0.5 atm. of air) at 37°C. Other samples were measured at 37°C and in air.

Table 2. Pseudo-first order rate constant (s^{-1}) for the reduction of galvinoxyl radical with ascorbic acid, cysteine, and glutathione in various media at 37°C

	Homogeneous solution[a]	Micelle[b]	Liposome[c]
Ascorbic acid	too fast	1.7×10^{-4}	too fast
Cysteine	1.0×10^{-3}	4.5×10^{-5}	6.5×10^{-4}
Glutathione	1.8×10^{-4}	1.7×10^{-5}	2.0×10^{-4}

[a] In acetone/water (9/4 by v/v).
[b] Galvinoxyl dissolved in ethyl palmitate in 10 mM Triton X-100 aqueous dispersions.
[c] Galvinoxyl incorporated into dimyristoyl phosphatidylcholine liposome in 0.1 M NaCl aqueous dispersions.

ascribed to hydrogen bonding between solvent and α-tocopheroxyl radical. The coupling constants in PC liposomes were similar to those obtained in hexane and acetonitrile, while those in SDS micelles were similar to those in water or in methanol. These results imply that α-tocopheroxyl radical does not stick its phenolic group out of the membrane into the aqueous phase, but rather is situated in the lipophilic domain of the membrane. On the other hand, the phenolic oxygen of α-tocopheroxyl radical in the micelle is suggested to be located in or closer to the water phase.

As described later, the reduction of phenoxyl radical to regenerate phenolic antioxidant is an important step in the synergistic inhibition of oxidation. For example, the reduction of α-tocopheroxyl radical by ascorbate is well known (Packer, 1992). The relative activities of ascorbate, cysteine and glutathione in the reduction of galvinoxyl have been measured in different media by following the decay of the ESR signal of galvinoxyl. As shown in Table 2, the reactivities decreased in the order ascorbate > cysteine > glutathione (Tsuchiya et al., 1985). The reduction proceeded rapidly in homogeneous solution for any reductant, slower in the liposome system and most slowly in the micelle system using Triton X-100 as a surfactant.

Furthermore, this reaction can be used for measuring the mobility of antioxidant between the membranes. The interaction of galvinoxyl incorporated into dimyristoyl phosphatidylcholine liposomal membranes and chromanols En with various side chain lengths incorporated into different dimyristoyl phosphatidylcholine liposomal membranes, was studied by following the decay of the ESR signal of galvinoxyl. The rate decreased with increasing length of side chain of chromanols. In agreement with these results, the antioxidant efficiency against free radical-induced oxidative hemolysis of En incorporated into dimyristoyl phosphatidylcholine liposomes decreased with increasing length of the side chain (Niki et al., 1988).

Studies using spin label

A spin label technique, first developed in a study using the chlorpromazine cation as a probe of drug-DNA interactions (Ohnishi and McConnell, 1965), has been applied to the structure-function problem in biological systems (Smith et al., 1976). We have also used N-oxyl-4,4'-dimethyloxazolidine derivatives of stearic acid NS as spin labels in the study of the dynamics of the action of antioxidant in membranes and lipoproteins.

One piece of information obtained from the spin probes incorporated into the membranes is the physical property of the microenvironment. For example, the ESR spectra of 5-NS, 1 2-NS and 1 6-NS in homogeneous solution are indistinguishable from each other, but they give different spectra when incorporated into membranes or low density lipoprotein (LDL) due to the different fluidity of the environment. α-Tocopherol, like cholesterol, decreases the membrane fluidity but its effect is quite small at physiological concentrations (Takahashi et al., 1988).

Ascorbic acid reduces nitroxide quite rapidly. When the spin probe NS is incorporated into the

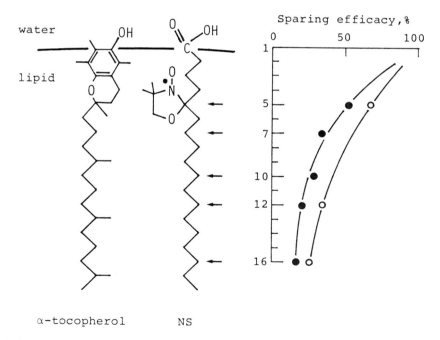

Figure 2. Sparing efficiency by α-tocopherol of NS spin probe having a nitroxide group at different positions along the carbon chain of stearic acid incorporated into soybean phosphatidylcholine (○) or dimyristoyl phosphatidylcholine (●) liposomal membranes in the presence of lipophilic azo radical initiator. α-Tocopherol, NS and AMVN were incorporated into liposomal membranes simultaneously and the reaction was induced at 37°C in air.

liposomal membranes, the rate of reduction by ascorbate decreases as the nitroxide radical goes deeper into the interior of the membrane (Schreier-Mucillo et al., 1976); Takahashi et al., 1988). Similar results are obtained with LDL. Interestingly, cholesterol ester of 16-NS incorporated into LDL, which is assumed to be located in the LDL core, is not reduced by ascorbate.

The spin label NS acts as an antioxidant and suppresses the oxidation of liposomal membranes in which it is incorporated, although the antioxidant activity is much smaller than that of α-tocopherol (Takahashi et al., 1989). It is consumed as the oxidation proceeds and when it is completely depleted, a fast oxidation takes place at a similar rate to that in the absence of the spin label. When both NS and α-tocopherol are incorporated into the same liposomal membranes, they compete in scavenging radicals, but α-tocopherol is much more potent and spares NS at the expense of itself. The sparing efficiency by α-tocopherol is dependent on NS. It spares 5-NS more efficiently than 16-NS. It also depends where the radicals are formed initially. The sparing efficiency is less efficient when the radicals are formed initially within the membranes by the lipophilic azo compound AMVN than when the radicals are formed in the aqueous phase by water-soluble AAPH. Furthermore, the sparing efficiency by α-tocopherol for different NS is higher in soybean phosphatidylcholine liposomes than in dimyristoyl phosphatidylcholine liposomes in the oxidations induced by AMVN (Fig. 2) (Takahashi et al., 1989). All these data suggest that the active site of α-tocopherol is located at or near the membrane surface and that the efficacy for radical scavenging by α-tocopherol decreases as the radicals go deeper into the membranes.

Synergistic inhibition of oxidation by antioxidants

The antioxidants act not only individually but also cooperatively or even synergistically. Above all, the interactions between α-tocopherol (vitamin E) and ascorbate (vitamin C) have received much attention (Packer, 1992). This reduction of α-tocopheroxyl radical by ascorbate was first proved experimentally by Packer et al. (1979) and later by us using ESR (Niki et al., 1982). α-Tocopherol, the most active and abundant lipophilic antioxidant, scavenges lipophilic lipid-derived peroxyl radical to break the chain propagation. Ascorbate reduces α-tocopheroxyl radical to regenerate α-tocopherol, although it cannot scavenge lipophilic radical within the membranes efficiently (see above). Thiols may also contribute to the reduction of α-tocopheroxyl radical. In fact, the reduction of α-tocopherol by cysteine (Motoyama et al., 1989) and glutathione (Niki et al., 1982) has been observed.

It has been found that similar interactions can take place between ascorbate and α-tocopherol in LDL (Sato et al., 1990; Kagan et al., 1992; Kalyanaraman et al., 1992). That α-tocopherol is located primarily at the surface of LDL is also supported by a facile interaction between endogenous α-tocopherol in LDL and copper (II) ion, by which the formation of α-tocopheroxyl radical is confirmed by ESR along with the formation of copper (I) ion (Yoshida et al., 1994).

References

Iwatsuki, M., Tsuchiya, J., Komuro, E., Yamamoto, Y. and Niki, E. (1994) Effects of solvents and media on the antioxidant activity of α-tocopherol. *Biochim. Biophys. Acta* 1200: 19–26.

Kagan, V.E., Serbinova, E.A., Forte, T., Scita, G. and Packer, L. (1992) Recycling of vitamin E in human low density lipoproteins. *J. Lipid Res.* 33: 385–397.

Kalyanaraman, B., Darley-Usmar, V.M., Wood, J., Joseph, J. and Parthasarathy, S. (1992) Synergistic interaction between the probucol phenoxyl radical and ascorbic acid in inhibiting the oxidation of low density lipoprotein. *J. Biol. Chem.* 267: 6789–6795.

Motoyama, T., Miki, M., Mino, M., Takahashi, M. and Niki, E. (1989) Synergistic inhibition of oxidation in dispersed phosphatidylcholine liposomes by a combination of vitamin E and cysteine. *Arch. Biochem. Biophys.* 270: 655–661.

Niki, E., Tsuchiya, J., Tanimura, R. and Kamiya, Y. (1982) Regeneration of vitamin E from α-chromanoxyl radical by glutathione and vitamin C. *Chem. Lett.* 789–792.

Niki, E., Komuro, E., Takahashi, M., Urano, S., Ito, E. and Terao, K. (1988) Oxidative hemolysis of erythrocytes and its inhibition by free radical scavengers. *J. Biol. Chem.* 263: 19809–19814.

Ohnishi, S. and McConnell, H.M. (1965) Interaction of the radical ion of chlorpromazine with deoxyribonucleic acid. *J. Am. Chem. Soc.* 87: 2293.

Packer, J.E., Slater, T.F. and Willson, R.L. (1979) Direct observation of a free radical interaction between vitamin E and vitamin C. *Nature* 278: 737–738.

Packer, L. (1992) Interactions among antioxidants in health and disease: vitamin E and its redox cycle. *Proc. Soc. Expt. Biol. Med.* 200: 271–276.

Sato, K., Niki, E. and Shimasaki, H. (1990) Free radical-mediated chain oxidation of low density lipoprotein and its synergistic inhibition by vitamin E and vitamin C. *Arch. Biochem. Biophys.* 279: 402–405.

Schreier-Mucillo, S., Marsh, D. and Smith, I.C.P. (1976) Monitoring the permeability profile of lipid membranes with spin probes. *Arch. Biochem. Biophys.* 172: 1–11.

Smith, I.C.P., Schreier-Muccillo, S. and March, D. (1976) Spin labeling. *In:* W.A. Pryot (ed.): *Free Radicals in Biology.* Vol. I., Academic Press, New York, pp 149–197.

Takahashi, M., Tsuchiya, J., Niki, E. and Urano, S. (1988) Action of vitamin E as antioxidant in phospholipid liposomal membranes as studied by spin label technique. *J. Nutr. Sci. Vitaminol.* 34: 25–34.

Takahashi, M., Tsuchiya, J. and Niki, E. (1989) Scavenging of radicals by vitamin E in the membranes as studied by spin labeling. *J. Am. Chem. Soc.* 11: 6350–6353.

Tsuchiya, J., Yamada, T., Niki, E. and Kamiya, Y. (1985) Interaction of galvinoxyl radical with ascorbic acid, cysteine, and glutathione in homogeneous solution and in aqueous dispersions. *Bull. Chem. Soc. Jpn.* 58: 326–330.

Yoshida, Y., Tsuchiya, J. and Niki, E. (1994) Interaction of α-tocopherol with copper and its effect on lipid peroxidation. *Biochim. Biophys. Acta* 1200: 85–92.

Bioradicals Detected by ESR Spectroscopy
H. Ohya-Nishiguchi & L. Packer (eds)
© 1995 Birkhäuser Verlag Basel/Switzerland

In vivo EPR spectroscopy

H.M. Swartz, G. Bacic, B. Gallez, F. Goda, P. James, J. Jiang, K.J. Liu, K. Mäder,
T. Nakashima[1], J. O'Hara, T. Shima[1] and T. Walczak

Department of Radiology, Dartmouth Medical School, H.B. 7252, Hanover, NH 03755-3863, USA
[1]Faculty of Third Department of Internal Medicine, Kyoto Prefectural University of Medicine, Kyoto 602, Japan

Summary. This chapter is intended to provide a brief overview of the principles of electron paramagnetic resonance (EPR, or completely equivalently electron spin resonance, ESR) spectroscopy applied to living animals. It attempts to indicate especially those areas in which this approach is likely to be of value because it can provide useful information that cannot be provided as well by other approaches. As a matter of convenience the descriptions are drawn principally from the authors' laboratory but it should be noted that there are a number of laboratories around the world, especially in Japan, which are also actively pursuing these developments. Because of the need for brevity in this volume, the coverage is illustrative rather than comprehensive but this fits well with the aim of the book which is to provide a review that will be useful for the longer term rather than only a review of the current state of development.

Introduction

Definition of in vivo *EPR spectroscopy*

It should be emphasized that the key term is spectroscopy; that is, obtaining well-resolved EPR spectra from defined regions of living animals. This is quite different from *imaging* which provides an indication of the spatial distribution of the intensity of the EPR spectra and not the spectra themselves. There is a technique, termed spectral-spatial imaging, which can provide both types of information simultaneously but this requires a sensitivity that will be very difficult to achieve *in vivo*; it has been applied successfully *in vitro* (Woods et al., 1989).

The volumes from which the spectra are obtained in principle can be any size but in practice are likely to be of the order of 1 cm^3 for diffuse paramagnetic species and can be much less than 1 mm^3 for some particulate paramagnetic species such as lithium phthalocyanine (LiPc) and fusinite (Liu et al., 1993; Vahidi et al., 1994). With the latter types of paramagnetic species it is also quite feasible to obtain spectra from several sites simultaneously (Smirnov et al., 1993).

Potential advantages and limitations of this approach

The principal advantages of obtaining spectra instead of imaging are the much richer information content available in spectra *vs.* imaging and the much better signal/noise ratio achievable with spectroscopy. As is the situation *in vitro*, the EPR spectra of paramagnetic species *in vivo* can be very sensitive indicators of their environment, with changes in the shape and splitting of the spectral lines reflecting parameters such as molecular motion, pH, temperature, and, especially important, the concentration of oxygen (Swartz, 1990).

The potential limitation of spectroscopy is the inability to obtain the full distribution of intensities that theoretically can be obtained by imaging although, as indicated above, it is usually very difficult to obtain adequate signal/noise for well-resolved images.

The key question to consider, in regard to what is the appropriate technique to be used for an *in vivo* study, is: What type of information is needed? In many, perhaps most cases, the information contained in the spectra is what is needed to resolve the experimental question that is being considered and therefore spectroscopy will be the approach of choice.

Technical aspects of *in vivo* EPR spectroscopy

Potential constraints in carrying out EPR in vivo

The principal special technical considerations in the development of *in vivo* EPR are:
1) The large amount of water in living systems, leading to potential severe non-resonant loss of the exciting radiation;
2) Potential physical constraints on getting the volume of interest in the subject properly positioned in the spectrometer;
3) The occurrence of physiological and voluntary motions of the subject;
4) The limited amounts of naturally occurring paramagnetic species which are suitable for EPR studies.

These potential constraints have significantly shaped the technical developments and applications of *in vivo* EPR techniques. As a consequence of the efforts of several different groups, most of the potential problems have been surmounted or the means to do so seem achievable.

EPR instruments for in vivo *EPR*

Types of detectors
The instruments used for *in vivo* EPR spectroscopy do not necessarily differ from conventional instruments, except in regard to the detectors and the frequency. Depending on the type of study being pursued, different types of detectors may be optimal. Fortunately, a number of different alternatives for detectors have been developed and continue to be improved. These include surface detectors, structures into which the sample is inserted, configurations in which the detector fits around the sample (e.g., as a loop), and detectors which can be inserted as a needle or a catheter.

Choice of frequency
Instruments are being developed and/or used for *in vivo* EPR at frequencies ranging from 300–9,500 MHz (Swartz and Walczak, 1993; Koscielniak and Berliner, 1994; Zweier and Kuppusamy, 1988; Halpern et al., 1989; Ishida et al., 1992; Alecci et al., 1992). The higher the frequency, the higher the sensitivity but, also, the greater the nonresonant loss of the exciting radiation with consequent loss of penetration. At 9,500 MHz the microwave penetrates less than 1 mm while at 300 MHz the penetration is more than 10 cm. Consequently studies at the higher frequency are limited to objects such as tails of mice or thin sections of skin, while the lower frequencies can potentially be used to study deep structures in humans. Our current approach is to use 1.1 GHz, which provides an effective compromise between sensitivity and depth of penetration. This frequency provides sufficient sensitivity for many important applications with existing technology, and can study regions which are within 10 mm of the surface or accessible by catheters or needles.

Determining volume of interest
The ability to define the volume from which the spectra are obtained is one of the keys to the usefulness of *in vivo* EPR spectroscopy. Definition and localization of the volume that is studied can be determined by instrumental factors and/or the nature of the paramagnetic material that is used. The depth of penetration of effective amounts of the exciting radiation (300–9,500 MHz) defines the potential volume that can be studied. The distribution of the field depends on the frequency and type of detector. For example, the field from the loop-gap resonator which we use as a surface detector (Nilges et al., 1989) falls off exponentially with depth while the sensitive volume of a resonant cavity extends throughout the cavity with a distribution that is typical for the mode of the cavity. The sensitive volume then is further defined by the nature of the material within it (especially as the material leads to dielectric losses and eddy currents); the distribution of the modulation field; and the distribution of the paramagnetic material.

When the paramagnetic material has a limited distribution, as is the case with the use of many particulate materials and large encapsulation devices, this then becomes the dominant factor in the delineation of the sensitive volume. This can have some important experimental advantages. The placement of the paramagnetic material provides precise delineation of the volume that is being studied. Under this circumstance it is relatively easy to obtain high quality spectra from several sites simultaneously by placing the particles at the desired sites and applying external magnetic field gradients to modify the value of the main external field at which resonance is achieved (Smirnov et al., 1993). The distance that can be resolved between points depends on the value of the gradient(s) that is applied and the line width of the spectra and whether some distortion of the EPR spectra is acceptable; points separated by 0.3 mm should be readily distinguished. In principle it also should be possible to resolve several regions simultaneously when using diffused rather than particulate paramagnetic materials, by the use of spectral-spatial imaging techniques with large voxels.

Types of paramagnetic materials used for in vivo *EPR spectroscopy*

In vivo EPR spectroscopy can be used to study both naturally occurring and introduced paramagnetic materials. The number of naturally occurring paramagnetic materials that are present in sufficient quantities to be observed are limited but do offer some very interesting possibilities. Currently melanin and, perhaps, the ascorbyl radical are the only naturally occurring free radicals that can normally be observed by this approach. The administration of some substances such as drugs or metal ions may result in detectable levels of paramagnetic substances which also might be considered to be naturally occurring inasmuch as they become paramagnetic as a result of normal biochemical and/or metabolic processes (Fujii et al., 1994; Liu et al., 1994).

Many of the most productive studies have used the administration of stable paramagnetic materials, especially particulate materials and nitroxides. The particulates, which have a high sensitivity to the concentration of oxygen, high spin concentrations, and chemical and biological inertness, have the potential for becoming the basis of the widespread use of *in vivo* EPR for oximetry of tissues in experimental animals and perhaps in patients. The particulates can offer the additional advantages of being localized so that the site from which they are reported can be determined precisely and multiple sites can be studied simultaneously.

Because of their chemical versatility the nitroxides can be used for a number of different purposes including as spin labels for various types of molecules or structures (e.g., liposomes), as redox indicators, for biophysical studies (e.g., for motion in membranes), and as indicators of the concentration of oxygen. The use of nitroxides in viable biological systems can have the additional complication that they tend to be metabolized from and back to the paramagnetic state but this complexity can be exploited, especially for the study of redox metabolism (Swartz, 1987). Paramagnetic materials, especially the nitroxides, also can be administered in structures such as liposomes (Bacic et al., 1989), albumin microspheres (Liu et al., 1994), and macroscopic plastic

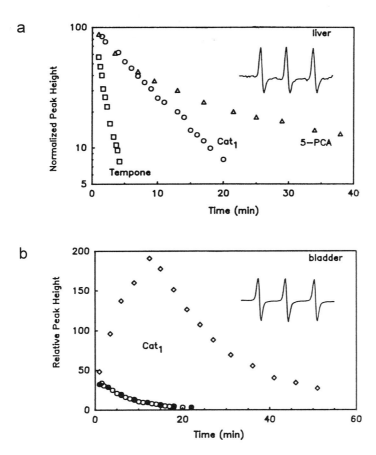

Figure 1. Measurement of pharmacokinetics in tissues by *in vivo* EPR spectroscopy.(a) Decrease in the intensities of the EPR signals of nitroxides from the liver region of mice following the IV injection of three different nitroxides: Tempone, Cat_1, and PCA at doses 0.10–0.15 mmol/kg body wt. The signals of all nitroxides were normalized to the signal intensity at the termination of injection. The temperature of the mice was maintained at 37°C. The insert shows the *in vivo* spectrum of PCA obtained 5 min after injection. (b) Observation of the excretion of the nitroxides by measuring the intensity of the EPR spectra of Cat_1 from the bladder region under the same experimental conditions. The decrease of the signal from the vascular pool, measured by the placement of the surface probe over regions without localized uptake, is also shown (solid and open circles indicate points from two different experiments). (Reproduced by permission of Academic Press from Bacic, G., Nilges, M.J., Magin, R.L., Walczak, T. and Swartz, H.M. (1989) *Magn. Reson. Med.* 10: 266).

containers (Subczynski et al., 1986); these approaches decrease the metabolism and/or the biodistribution of the nitroxides and have led to some potentially valuable applications.

Illustrative results

A large and rapidly growing number of studies have been carried out with *in vivo* EPR spectroscopy. These provide the strongest evidence and arguments for the value and importance of this experimental approach. In this manuscript we cannot provide a complete description with the details of all of the studies which have used *in vivo* EPR spectroscopy in a useful manner, but instead we attempt to provide an indication of what types of studies are feasible.

Pharmacokinetics and measures of redox metabolism

Many of the studies that have been reported can be considered within this category and it seems likely that this will be an important area of application of *in vivo* EPR spectroscopy. Many of the usual principles of any study using labeled substances apply to these experiments. The potential advantages of using paramagnetic labeling include: high sensitivity; the avoidance of radioactivity; a very wide range of substances that potentially can be labeled because of the availability of spin labeled compounds which are analogs of virtually any type of substance; and the ability to obtain additional information simultaneously due to the metabolism of the compound (which reflects redox states) and/or the response of the paramagnetic material to biologically important environmental factors. The latter can include the concentration of oxygen, pH, electrostatic potentials, and molecular motion.

Using techniques which obtain spectra from the whole body, the metabolism of nitroxides to and from nonparamagnetic states can be followed quantitatively (Fig. 1a). The use of spectroscopy from the whole animal has been especially useful in delineating and exploiting the principles of the metabolism of nitroxides, especially structure-function relationships and factors which affect the rate of metabolism, such as the concentration of oxygen.

More specific information on the sites of metabolism and the distribution of the spin labeled substances has been obtained by localized spectroscopy. Excretion (e.g., into the bladder) as well as local redox metabolism can be followed by this approach (Fig. 1b). If the paramagnetic substance is attached to a molecule or structure (e.g., a liposome) then localized spectroscopy can be used to follow the distribution of the labeled material and under favorable conditions, the metabolism of the labeled material. For example, Figure 2 indicates how, by using readily bioreduced nitroxides incorporated in liposomes, both the distribution and the integrity of the liposomes could be followed by *in vivo* EPR spectroscopy (Bacic et al., 1989).

Still more information may be obtained when the additional variable of perfusion of tissues is considered. In this type of study the paramagnetic substance (usually a nitroxide) is injected into

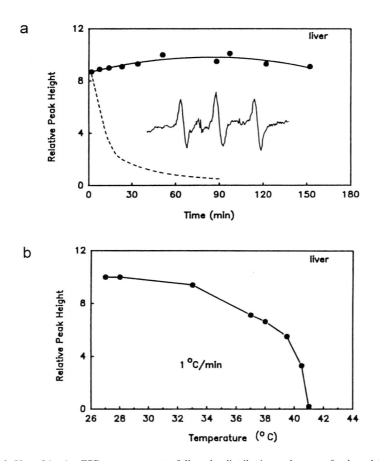

Figure 2. Use of *in vivo* EPR spectroscopy to follow the distribution and status of a drug delivery system. The intensity of the EPR signal of a nitroxide from the liver region of a mouse following the IV injection of Cat$_1$ encapsulated in DPPC/DPPG liposomes was monitored by *in vivo* EPR spectroscopy. The injected dose was 0.03 mmol/kg of nitroxide. (a) Changes in the signal intensity *vs.* time after injection. The temperature of the mouse was maintained at 30°C i.e., below the phase transition point of liposomes. The insert is a typical spectrum obtained 30 min after injection. The dashed line shows a typical blood clearance curve for these liposomes. (b) Changes in the signal intensity *vs.* temperature of the mouse. Heating of the mouse by a warm air flow (1°C/min.) was started 30 min after injection of the liposomes. The contents of liposomes will leak out at the transition temperature and this is indicated by the loss in intensity of the EPR signal because the nitroxide that was used is rapidly bioreduced and eliminated in tissues. (Reproduced by permission of Academic Press from Bacic, G., Nilges, M.J., Magin, R.L., Walczak, T. and Swartz, H.M. (1989) *Magn. Reson. Med.* 10: 266).

the tissue of interest (e.g., a tumor) and then the metabolism (especially the rate of reduction to the hydroxylamine) and the perfusion of the tissue can be measured quantitatively by observing the rate of disappearance in the presence and absence of temporary blockage of perfusion (Fig. 3). Analogous experiments can measure phenomena such as the rate of crossing the blood-brain barrier (Ishida et al., 1992); under suitable conditions this could be used to measure the integrity of the blood-brain barrier as well as the normal distribution of the labeled substance.

Spin trapping

Recently we have demonstrated that under favorable circumstances it is possible successfully to pursue one of the principal reasons for the development of *in vivo* EPR techniques: the ability to follow very reactive species by means of spin trapping. This has been demonstrated by the observation of spin adducts of 5,5-dimethyl-1-pyrroline n-oxide (DMPO) and sulfite radicals which were trapped and followed directly *in vivo,* Figure 4 (Jiang et al., 1995). While there remain very significant technical problems for the extension of the technique to other types of free radicals,

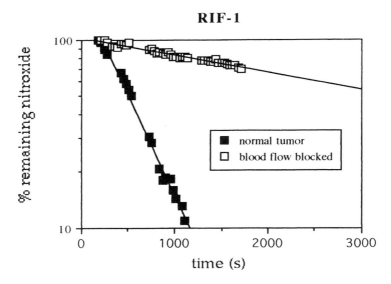

Figure 3. Measurement of perfusion and redox activity in tissues by *in vivo* EPR spectroscopy. The EPR spectra of the nitroxide Cat₁ were measured after the nitroxide was injected into a RIF-1 tumor; measurements were made before (solid squares) and after (open squares) restriction of the blood supply of the same tumor. The rate of decrease of the EPR signal in the blood-flow restricted conditions reflects the rate of redox-dependent reduction of the nitroxide. The rate of decrease of the EPR signal with normal blood flow reflects a combination of perfusion (wash out) and the redox-dependent reduction of the nitroxide.

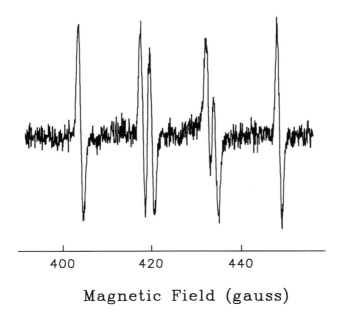

400   420   440

Magnetic Field (gauss)

Figure 4. Measurement of spin trapped free radicals (spin adduct) in tissues by *in vivo* EPR spectroscopy. The EPR spectrum, typical for a DMPO-SO$_3$ spin adduct, was obtained from an anesthetized mouse after intravenous injection of DMPO, sodium sulfite, and sodium dichromate.

especially in regard to the availability of spin traps with the required stability for adducts of oxygen-centered free radicals, the results with sulfite radicals indicate that success is possible.

Metal ions

While the biological and spectral characteristics of most metal ions make it unlikely that they will be able to be observed *in vivo*, it has recently been shown that under favorable conditions some paramagnetic metal ions can be observed and followed quantitatively by *in vivo* EPR spectro-scopy, Figure 5 (Liu et al., 1994). The figure demonstrates the production of paramagnetic chromium V, following the administration of chromium VI.

Free radicals

As noted previously, melanin and ascorbate are probably the only naturally occurring free radicals with sufficient concentrations to be detected in biological systems. It has recently been shown,

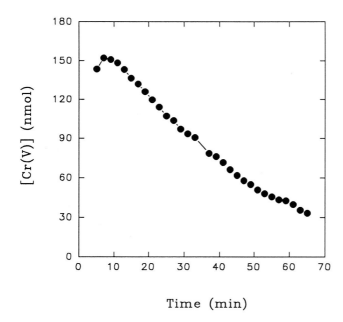

Figure 5. Measurement of paramagnetic metal ions in tissues by *in vivo* EPR spectroscopy. The Figure demonstrates the time course of the formation and decay of Cr(V) signal following intravenous injection of sodium dichromate into an anesthetized mouse. The EPR signal was scanned every minute immediately after the injection; time zero is the time of injection. The signal intensity was measured from the peak-to-peak height of the signal.

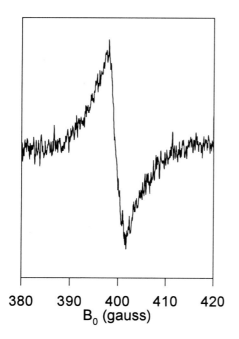

Figure 6. Detection of a free radical intermediate of a therapeutic drug in tissues by *in vivo* EPR spectroscopy. The skin of a SKHl mouse was treated for six hours with a typical therapeutic preparation of anthralin (1% anthralin in a lipophilic aqua base). This spectrum, with a g-factor of 2.0036 and a line width of 6 gauss, was recorded 24 h after the first treatment. It is probably due to free radicals in a metabolic product of anthralin, a polymeric structure termed "anthralin brown".

however, that *in vivo* EPR spectroscopy can be used to follow the generation of free radical inter-mediates from xenobiotics (Fujii et al., 1994) or drugs. Figure 6 illustrates the occurrence of a free radical intermediate of a therapeutic drug which was readily detectable by *in vivo* EPR.

Oximetry

Perhaps the most promising, and already the most widespread, use of *in vivo* EPR spectroscopy is to measure the concentration of oxygen in tissues *in vivo*. This has occurred because of the experimental and clinical importance of such measurements and the lack of other methods to make these measurements with the capabilities demonstrated by *in vivo* EPR oximetry. These advantages include sensitivity, accuracy, repeatability, and non-invasiveness. The method is already being used extensively to obtain measurements of the concentration of oxygen in tissues with an ease and accuracy that has not been possible previously. It also appears quite possible

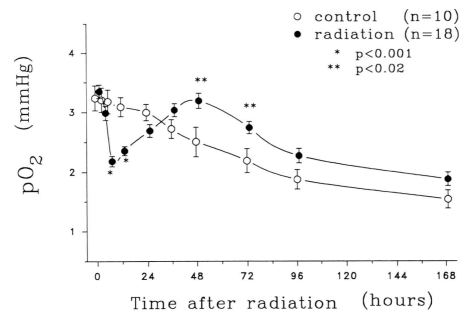

Figure 7. Demonstration of the capability of *in vivo* EPR spectroscopy to resolve differences of pO$_2$ of less than 1 Torr. The pO$_2$ in MTG-B tumors (average size 200 mm^3) was measured in unanesthetized mice using India ink injected into the tumors 24 h prior to the administration of 20 Gy of X-rays. The figure shows the average values and standard errors of the means at the various time points at which the measurements were made. Note the ability to measure significant differences of less than 1 Torr and the capability of making repeated measurements in the same animal at the same site throughout the duration of the experiment.

that the technique will become a widely used clinical tool. The development of EPR oximetry has been catalyzed by the development of several different paramagnetic species which combine high sensitivity to the pO_2 with a high degree of inertness in biological systems. Figures 7–10 illustrate the types of data that have been obtained with *in vivo* spectroscopy using our 1.1 GHz EPR spectrometer. Figure 7 illustrates the ability of the technique to follow changes in pO_2 in tumors, resolving significant changes of less than 1 Torr. Figure 8 illustrates the capability of particles of India ink to provide selective measurements of the pO_2 within Kupffer cells *in vivo*. Figure 9 illustrates the possibility of using *in vivo* EPR to measure the effects of anesthesia on the pO_2 in the brain. Figure 10 illustrates the ability to measure pO_2 simultaneously in two different sites.

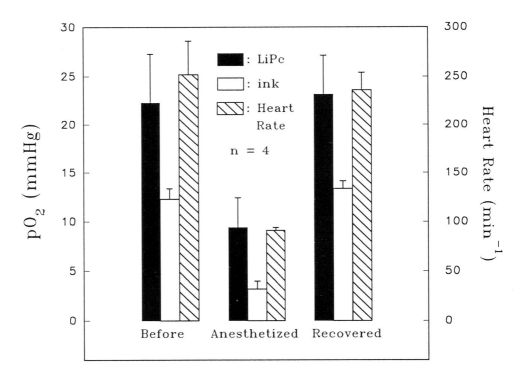

Figure 8. Demonstration of the capability of *in vivo* EPR spectroscopy to measure the pO_2 at physiologically different locations in the same organ (liver). The pO_2 in the livers of mice was measured by two different paramagnetic oxygen sensitive materials, India ink and lithium phthalocyanine (LiPc). India ink is taken by Kupffer cells and therefore reflects the pO_2 in these cells, while the macroscopic crystal of LiPc reflects the overall pO_2 in the liver. The experiments were carried out with live mice and effects of the anesthetic (pentobarbital) on pO_2 were demonstrated. In these experiments the pO_2 was the same in both physiological locations, but in other studies apparently significant differences have been found.

Conclusions

The feasibility and usefulness of *in vivo* EPR spectroscopy has now been well demonstrated with experimental studies in small animals. There do not appear to be any significant barriers to the extension of the technique to large animals and, eventually to patients. There are a large number of potentially important areas of applications of this technique which meet the essential requirement that the EPR approach provides significant advantages over other techniques aimed at obtaining the same kinds of information. The most promising and important area appears to be for the measurement of the concentration of oxygen in tissues *in vivo* but there are a number of other areas which also seem very likely to be quite important. These includes probes of redox

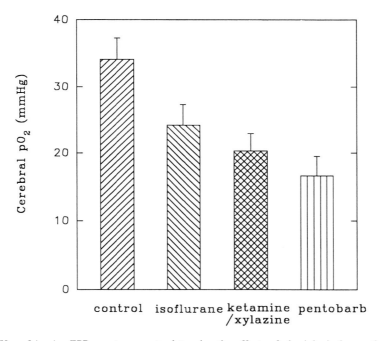

Figure 9. Use of *in vivo* EPR spectroscopy to determine the effects of physiological or pathophysiological processes on cerebral oxygenation. The figure demonstrates the effect of anesthetics on the cerebral pO_2. Ketamine/xylazine mixture (100/10 mg/kg) and sodium pentobarbital (50 mg/kg) were injected into the rats intraperitoneally; isoflurane (1%, flow rate 3 ml/min.) was delivered to the animal by a gas system. The values are expressed as mean ± SEM (n = 3). Such data, which cannot readily be obtained by other methods, provide an approach to understand critical physiological and pathophysiological processes and to determine the effects of agents or procedures designed to treat them. In the study presented here, the different effects of different anesthetics on cerebral pO_2 are clearly shown.

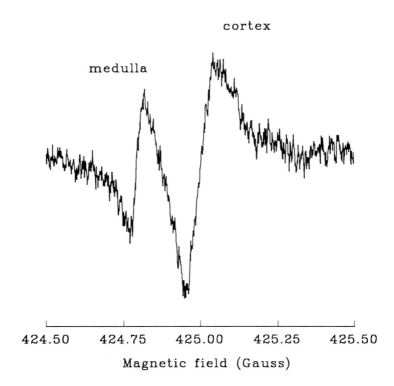

Figure 10. Demonstration of the capability of *in vivo* EPR spectroscopy to measure the pO_2 at two different anatomical locations at the same time. Crystals of LiPc were placed into the cortex and medulla of the kidney of a live mouse (crystals inserted at laparotomy, with the separation of the locations of the crystals being 2.8 mm), and EPR spectra obtained. The figure shows a typical spectrum (20 seconds scan) with the signal from both cortex and outer medulla recorded simultaneously. Using such spectra and appropriate calibration curves it was determined that the average pO_2 in the cortex was 22.5 Torr and in the medulla 15.2 Torr (corresponding line widths of LiPc were 141 mG and 95 mG).

metabolism, a variety of new approaches for pharmacokinetics, the observation *in vivo* of reactive intermediates with unpaired electrons, and the study of redox metabolism.

References

Alecci, M., Della Penna, S., Sotgiu, A., Testa, L. and Vannucci, I. (1992) Electron paramagnetic resonance spectrometer for three-dimensional *in vivo* imaging at very low frequency. *Rev. Sci. Instrum.* 63: 4263–4270.
Bacic, G, Nilges, M.J., Magin, R.L., Walczak, T. and Swartz, H.M. (1989) *In vivo* localized spectroscopy reflecting metabolism. *Magn. Reson. Med.* 10: 266–272.
Fujii, H., Zhao, B., Koscielniak, J. and Berliner, L.J. (1994) *In vivo* EPR studies of metabolic fate of nitrobenzene in the mouse. *Magn. Reson. Med.* 31: 77–80.

Halpern, H.J., Spencer, D.P., vanPolen, J., Bowman, M.K., Nelson, A.C., Dowey, E.M. and Teicher, B.A. (1989) Imaging radio frequency electron-spin-resonance spectrometer with high resolution and sensitivity for *in vivo* measurements. *Rev. Sci. Instrum.* 60: 1040–1050.

Ishida, S., Matsumoto, S., Yokohama, H., Mori, N., Kumashiro, H., Tsuchihashi, N., Ogata, T., Yamada, M., Ono, M., Kitajima, T., Kamada, H. and Yoshida, E. (1992) An ESR-CT imaging of a living rat receiving an administration of a nitroxide radical. *Magn. Reson. Imag.* 10: 109–114.

Jiang, J., Liu, K.J., Shi, X. and Swartz, H.M. (1995) Detection of short-lived free radicals by low frequency EPR spin trapping in whole living animals: evidence of generation of the sulfur trioxide anion free radical *in vivo*. *Arch. Biochem. Biophys.* 319: 570–573.

Koscielniak, J. and Berliner, L.J. (1994) Dual diode detector for homodyne ESR microwave bridges. *Rev. Sci. Instrum.* 65: 2227–2230.

Liu, K.J., Gast, P., Moussavi, M., Norby, S.W., Vahidi, N., Walczak, T., Wu, M. and Swartz, H.M. (1993) Lithium phthalocyanine: A probe for EPR oximetry in viable biological systems. *Proc. Natl. Acad. Sci. USA* 90: 5438–5442.

Liu, K.J., Jiang, J., Swartz, H.M. and Shi, X. (1994) Low frequency EPR detection of chromium(V) formation by chromium(VI) reduction in whole mice. *Arch. Biochem. Biophys.* 313: 248–252.

Liu, K.J., Grinstaff, M.W., Jiang, J., Suslick, K.S., Swartz, H.M. and Wang, W. (1994) *In vivo* measurement of oxygen concentration using sonochemically synthesized microspheres. *Biophys. J.* 67: 896–901.

Nilges, M.J., Walczak, T. and Swartz, H.M. (1989) 1 GHz *in vivo* ESR spectrometer operating with a surface probe. *Phys. Med.* 5: 195–201.

Smirnov, A.I., Norby, S.W., Clarkson, R.B., Walczak, T. and Swartz, H.M. (1993) Simultaneous multi-site EPR spectroscopy *in vivo*. *Magn. Reson. Med.* 30: 213–220.

Subczynski, W.K., Lukiewicz, S. and Hyde, J.S. (1986) Murine *in vivo* L-band ESR spin label oximetry with a loop-gap resonator. *Magn. Reson. Med.* 3: 747–754.

Swartz, H.M. (1987) Use of nitroxides to measure redox metabolism in cells and tissues. *J. Chem Soc., Faraday Trans. 1*, 83: 191–202.

Swartz, H.M. (1990) The use of nitroxides in viable biological systems: An opportunity and challenge for chemists and biochemists. *Pure & Appl. Chem.* 62: 235–239.

Swartz, H.M. and Walczak, T. (1993) *In vivo* EPR: Prospects for the '90s. *Phys. Med.* 9: 41–48.

Vahidi, N., Clarkson, R.B., Liu, K.J., Norby, S.W., Wu, M. and Swartz, H.M. (1994) *In vivo* and *in vitro* EPR oximetry with fusinite: A new coal-derived, particulate EPR probe. *Magn. Reson. Med.* 31: 139–146.

Woods, R.K., Hyslop, W.B, and Swartz, H.M. (1989) Mapping oxygen concentrations with 4D electron spin resonance spectral-spatial imaging. *Phys. Med.* 5: 121–137.

Zweier, J.L. and Kuppusamy, P. (1988) Electron paramagnetic resonance measurements of free radicals in the intact beating heart: A technique for detection and characterization of free radicals in whole biological tissues. *Proc. Natl. Acad. Sci. USA* 85: 5703–5707.

Bioradicals Detected by ESR Spectroscopy
H. Ohya-Nishiguchi & L. Packer (eds)
© 1995 Birkhäuser Verlag Basel/Switzerland

Pharmaceutical aspects of ESR investigations on drug delivery systems, tissues and living systems

R. Stösser[1], K. Mäder[2], H.-H. Borchert[2], W. Herrmann[3], G. Schneider[4] and A. Liero[4]

[1]Humboldt-University, Institute of Chemistry, Hessische Str. 1-2, D-10115 Berlin, Germany; [2]Humboldt-University, Institute of Pharmacy, Berlin, Germany; [3]Federal Institution of Material Research and Testing, Rudower Chaussee 5/Haus 18.12, D-12489 Berlin, Germany; [4]Magnettech GmbH, Rudower Chaussee 6, D-12489 Berlin, Germany

Summary. Examples are given for the application of S- and X-band ESR in the fields of pharmacy and medicine respectively. Drug delivery systems, frozen tissue inclusive subsystems and in vivo ESR on fungi and mice are included. Endogeneous paramagnetic species like Fe(III), Mn(II), Cu(II) in various complex environments located in the samples are used to characterize the drug delivery as well as the in vitro systems like liver and kidney tissues, blood and others. By means of spin labeling techniques it was possible to determine microviscosity, micropolarity, pH-values and partial pressure of oxygen. Pharmaceutical aspects of ESR imaging as well as in vivo ESR are discussed, especially the liberation, distribution and biotransformation of the applied paramagnetic model drugs.

Introduction

The method of electron spin resonance (ESR or EPR) represents today – fifty years after its discovery by Zavoiskij (1944) in Kazan – a useful tool which is based on a well-developed theory, and which can in general be carried out with highly sophisticated experimental equipment. Applications of ESR are reported from all sciences applying physical methods, and even in the field of pharmacy some review papers now exist (Chingnell, 1979; Mäder et al., 1994). For the present contribution rather different examples were selected with respect to pharmaceutical questions and unresolved problems. The power of ESR should give a deeper insight into the processes occuring in drug delivery systems and the characterization of paramagnetic species formed or changed during metabolic processes in tissues and subsystems (e.g., mitochondria and enzymes) can be demonstrated. In terms of experimental-methodological aspects it was of interest to observe the ESR response in the systems mentioned above as a function of parameters like temperature, concentration, pressure and microwave frequency.

Finally, the findings of in vivo ESR will be discussed in terms of the biotransformation of paramagnetic drugs in mammals and in fungi.

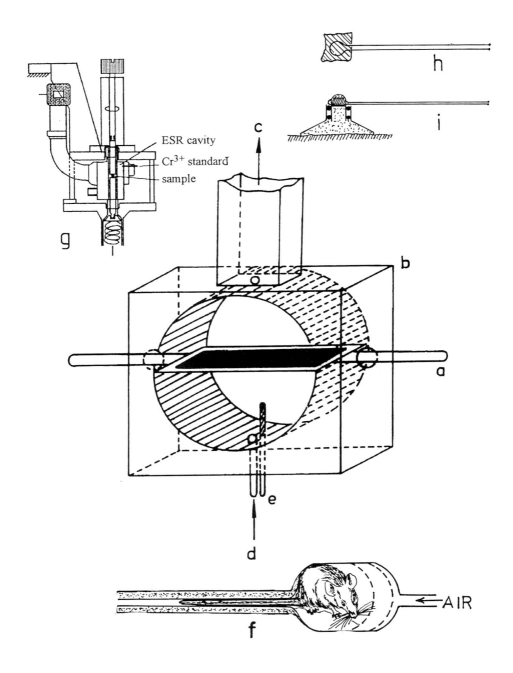

ESR cavity
Cr^{3+} standard
sample

Materials and methods

The ESR investigations were performed in X-band with spectrometers E4 (Varian,USA) and ERS 300 (Centre of Scientific Instruments, Berlin-Adlershof, Germany). Special applications could be carried out on the spectrometer E4 by equipping it with an adapted pressure device (Fig. 1a, g); (Stösser et al., 1990) and on the ERS 300 by combination with a ESR tomography unit (Centre of Scientific Instruments, Berlin-Adlershof, Germany).

For the S-band measurements the spectrometer containing a microwave bridge MWS-4G (Magnettech GmbH, Berlin, Germany) was used. The measurements on mouse skin were done with the help of a surface coil (Fig. 1h and i). In general this microwave unit could be combined either with a stationary laboratory magnet or with an easily transportable permanent split-ring magnet of ~15 cm diameter. In the last configuration even studies on human skin could be performed (Stösser et al., 1994b). In any case the temperature and the partial pressures of N_2, H_2O and O_2 were kept constant by a suitable gas flow.

The best results in the preparation of tissue and blood samples were obtained by a rapid freezing technique. Rapid freezing was achieved either by injection of the sample into a low temperature bath, or by dispersion of a fluid sample with a set of nozzles and collecting the drops on the surface of a fast-rotating cryostat. The transfer from 77 K to the ESR-He-flow cryostat (APD Cryogenics, Inc., USA) with $T \geq 2.6$ K could be achieved without intermediate warming up of the samples.

While the *in vivo* ESR could be realized without problems in S-band, for the X-band spectrometers only the tails of the mice could be used for the ESR measurements. A special teflon-support was manufactured and located in the horizontally arranged TM 110 cavity (Fig. 1a−f). This cavity arrangement appeared to be useful for the investigations of polymers and skin samples because it allows the samples to be fixed and the desired conditions (e.g., pH_2O) realized in an easy way (Mäder et al., 1993; Stösser et al., 1995).

The stable nitroxyl radicals were purchased from Aldrich Chemie GmbH, Steinheim, Germany and the pH sensitive probes (Kroll et al., 1995) were obtained from the group of Dr. Volodarsky / Dr. Grigoriev, Institute of Organic Chemistry, Russian Academy of Sience, Novosibirsk, Russia. From the solids used in pharmacy as drug delivery systems or for other therapeutic purposes

Figure 1. Experimental equipment for X- and S-band. (a) Support for skin or polymer samples in a horizontal arrangement of the TM110 X-band cavity (b); (c) Wave guide to the micro wave bridge; (d) Inlet for the gas mixture of the desired temperature T; (e) MgO/Cr^{3+} standard for the determination of intensity and the g-values; (f) special teflon made support for *in vivo* ESR on mice which can be inserted instead of (a); (g) high pressure device for X-band; (h) S-band surface coil above a piece of skin, treated with nitroxyl radicals; (i) gas impermeable arrangement located above the skin. The S-band coil encloses a 6 mmm quartz tube filled on the top with a material (e.g., fusinite or $CaCO_3/Mn^{2+}$) which is sensitive to the partial pressure of oxygen or water.

kaolinites ($Al_2O_3 \cdot SiO_2 \cdot H_2O$) and montmorillonites($Al_2O_3 \cdot 4SiO_2 \cdot H_2O$) were selected and obtained from different natural deposits as well as from pharmaceutical sources. ZnO was of pharmaceutical origin.

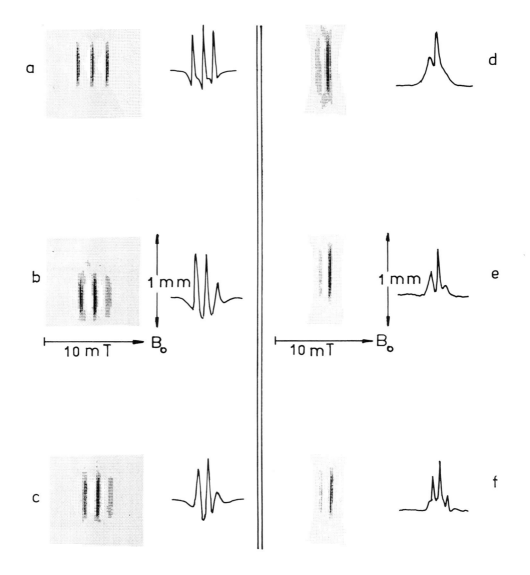

Figure 2. Spatial-spectral resolved ESR tomograms and corresponding spectra (second derivation of solutions of TEMPOL in water (a), glycerol (b), polyethelenglycol 3000 (c); (d) TEMPOL incorporated in polylactide foils (dry); (e) after 2h and (f) after 23h penetration of water.

The preparation of the microemulsions is described in Mäder (1993) and the preparation and reactions of the subsystems (mitochondria and CuZnSOD) conditions are summarized in Henke et al. (1993) and Haberland et al. (1993).

Results and discussion

Drug delivery and corresponding model systems

Some drug delivery systems contain paramagnetic species *per se*, for example Fe^{3+}- or Mn^{2+} ions are detectable in minerals (kaolinite and montmorillonite) and also in organic products of natural origin (alginic acid). By means of ESR it is possible to get information concerning the concentration and the environment of these paramagnetic compounds. Temperature or high pressure treatments can lead to structural changes in the system, which can be detected by the ESR spectra of these endogenous probes.

In the next part three applications of X-band ESR will be demonstrated. The method was used to study the structure and dynamics of microemulsions, to describe the primary liberation steps in biodegradable polyesters, and to show what kind of pressure-induced effects can be expected in drug delivery systems.

Microemulsions and polymers

Microemulsions are generally regarded as optically isotropic and thermodynamically stable mixtures of lipids, water and tensides. Because they exhibit high penetration and solubilization properties, they are interesting drug delivery systems. Although much work has been done to understand their peculiarities, different and contrary models still exist. Here, the influence of the water content on dynamics and structures of a system containing Tween 80, dodecanol, cera perliquida and propyleneglycol is discussed. Measurements of the macroscopic viscosity, conductivity and microwave absorption were carried out (Mäder, 1993). The reorientation time, the magnitude of hyperfine coupling as an indicator of molecular polarity and the saturation behaviour of different spin probes were measured by ESR. Furthermore, spin exchange measurements were applied for the determination of dynamic magnetic interactions between paramagnetic compounds. The results obtained indicate differences between macroscopic and molecular viscosity. The thermodynamic stability of microemulsions demands the binding of water. We conclude the existence of dynamic structures with a size lower than 100 nm.

ESR was also used to prove the penetration of water into foils and microparticles made from copolymers of hydroxyacetic acid and lactic acid (Mäder et al., 1991a, b). The spectral and dynamic changes induced by the penetration of water indicate a fast diffusion of water, whereas other molecules (ascorbate, Mn^{2+}-ions, TEMPOL) are not able to penetrate to a comparable extent. Polar regions with high mobility of the spin probe were formed, but the liberation of the spin probe was quite low. CW-ESR imaging including spatial-spectral resolution allows description of the mobility of paramagnetic substances (Fig. 2a–f) as a function of their localization (Mäder et al., 1991a, b). The mobility of the spin probe is a function of the polymer matrix, the penetration time and localization.

On the influence of high pressure and other mechanical impacts on drug delivery systems
From a pharmaceutical point of view the application of high pressures covers a field ranging from the manufacture of pharmaceutical dosage forms (including the densification of materials and the accompanying structural and chemical changes, Parrot, 1990) up to pressure sterilization methods. The influence of the pressure treatment on the processes of liberation, absorption, distribution, metabolism and excretion (LADME) is of special interest. With respect to the pressure range usually applied in this field ($p \leq 1$ GPa) one can expect small but significant irreversible changes in the system investigated. Such changes are hard to detect by macroscopic methods.

As the first attempts in this field (Stösser et al., 1994a) show, ESR is sensitive enough to detect even small pressure effects and can therefore be used for this purpose.

Although pressure treatments are common in pharmacy, e.g., to increase the density of tablets and diminish their porosity and specific surface, up to now pressure effects have been more or less uninvestigated in this field.

Here spectroscopic evidence should be given for the activation of solids after a pressure or grinding procedure using the intrinsic paramagnetic centres as well as typical spin probes and traps as reporters. Kaolinites, montmorillonites as well as zinc oxide were used as models for drug delivery systems.

Two kinds of intrinsic paramagnetic species are contained in the kaolinite investigated: Fe^{3+} high spin ions substituting Al^{3+} ions in the lattice. Furthermore, the defects present, $O_2^{-\cdot}$ and trapped holes (h^+), are paramagnetic (Fig. 3a, c, e, f, $g' \sim 2$ region). Additionally, the impurity phase $FeCO_3$ is present which decomposes during the pressure treatment forming magnetically ordered particles of a Fe_3O_4-like composition by subsequent reaction of the primary produced FeO with oxygen. The ordered phase gives a large broad line contributing to the ESR spectrum displayed in Figure 3c. Further pressure treatment causes a transformation of the Fe-O species into Fe_2O_3. This reaction was demonstrated by separate treatments of the $FeCO_3$ and Fe_3O_4 respectively. Parallel to the chemical reactions and distortions of the magnetic order the local symmetry of the Fe^{3+} ions localized on Al^{3+} sites is lowered as indicated by the spectral changes in the low field regions (Fig. 3a and c). Therefore, the essential effect of the pressure treatment results in an activation of the system.

In contrast to the layer silicates, in ZnO the formation of trapped holes proceeds even at low pressures. The pharmaceutical consequences of this behaviour are under investigation.

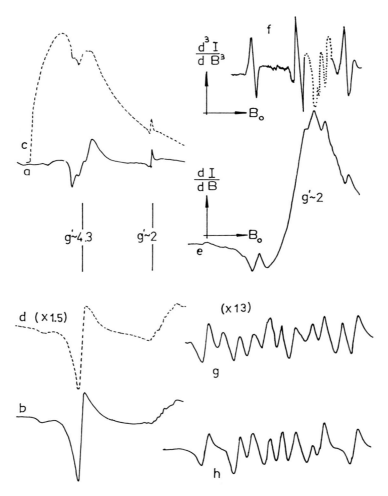

Figure 3. a–d: ESR spectra of untreated kaolinite (Caminau) and montmorillonite (Jelsovy Potok) before and after pressure treatment in the GPa range. (e) ESR spectrum (O_2^- and $RR'NO\cdot$) of a pressure-treated kaolinite sample intercalated with DMPO; (f) second derivative of (e) to suppress the spectrum of O_2^-; (g) ESR spectra obtained by heterogeneous reaction of a deoxygenated solution of DMPO in benzene with mechanically activated montmorillonite; (h) reaction of solution of (g) with DPPH.

The pressure response of the three-layer silicate montmorillonite differs markedly from that of kaolinite. There is a pressure-dependent interplay between the Fe^{3+} ions localized on rhombically distorted sites (g´ = 4.3; Fig. 2b and d) and Fe^{3+} ions coupled with other Fe ions in clusters or precursors of autonomous phases (g = 2). The irreversible structural effects can be reduced by the intercalation of surfactants and related molecular species. As the reactions of pressure-treated kaolinite and montmorillonites with solutions of the spin trap DMPO (Fig. 1g and h) indicate, there is an activation in both systems. But for the direct reaction with the intercalated trap the pressure range applied here was sufficient only for the kaolinite (Fig. 1e and f). Although the FeO_x content is much higher in montmorillonite, the mechanical activation of the reaction of the

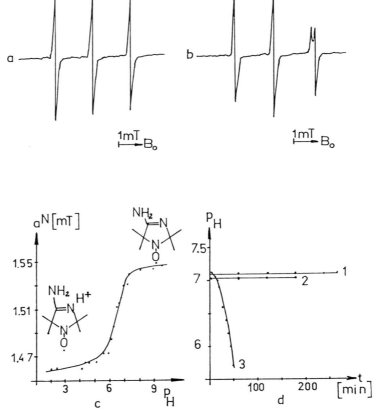

Figure 4. (a) ESR spectrum of the spin probe 2,2,5,5-tetramethyl-4-amino-3-imidazoline-1-yloxy in water at pH 8.0; (b) as (a) at pH 6.4; (c) dependence of the hyperfine coupling constant a^N on the pH-value in aqueous solutions; (d) monitoring of the pH change of the water phase of W/O EUCERINIUM ointments; 1: 25°C without addition; 2: addition of benzocain, 40°C; 3: addition of acetylsalicylic acid, 40°C.

FeCO$_3$ impurity phase in kaolinite is favoured. The internal reaction of montmorillonite with DMPO was achieved by grinding the intercalated system in a corundum mortar.

The state of activation of all solids investigated was proved by heterogeneous reactions using polar and non-polar solutions of the spin trap DMPO. By ESR paramagnetic species could be observed which are derived from DMPO by complex redox reactions under participation of oxygen. The reaction starts with the electron transfer from DMPO to the transition metal ions contained in the delivery systems. Primary reactions with free radicals could here be excluded.

The observed mechanical activation processes in layer silicates are in direct correspondence to mechanically-treated pharmaceutical drug delivery systems. Such processes have to be taken into account if less stable drugs are incorporated or long time reactions are of importance. (See also Kuzuya, 1993).

On the determination of pH value-proportional quantities by ESR spectroscopy
The determination of pH values is of basic interest in the field of pharmacy. Important applications include the adjustment of the pH value in various drug delivery systems such as eyedrops and infusions, and the monitoring of degradation processes in drugs. In the context of advantages and disadvantages of pH determining methods, the ESR can directly observe pH value-proportional quantities in a continuous way without any further disturbance. The basis here is the use of pH-sensitive spin probes (Khramtsov and Weiner, 1988; Kroll et al., 1995). From a pharmaceutical point of view the commonly used drugs acetylsalicylic acid and benzocain were chosen as model compounds with hydrolyzable ester groups. The change in the pH value by the decomposition of these compounds was monitored directly and continuously in water/oil (W/O) systems by the ESR of the imidazoline-derived spin probes (Fig. 4a–d). This nondestructive method makes it possible to perform measurements in nontransparent W/O systems without the need for any preparatory steps. While the hydrolysis of acetylsalicylic acid caused a rapid decrease of the pH value over five hours at 25°C within the emulsion, the benzocain exhibited a relatively higher stability (Fig. 4d). Further investigations performed on tablets and polymer foils demonstrated the ability of this ESR method to follow directly the biodegradation of the materials by registration of the change of the pH value.

ESR investigations on tissues and subsystems

S- and X-band ESR on mouse skin
The percutanous administration of drugs is used for local and systemic treatments as well. For a deeper understanding of the function of the drug delivery systems with respect to the therapeutic

aim, it is necessary to investigate the interaction between drug, drug delivery system and skin at a microscopic level. With respect to spin labeled drugs, stable nitroxyl radicals were used as model compounds. Since the reduction of the spin probes takes place exclusively outside the delivery systems, it is possible to use the change in the ESR intensity caused by the skin as a parameter to characterize the combined action of the penetration and reduction processes. The experimental findings obtained in ESR S-band are represented in Figures 5 and 6.

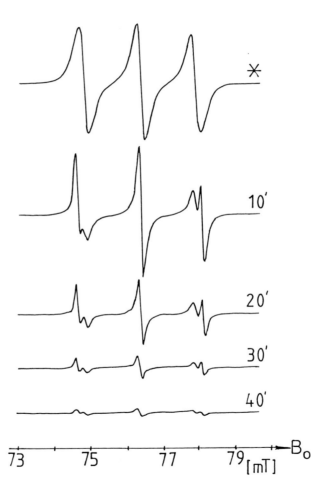

Figure 5. ESR spectra (2.4 GHz) of an aqueous solution of DTNB (*) showing penetration through mouse skin as a function of time.

The larger penetration depth of the microwaves on the one hand and the relatively small g-shifts on the other hand are responsible for the spectral pattern in S-band. While the properties of the delivery systems Miglycol (neutral oil) essentially govern the hyperfine splitting as well as the signal decay under the influence of the skin (Stösser et al., 1995) a distinct splitting and a faster

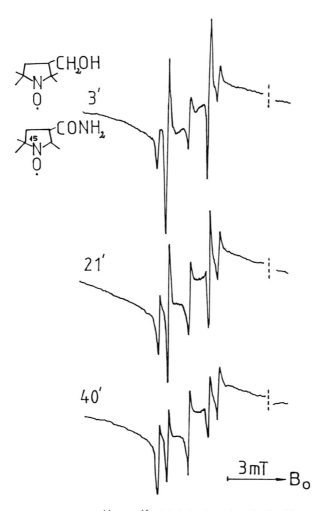

Figure 6. Simultanous detection of the ^{14}N and ^{15}N labeled spin probes dissolved in water in the penetration of mouse skin at 9.2 GHz ESR. On addition of K$_3$[Fe(CN)$_6$] no ESR signal appears because of strong magnetic interactions.

decay of the narrow signals are observed if aqueous solutions are used (Fig. 5). Here the spin probe is distributed between hydrophobic and hydrophilic local environments in the skin. The last fact is responsible for the larger splitting because the more polar structure of the probe produces a larger spin density at the nitrogen atom of the probe.

Similar distribution phenomena were observed using DMSO/water mixtures as carrier systems and TEMPOL as a spin probe. Figure 6 shows the effect of the application of a mixture of model compounds where the ^{15}N labelling allowed the two species to be detected simultaneously. The addition of a $K_3[Fe(CN)_6]$ solution to the starting mixture abolishes the signal outside the skin completely and only the interaction with the skin could be followed. As Figure 6 indicates, the polar CMP (^{15}N) is reduced more rapidly by the penetration through the skin. The phenomenon of spatial distribution of stable spin probes was also investigated by ESR imaging techniques (Mäder, 1993). From the results obtained by S- and X-band ESR carried out *in vitro* it can be concluded that the effect of the carrier or drug delivery system on the liberation and reactions of spin probes of different structure can be clearly demonstrated and quantified.

Monitoring metabolic processes in biological systems using ESR freezing methods at T ≥ 4 K
Large anisotropies of the coupling tensors, short relaxation times and the usually small native concentrations of the paramagnetic species of interest reduce the general advantages of the cw-ESR method for applications in the field of biology. To overcome these problems rapid freezing experiments are commonly used. Not only are spin relaxation times and the signal to noise ratio enlarged but also the actual state of the biological system can be conserved in some sense; therefore one can detect metabolic processes without further preparatory steps. Rapid freezing means cooling the specimen down in as short a time as possible. The first step of the process is done by immersion in liquid nitrogen (77 K) because of the higher heat capacity of nitrogen compared to that of helium.

The advantages of the method can be demonstrated on the basis of examples given in Figure 7a–k. Signals of catalase and low-spin cytochrome P 450, the reduced form of mitochondrial iron-sulfur proteins, Mo^{5+} as well as free radical species were detected at 77 and 4 K respectively (Mäder, 1993; Mäder et al., 1994; Stösser, 1993).

As examples of metabolic processes the ESR signals obtained after intraperitoneal administration of $Na^{15}NO_2$ and other nitric oxide-releasing systems were investigated (Fig. 7j). While the signals of cytochrome P 450 decrease in a dose-dependent manner, the signals caused by the iron sulfur proteins remain unaffected. At the same time the ESR response from ^{15}NO-iron porphyrine species appear (Fig. 7k) indicating a certain role of cytochrome P 450 in the metabolism of nitrite. To get an unequivocal assignment of the ESR transitions observed the ESR spectra were simulated on the basis of an appropriate spin-Hamiltonian.

The experiments at 4 K must be done with care in closed ampoules. Because of the ampoules possible condensation of oxygen in the sample from the environment, depending on the concentration a spectral contribution of the S = 1 system of $^3O_2$ can occur which obscures the spectra of other, e.g., iron species (Simoneau et al., 1971).

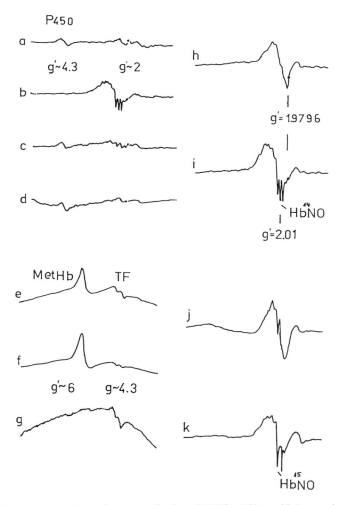

Figure 7. (a) ESR spectra of rat liver after i.p. application of $NaNO_2$ (170 µmol/kg):control; (b), (c), (d): 5, 60 and 180 min after application; (e) low-field region of the ESR spectra of rat blood 5 min after application of $NaNO_2$ (170 µmol/kg); (f), (g): 30 and 150 min after application; (h) ESR spectrum of rat heart 5 min after application of $NaNO_2$ (170 µmol/kg); (i) as (h) for rat liver; (j) ESR spectrum of mouse heart 5 min after i.p. application of $Na^{15}NO_2$ (57 µmol/kg); (k) as (i) for mouse liver

The advantage of the ESR application in this field results essentially from two facts: no further preparation or separation techniques are necessary and the structure and dynamic of the paramagnetic species can be determined directly.

Selected subsystems

As examples of selected subsystems, isolated mitochondria and isolated enzymes (CuZnSOD and LOX) will be discussed.

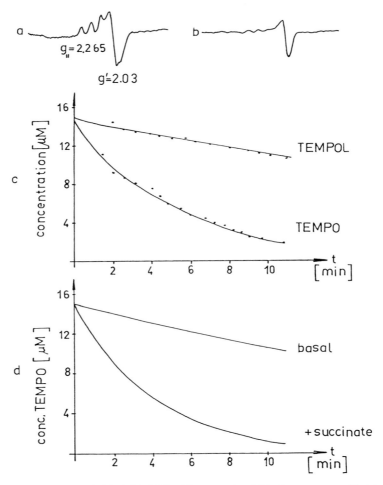

Figure 8. (a), (b) ESR spectra of CuZnSOD (0.22 mM) untreated and 5 h after treatment with 4.4 mM H_2O_2. (c) Breakdown of TEMPO and TEMPOL by rat kidney mitochondria measured by ESR; stimulation by 20 mM succinate. (d) Stimulation of the basal TEMPO breakdown by 20 mM succinate.

Mitochondria. The application of nitroxide spin probes and labels to biological systems and subsystems requires information about their metabolic interactions. This question was studied (Henke et al., 1993) on isolated mitochondria of rat kidney cortex applying the spin probes TEMPO and its 4-hydroxy derivative TEMPOL.

As the ESR studies show (Fig. 8c and d), TEMPO reacts with the mitochondria whereas TEMPOL is of little effect. TEMPO, using succinate as respiratory substrate, increases state four of respiration of isolated mitochondria and is without effect on state or on three uncoupled respiration.

Measuring the NADH-linked respiration with glutamate/malate TEMPO inhibits the rate of state three and uncoupled respiration. State four respiration is raised by TEMPO. TEMPOL is without effect on respiratory functions. TEMPO is not only metabolised by renal mitochondria but also induces mitochondrial dysfunctions (Henke et al., 1993).

Enzymes. CuZn superoxide dismutase (CuZnSOD) contributes to the regulation of the steady-state concentration of reactive oxygen species in cells and minimizes pathological consequences of these reactive oxygen species. During increased formation of reactive oxygen species, often resulting from an activation of phagocytic cells, CuZnSOD is administered with a therapeutic purpose. But inhibition of the endogenous or administrated CuZnSOD by products generated during the process of formation of reactive oxygen species (H_2O_2, HOCl, OH, products of lipid peroxidation) might intensify cell damage. Here the influence of malondialdehyde (MDA, a highly reactive molecule formed in lipid peroxidation) and H_2O_2 (known to inhibit CuZnSOD) on bovine CuZnSOD will be discussed. It was established that MDA reacts with CuZnSOD. The reaction was found to be both concentration- and time-dependent, which was demonstrated by the formation of fluorophors (Haberland et al., 1993). ESR spectroscopy revealed that this reaction should have no influence on the activity of CuZnSOD since the catalytic centre of the CuZnSOD was not affected by MDA. In contrast, H_2O_2 modified the catalytic centre which caused an activity decrease (Fig. 8a and b).

Examples of in vivo *ESR*

The progress of *in vivo* ESR is reflected in the increase of the number of publications during the last few years (Mäder et al., 1994; Colacicchi et al., 1992; Swartz et al., 1991). The main problem of ESR measurements in whole animals is the great nonresonant dielectric loss caused by the high water content of biological tissue. This loss increases with the microwave frequency and with the size of the sample. Microwaves with a frequency of 10 GHz penetrate about 1 mm, 1 GHz about 16 mm and at 250 MHz 55 mm into muscle tissue. Therefore it is convenient to use frequencies lower than the conventional X-band (9–10 GHz). However, the decrease in the frequency is accompanied by a decrease in sensitivity due to the lower splitting of the energy levels. The most commonly used frequency is therefore 1–2 GHz but good results have been reported from experiments carried out in the MHz range (Colacicchi et al., 1992).

X-band ESR on living mice

X-band has been used for the noninvasive measurements of systemic nitroxide levels by putting the tails of mice into the cavity (Fig. 9a and b). The successful detection of nitroxide radicals after

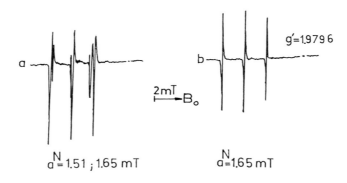

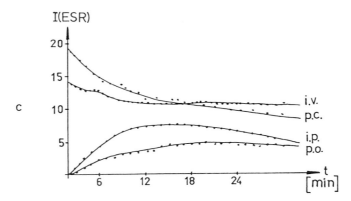

Figure 9. (a) ESR spectrum of a mouse tail recorded 30 min after inhalation of DTNB. (b) ESR spectrum of blood collected from mouse decapitated 30 min after inhalation of DTNB. (c) Time evolution of the ESR signal intensity (I) after i.v. (intravenous), p.c. (percutaneous, tail), i.p. (intraperitonal) and p.o. (per os) application CPM.

a dosage of 10 µmol/kg demonstrates the high sensitivity of *in vivo* X band EPR (Mäder et al., 1993b).

The *in vivo* experiments in X-band using the mouse tail (Fig. 1f; Mäder et al., 1993b) gave some further insight into pharmaceutical processes. The influence of the partial pressure of oxy-

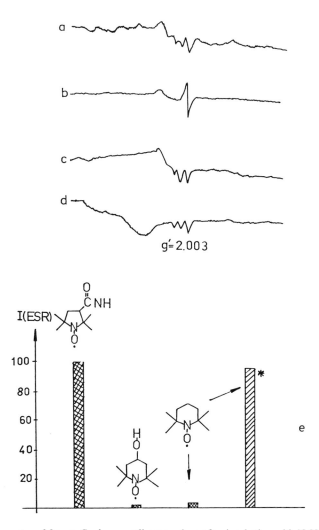

Figure 10. (a) ESR spectra of frozen *C. elegans* cell suspensions after incubation with NaNO$_2$ (0.7 mM, 25°C, 1h); (b), (c) and (d) after 2, 16 and 24h. (e) Reduction of PCM, TEMPOL and TEMPO (1 mM) in cell suspensions (1h, 28°C) of *Cunninghamella elegans* ATCC 361 12; *: the suspension was heated (5 min, 90°C) before incubation with TEMPO.

gen during respiration of the mouse on the decay rate of the ESR signal was clearly demonstrated. In the same experimental arrangement (Fig. 1f) it was possible to follow the oxidation of the hydroxylamine applied to the mouse. As a result the corresponding nitroxyl radical was detected *in vivo*. These findings represent valuable information for the discussion of the metabolic pathways of administrated drugs. Nitroxide radicals were administrated per os, intravenously and intraperitoneally (Fig. 9c). The signal intensity and form changed temporally as a function of these different types of application. Only signals from the applied probes were detectable. All probes with a 2,2,6,6-tetramethylpiperidine structure were observable for only short time. In contrast, the DTBN- and 2,2,6,6-tetramethylpyrrolidine intensities decreased slowly. Even 25 min after 1×10^{-5} mol/kg i.p. injection of carbamoylproxyl the well-known spectrum was still detectable. The ring structure influences the biological halflife more powerfully than substitutions on the ring. DTNB was also applied by inhalation. In any case after DTNB application (inhalation, i.p., p.o.) a superimposed EPR spectrum consisting of a polar species (g = 2.0057; a = 1.65 mT) and a less polar paramagnetic species (g = 2.0061; a = 1.51 mT) was detected in the mouse tail. Collected blood exhibits only the polar spectrum (Fig. 9a and b).

ESR on fungi

Microbial models of mammalian drug metabolism exhibit certain characteristics which make them attractive as alternatives or supplements to the more classical *in vitro* and *in vivo* animal models in preclinical drug research (Borchert et al., 1995). Based upon this background, recently the coronary active drug trapidil and other substrates of cytochrome P450-dependent monooxygenase were examined for metabolic parallels between mammalian and selected fungal systems. The findings of these structures suggest that the fungal strain *Cunninghamella elegans* ATCC36112 appears to be particularly promising with regard to the broad range of phase I biotransformations observed in mammals. Based on ESR investigations one can state that there are parallels between the fungal strain *Cunninghamella elegans* and mammalian systems with regard both to cytochrome P450-dependent drug metabolism and metabolism of free radicals. In agreement with the data obtained with mammalian systems, cells of *Cunninghamella elegans* are able to carry out the bioreduction of nitroxyl radicals to the diamagnetic hydroxylamines. The pyrrolidine derivative PCM was reduced at rates much slower than the piperidines TEMPO and TEMPOL. Both enzymatic and nonenzymatic reductions mechanisms can be assumed. However, the heat-resistant reducing activity was low (Fig. 10e). Furthermore the interest was focussed on nitric oxide-associated metabolic processes.

During the incubation of *Cunninghamella elegans* cell suspensions with precursors of nitric oxide such as glycerolnitrate, sodium nitroprusside and sodium nitrite Fe(II)-NO compexes are produced which show a peculiar g´ = 2.03 ESR signal of porphyrin-Fe(II)-NO complexes formed in mammalian tissues. Corresponding ESR spectra are shown in Figure 10a–d.

Conclusions

Typical parts of LADME processes were investigated by ESR using endogeneous paramagnetic ions as well as spin probes of different structure as reporters for structural and dynamic changes in the systems. Unique information is obtained about the local enviroment of the probes depending on the variation of the other parameters temperature, pressure and concentration of xenobiotics. The *in vivo* experiments clearly demonstrated the influence of the kind of application and of the delivery system used for the paramagnetic probes.

Acknowledgements
We thank C. Kroll, A. Haberland and W. Henke for providing samples, assistance at the measurements and for helpful discussions.

References

Borchert, H.-H., Mäder, K., Stösser, R. and Kroll, C. (1995) Free radical intermediates in fungal metabolism as a model of mammalian drug metabolism. *Magnetic Resonance in Medicine* 6: 375–377.
Chingnell, C.F. (1979) Spin labeling in pharmacology. *In*: L.R. Berliner (ed.): *Spin Labeling II,* Academic Press, New York.
Colacicchi, S., Ferrari, M. and Sotgin, A. (1992) *In vivo* electronparamagnetic resonance spectroscopy/imaging: first experiences, problems and perspectives. *Int. J. Biochem.* 24: 205–214.
Haberland, A., Mäder, K., Stösser, R. and Schimke, I. (1993) Comparison of malondialdehyde and hydrogen peroxide modified CuZn SOD by EPR spectroscopy. *Agents and Actions* 40: 166–170.
Henke, W., Mäder, K., Nickel, E. and Stösser, R. (1993) Effects of spin labels on respiration of renal cortical mitrochondria. Intern. Conference of Critical Aspects of Free radicals in Chemistry, Biochemistry and Medicine, Vienna, Austria, *Book of Abstracts*, p. 100.
Khramtsov, V.V. and Weiner, L.M. (1988) Proton exchange in stable nitroxyl radicals: pH-sensitive spin probes. *In*: L.B. Volodarsky (ed.): *Imidazoline Nitroxides Vol. II,* CRC Press, Boca Raton, pp 37–80.
Kroll, C., Mäder, M., Stösser, R. and Borchert, H.-H. (1995) Direct and continous determination of pH-values in nontransparent W/O systems by means of EPR-spectroscopy. *Europ. J. of Pharm. Sci.* 3: 21–26.
Kuzuya, M., Kondo, S. and Murase, K. (1993) A novel single electron transfer in solid-state organic compounds: mechanically induced reduction of dipyridinium salts. *J. Phys. Chem.* 97: 7800–7802.
Mäder, K., Stösser, R., Borchert, H.-H., Mank, R. and Nerlich, B. (1991a) ESR-Untersuchungen zur Wasser-penetration in Polymerfolien und Mikropartikeln auf der Basis von biologisch abbaubaren Polyestern. *Pharmazie* 46: 342–345.
Mäder, K., Borchert, H.-H., Stösser, R., Groth, N. and Herrling, T. (1991b) Modelluntersuchungen zur Lokalisation und Mobilität von Arzneistoffen in Polymerfolien mit Hilfe der ESR-Tomographie. *Pharmazie* 46: 439–442.
Mäder, K. (1993a) *Application of ESR and ESR Tomography in Biopharmacy*. Thesis, Humboldt-University, Berlin.
Mäder, K., Stösser, R. and Borchert, H.-H. (1993) Detection of free radicals in living mice after inhalation of DTNB by X-band ESR. *Free Radical Biology & Medicine* 14: 339–342.
Mäder, K., Swartz, H.M., Stösser, R. and Borchert, H.-H. (1994) The application of EPR-spectroscopy in the field of pharmacy. *Pharmazie* 49: 97–101.

Parrot, E.L. (1990) Compression. *In:* H.A. Liebermann, L. Lachmann and J.B. Schwartz (eds)*: Pharmaceutical Dosage Forms:Tablets,* Vol 2, Marcel Dekker, pp 201–203.

Simoneau, R., Harvey, J.S.M. and Graham, G.M. (1971) EPR of impurity in solid N_2, Co, Ar and CD_4 prepared from vapor. *J. chem. Phys.* 54: 4819–4824.

Stösser, R., Nofz, M., Brenneis, R., Klein, J. and Rericha, A. (1990) Mechanical and magnetic properties of glassy-crystalline $MeOSiO_2Al_2O_3$. *Materials Science Forum* 62–64: 279–280.

Stösser, R. (1993) Monitoring metabolic processes in biological systems using rapid freezing ESR at T ≥ 4 K. *Workshop "In vivo EPR and EPR studies of viable biological systems".* Dartmouth-Hitchcock Medical Center, Hanover, USA, Oct. 1993.

Stösser, R., Mäder, K., Borchert, H.-H. and Lück, R. (1994a) High pressure treatment of pharmaceutical drug delivery and related systems. *High Pressure Research* 13: 29–33.

Stösser, R., Mäder, K., Borchert, H.-H., Schneider, G. and Herrmann, W. (1994b) *In vivo* measurement of the oxygen tension at human skin, aspects of dermatology. Intern. Conf. on Bioradicals Detected by ESR Spectroscopy. *Book of Abstracts, Yamagata, June 12–16, Japan.*

Stösser, R., Mäder, K., Borchert, H.-H., Herrmann, W., Schneider, G. and Liero, A. (1995) *In vitro* and *in vivo* ESR in S- and X-band to monitor biopharmaceutical processes in mammalian skin. *Magnetic Resonance in Medicine* 6: 349–351.

Swartz, H.M., Boyer, S., Gast, P., Glockner, J.F., Hu, H., Liu, K.J., Moussavi, M., Norby, S.W., Vahidi, N., Walczak, T., Wu, M. and Clakson, R.B. (1991) Measurements of pertinent concentration of oxygen *in vivo. Magnetic Resonance in Medicine* 20: 333–339.

Zavoiskij, E. (1944) Paramagnetic absorption of a solution in parallel fields. *J. Phys.* 8: 377–380.

Bioradicals Detected by ESR Spectroscopy
H. Ohya-Nishiguchi & L. Packer (eds)
© 1995 Birkhäuser Verlag Basel/Switzerland

In vivo ESR measurement of free radical reactions in living animals using nitroxyl probes

H. Utsumi and K. Takeshita

Faculty of Pharmaceutical Sciences, Kyushu University, Higashi-ku, Fukuoka 812-82, Japan

Summary. This chapter covers the *in vivo* ESR measurement methods to estimate free radical reactions in living mice using nitroxyl radicals as probes. One of the following nitroxyl radicals, 2,2,6,6-tetramethylpiperidine-1-oxyl (TEMPO), 2,2,5,5-tetramethylpyrrolidine-1-oxyl (PROXYL), 4,4-dimethyloxazolidine-3-oxyl (DOXYL), and their derivatives, was dissolved in isotonic buffer and was intravenously, intramuscularly, transtracheally or intraperitoneally injected into female ddY mice. The ESR signal of nitroxyl radical in living mice decreased gradually by reducing to the corresponding hydroxylamine. The reduction rate depended on physiological and pathological conditions such as aging, γ-irradiation, and oxidative stress. Pre-treatment of antioxidants reduced the enhancement of signal decay by oxidative stress. Alveolar cell membrane had reducing system for nitroxyl radicals, and the activity was regulated by lipophilic SH modifying reagents. These results clearly demonstrate that *in vivo* ESR measurement with nitroxyl radical as a probe is very useful technique to estimate *in vivo* free radical reactions and to evaluate their relation to physiological and pathological phenomena.

Introduction

Free radicals such as active oxygen species and nitric oxide are believed to be very essential and functional compounds in various biological systems, and numerous studies have been made to detect active oxygen species and to clarify their role in biological phenomena. However, most reports were of *in vitro* experiments and few of *in vivo* investigations. Non-invasive measurement of *in vivo* free radical reactions is very important to understand the role of free radicals in the living body, since oxygen concentrations in tissues is much lower than in *in vitro* experiments and there are many substances and reaction pathways which influence *in vivo* radical reactions.

Recently, a low-frequency ESR spectroscopy has been developed and enabled non-invasive *in vivo* measurement of radicals in whole animals (Bacic et al., 1989; Ferrari et al., 1990; Ishida et al., 1989; Nishikawa et al., 1985; Subczynski et al., 1986; Utsumi et al., 1990; and also reviewed by Eaton et al., 1991). Despite its great potential, ESR measurement of *in vivo* radical generation in living animals has rarely been reported because of the poor sensitivity of the L-band ESR spectrometer. ESR signals of nitroxyl radicals are susceptible to oxygen concentration (Swartz, 1987), to active oxygens (Nilsson et al., 1989; Samuni et al., 1988) and to biological redox systems (Rauckman et al., 1984), indicating that combination of L-band ESR spectrometer with nitroxyl radicals as probes may provide valuable information about the biological function of free radical reactions including generation of active oxygen and activity of redox systems. Thus, with a high sensitivity *in vivo* ESR-CT apparatus which has been constructed by JEOL Co. Ltd at our request (Utsumi et al., 1990a,b, 1991), we have studied the *in vivo* free radical reactions using the nitroxyl radicals which have been administered intravenously (Utsumi et al., 1990b, 1991, 1992, 1993; Miura et al., 1992, 1995), intramuscularly (Masuda et al., 1991, 1992; Utsumi et al., 1993), intraperitoneally (Gomi et al., 1993), and transtracheally to mice (Takeshita et al., 1991, 1992, 1993). The results were reduction of nitroxyl radical and the revelation that the rate of reduction depends on the physiological and pathological conditions such as aging, oxidative stress, etc.

In this chapter, we review our recent results about non-invasive ESR measurement of *in vivo* free radical reactions in living mice using nitroxyl radicals as probes and discuss the utilization of this technique for clarifying their role in biological response.

Materials and methods

Chemicals

Table 1 demonstrate the major nitroxyl probes used in our laboratory. Most 2,2,6,6-tetramethyl-piperidine-1-oxyl (TEMPO) and 2,2,5,5-tetramethylpyrrolidine-1-oxyl (PROXYL) derivatives were purchased from Aldrich Chemical Co. Glutaramide- and methylglutaramide-TEMPO and 4,4-dimethyloxazolidine-3-oxyl (DOXYL) derivatives were synthesized in our laboratory. The water soluble probes were dissolved in isotonic solution, and the lipid probes were suspended with the other auxiliary lipids. Osmolarity of the solution was checked by freezing point depression. Pentobarbital sodium was purchased from Dainabot Co. Ltd. Other reagents were of the highest purity commercially available.

Animals

Female ddY mice (3–4 weeks old, 15–20 g body weight) were used throughout this study. Mice were anesthetized by intramusclar or intraperitoneal injection of pentobarbital (50 mg/kg) and fixed on a hand-made Teflon holder. Immediately after administration of isotonic solution of spin-probes through various routes, ESR spectra were measured.

Hypoxia and hyperoxia experiments were performed by exposing mice to an atmosphere of $N_2 - O_2$ mixture (12, 20 and 80% O_2 in N_2) for 45 min before ESR measurement. The isotonic solution of spin-probe was intravenously administered under the above atomosphere.

Table 1. Acronyms and structures of spin-probes used for *in vivo* ESR measurement

Acronym	Chemical name	R	Basic structure
1. TEMPO derivatives			
TEMPO	2,2,6,6-tetramethylpiperidine-1-oxyl	–H	
Hydroxy-TEMPO	4-hydroxy-2,2,6,6-tetramethylpiperidine-1-oxyl	–OH	
Amino-TEMPO	4-amino-2,2,6,6-tetramethylpiperidine-1-oxyl	–NH$_2$	
Carboxy-TEMPO	4-carboxy-2,2,6,6-tetramethylpiperidine-1-oxyl	–COOH	
Oxo-TEMPO	4-oxo-2,2,6,6-tetramethylpiperidine-1-oxyl	=O	
CAT-1	4-trimethylammonium--2,2,6,6-tetramethylpiperidine-1-oxyl iodide	–N$^+$(CH$_3$)I$^-$	
Phosphonooxy-TEMPO	4-phosphonooxy-2,2,6,6-tetramethylpiperidine-1-oxyl	–OPO$_3$H$_2$	
Glutaramide-TEMPO		–NHCO(CH$_2$)$_3$COOH	
Methylglutaramide-TEMPO		–NHCO(CH$_2$)$_3$COOCH$_3$	
2. PROXYL derivatives			
Carboxy-PROXYL	3-carboxy-2,2,5,5-tetramethylpyrrolidine-1-oxyl	–COOH	
Carbamoyl-PROXYL	3-carbamoyl-2,2,5,5-tetramethylpyrrolidine-1-oxyl	–CONH$_2$	
Aminomethyl-PROXYL	3-aminomethyl-2,2,5,5-tetramethylpyrrolidine-1-oxyl	–CH$_2$NH$_2$	
3. DOXYL derivatives			
2-DOXYL-butane	2-ethyl-2,4,4-trimethyloxazolidine-3-yloxy	R$_1$: –CH$_3$ R$_2$: –C$_2$H$_5$	
nSLS	n-(N-oxyl-4'4'-dimethyloxazolidine)-stearic acid		

Ischemia-reperfusion of mouse thigh was carried out by modifying the method of Oyanagui et al. (1988). Occlusion was done by tying the base of the femoral muscle with a thread for 20 min, and then followed by reperfusion. Spin probe was administered to femoral muscle of mice 1 min before reperfusion.

ESR measurement

ESR spectra from different domains from head to tail were obtained with an *in vivo* ESR spectrometer (JEOL, JES-RE-1L or -3L). The microwave frequency was 1.1–1.3 GHz and the power was 1.0–5.0 mW. The amplitude of the 100 kHz field modulation was 0.2 mT. The external magnetic field was swept at a scan rate of 5 mT/min.

Results and discussion

Characteristics of in vivo *ESR-CT system in our laboratory*

Figure 1a and 1b show the *in vivo* ESR-CT system and the fundamental structure of a loop-gap resonator of the system in our laboratory, respectively. The resonator has 4 gaps and its division is 33 mm i.d. and 5 or 24 mm long. The maximum amplitude of 100 kHz modulation is ca. 0.2 mT. The sensitivity of the *in vivo* ESR apparatus is about 1/40–1/300 of that of a conventional X-band ESR spectrometer, depending on the weight of animals and the observed domain. The spatial distribution of the sensitivity was fairly homogeneous in the resonator (Utsumi et al., 1990a, 1991).

ESR spectra of nitroxyl radicals in whole mice

Figure 2a and 2b show typical ESR spectra from the hepatic domain of a mouse with carbamoyl-PROXYL and hydroxy-TEMPO administered into the tail vein, respectively (Utsumi et al., 1990b). Three sharp lines were observed, with regular noise due to respiration. The hyperfine structure and the peak height ratios of carbamoyl-PROXYL coincided with those of the probe dissolved in saline at a concentration of less than 10 mM, suggesting that the spin probe should exist as a free monomer in veins at the hepatic domain. The ESR-CT imaging picture indicates the existence of carbamoyl-PROXYL in a bundle of the inferior vena cava, hepatic artery, and celiac aorta of mice after administration to the tail vein (Masumizu et al., 1991). Hydroxy-TEMPO in the hepatic domain of mice also gave triplet lines, but the intensity of the lines decreased gradually during field sweep. The quite similar spectra were also observed at head, chest, and lower abdomen after intravenous administration (Utsumi et al., 1990b, Miura et al., 1992). Intraperitoneal or intramusclar administration of spin probes also gave the same spectra as shown in Figure 2, but the period of signal appearance depended on the administration route (Masuda et al., 1991, 1992;

a   b

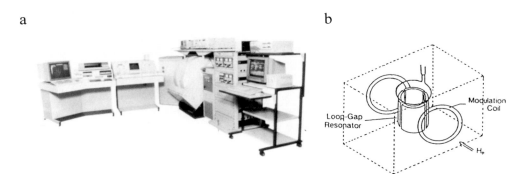

Figure 1. Picture of *in vivo* ESR-CT system (a) and structure of a loop-gap resonator (b) in our laboratory.

Gomi et al., 1993). Spin probes injected transtracheally into mouse lung showed the same triplet lines as those in water (Takeshita et al., 1991). Figure 3 demonstrates the ESR-CT imaging of carboxy-PROXYL injected into mouse lung. The scale and position of the image agreed well with those of the lung, suggesting that the spin probe distributes all over the lung (Takeshita et al., 1991).

Figure 4a and 4b show the spectra from the femur of mouse into which high concentrations of carbamoyl-PROXYL or amino-TEMPO (280 mM aq. solution) were intramuscularly adminis-tered (Masuda et al., 1991). The shapes of the spectra depended on the period after injection, and the signals seemed to consist of several components. Computer simulation was carried out by the

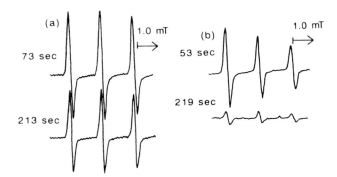

Figure 2. ESR spectra of carbamoyl-PROXYL (a) and hydroxy-TEMPO (b) in the hepatic domain of mice. One hundred microliters of an isotonic solution of spin probes (280 mOsM) were injected into a tail vein of a female ddY mouse and the ESR spectrum in the hepatic domain was observed with an L-band ESR spectrometer. The number at the left of each spectrum indicates the time after injection. (From Utsumi et al., 1990).

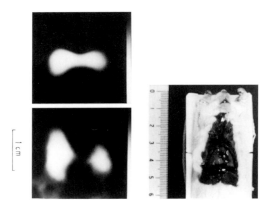

Figure 3. ESR image of nitroxyl-probe injected into mouse lung (a, b) and an anatomical picture of the mouse used for ESR image (c). ESR image was performed from two directions after transtracheal injection of carboxy-PROXYL (15 mM, 0.9 ml) into mouse lung; from the ventral side (a) and from tail along the body axis (b). (From Takeshita et al., 1991).

summation of three typical signals which were obtained with different concentrations of spin probes from 14 to 280 mM, and the simulated spectrum fitted well with the observed one (Fig. 4).

In Figure 5a and 5b are plotted the intensities of three components used for the simulation as a function of time after the injection. One minute after the injection of carbamoyl-PROXYL, a large amount of the 280 mM component was observed, and this component decreased gradually with an increase of 210 mM component. After 7 min, the 140 mM component appeared instead of the

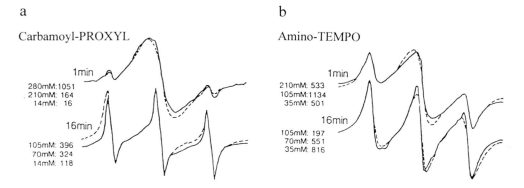

Figure 4. ESR spectra of carbamoyl-PROXYL (a) and amino-TEMPO (b) at the femur of female ddY mice after intramuscular injection of 50 microliters of an isotonic solution of spin-probes (280 mOsM). Solid and broken lines are the observed and simulated spectra, respectively. The number at the left of each spectrum indicates the intensity of the components used for the simulation. (From Masuda et al., 1991).

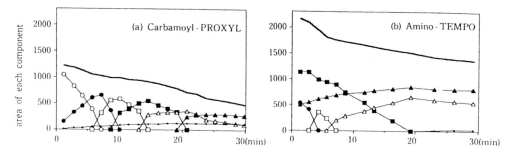

Figure 5. Time-course of occurrence of plural components in ESR spectra of carbamoyl-PROXYL (a) and amino-TEMPO (b) in the femur of female ddY mice after intramuscular injection. The concentration and intensity of each component was estimated from the corresponding computor-simulated spectrum shown in Figure 4. (From Masuda et al., 1991).

280 mM one, but it diminished after 15 min with the appearance of a lower one. Amino-TEMPO gave a different profile, that is 35 – 140 mM components were predominant for a long period. Thus, Figure 5 should indicate the diffusion of spin probes in femoral muscle (Masuda et al., 1991).

Pharmacokinetics of nitroxyl radicals and mechanism of reduction in living mice

It is established from numerous *in vitro* experiments that nitroxyl radicals are readily reduced to their corresponding hydroxylamines through enzymatic and non-enzymatic process, resulting in the loss of paramagnetism. In fact, ESR signals of nitroxyl radicals decrease gradually in living mice. Figures 6a and 6b respectively show semilogarithmic plots of the peak heights from the hepatic domain after intravenous administration of carbamoyl-PROXYL and hydroxy-TEMPO (Utsumi et al., 1990b). The plot was a straight line for at least 10 min, indicating that signal decay after i.v. injection should obey first order kinetics. Figure 6c shows a typical decay curve of carbamoyl-PROXYL in the head after intraperitoneal administration (Gomi et al., 1993). The peak height reached maximum immediately after intravenous injection, while intraperitoneal injection needed about 10 min to reach the maximum peak height level.

Table 2 demonstrates the decay constants of various spin probes in head and breast after intravenous injection (Sano et al., 1995). The decay constants for TEMPOs were several times larger than those for the PROXYLs, and the probes having large n-octanol/water partition coefficients were reduced more quickly. This tendency agreed well with that obtained by *in vitro* experiments. The decay constants at the head were significantly smaller than those at the breast.

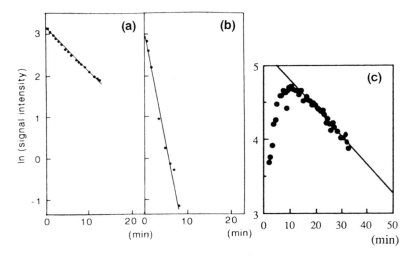

Figure 6. Decay curve of carbamoyl-PROXYL (a) and hydroxy-TEMPO (b) in the hepatic domain of mice after intravenous injection and that of of carbamoyl-PROXYL (c) in the head after intraperitoneal injection. (a and b; from Utsumi et al., 1990; c; from Gomi et al., 1993).

The probes shown in Table 2 cannot pass through the blood brain barrier. The discrepancy in the decay constants between head and breast may imply that the nitroxyl reduction occurs not only during blood circulation but also by interaction with tissues.

There should be several different mechanisms in spin clearance *in vivo*, including direct reduction to the corresponding hydroxylamine in blood circulation or cells, decomposition to other non-paramagnetic compounds, excretion through kidney or liver to urine or feces, binding to macromolecules, etc. Excretion of spin probes through kidney may be a major mechanism, since

Table 2. *In vivo* decay constants of ESR signals from head and breast of mice into which spin probes were intravenously injected. The decay constants (/min) are presented as mean ± S.D. The numbers of experiments are indicated in the parentheses. (From Sano et al., submitted)

	Breast	Head
Hydroxy-TEMPO	0.97 ± 0.33 (5)	0.55 ± 0.13 (5)
Amino-TEMPO	1.53 ± 0.54 (8)	0.92 ± 0.14 (5)
CAT-1	0.21 ± 0.08 (6)	0.07 ± 0.05 (5)
TEMPO-T	0.27 ± 0.10 (6)	0.12 ± 0.04 (6)
Carboxy-TEMPO	0.46 ± 0.10 (5)	0.25 ± 0.02 (5)
Carbamoyl-PROXYL	0.10 ± 0.05 (6)	0.05 ± 0.01 (5)
Carboxy-PROXYL	0.04 ± 0.01 (5)	0.04 ± 0.02 (5)

the accumulation of ESR signal of spin probes was observed in the bladder and urine (Bacic et al., 1989; Masuda et al., 1991). However, signal decay in the early stages may be considered to arise from the reduction of nitroxyl radicals to the corresponding hydroxylamine. A small amount of blood was collected from the mouse pre-loaded with spin probes, and the signal intensity in collected blood was compared before and after oxidation with ferricyanide. The result after oxidation was the stoichiometric restoration of signal intensity to the amount injected (Gomi et al., 1993). The presence of the corresponding hydroxylamine was also confirmed by thin-layer chromatography of the metabolic products in mouse lung (Takeshita et al., 1993).

In blood, there are many reductants of nitroxyl radicals, including ascorbic acid. The contribution of blood components after the collection to nitroxyl reduction must, however, be very small, since the decay constants of spin probes in the collected blood were about 0.1 times those observed from *in vivo* measurement (Utsumi et al., 1990b). Interaction of spin probes with blood vessels, endothelial cells or parenchymal cells of tissues may be important in the *in vivo* reduction of nitroxyl radicals.

Mouse lung also has a reduction system against nitroxyl radical, which cannot be washed out and is inactivated by homogenization (Takeshita et al., 1991, 1993). Nitroxyl radicals without any charges were reduced much more quickly than those with charges (Takeshita et al., 1992). The reduction was strongly inhibited with the membrane-soluble SH-blocker, N-ethylmaleimide, and amphiphilic ones whose maleimide groups locate within membrane hydrophobic regions, but not with the membrane-insoluble one, p-chloromercuriphenylsulfonic acid (Tab. 3, Takeshita et al.,

Table 3. Effect of sulfhydryl-blockers having different membrane permeabilities on reduction of nitroxyl radical in mouse lung. Hydroxy-TEMPO solution (5 mM) containing sulfhydryl-blocker (5 mM) was injected into mouse lung, and first-order reduction rate constants (k) were measured. (From Takeshita et al., 1993)

Sulfhydryl-blockers	Distance[a] (Å)	k[b] (min^{-1})		%
None		0.111 ± 0.013	(7)	100
Membrane permeable				
N-Ethylmaleimide		0.037 ± 0.001	(3)	33
Floating in membrane				
ε-Maleimidocaproic acid	10.8	0.040 ± 0.002	(3)	36
γ-Maleimidobutyric acid	8.3	0.076 ± 0.015	(4)	68
β-Maleimidopropionic acid	7.1	0.120 ± 0.011	(4)	108
Membrane-impermeable				
p-Chloromercuriphenylsulfonic acid		0.105 ± 0.023	(4)	95

[a]The distance is that from the reactive maleimide moiety to the carbon atom of the carboxyl group.
[b]Each value represents the mean ± S.D. The numbers in parentheses are numbers of animals.

Table 4. Effect of age and feeding on decay constants of carbamoyl-PROXYL in mouse head. (From Gomi et al., 1993)

Age (mo.)	Number of mice	Feeding	Spin clearance rate (min^{-1}) Mean ± SD
6	10	*ad libitum*	0.038 ± 0.006^1
30	4	*ad libitum*	$0.026 \pm 0.004^{1,2}$
39	4	restricted	0.035 ± 0.004^2

[1,2]Means are significantly different from each other at the p < 0.05 level as analyzed by Student's-test.

1993). The membrane-soluble reducing SH reagent, dithiothreitol, enhanced the reduction of nitroxyl radicals, although it showed no direct reactivity towards the radicals. These results indicate that the present reduction system in mouse lung is located within the membrane hydrophobic region and that SH-modifying lipophilic compounds regulate the activity of reduction.

Influence of aging and γ-irradiation on radical reduction

Table 4 represents the decay constants of carbamoyl-PROXYL in the head of 6- and 30-month-old, *ad libitum*-fed, and 39-month-old, food-restricted mice after intraperitoneal administration (Gomi et al., 1993). The decay constant for 30-month-old *ad libitum*-fed mice was significantly smaller than that for 6-month-old *ad libitum*-fed mice. It should be noted that the reduction for 39-month-old, food-restricted mice was significantly faster than that for 30-month-old *ad libitum*-fed mice and close to that for 6-month-old *ad libitum*-fed mice. These results suggest that reducing capacity in the cardiovascular system declines in old mice and that food-restriction prevents its age-dependent retardation.

Table 5. Decay constants for hydroxy-TEMPO and carbamoyl-PROXYL in whole mice under hypo- and hyperoxia (/min). (From Miura et al., 1992)

	Hydroxy-TEMPO		Carbamoyl-PROXYL	
	Abdomen	Head	Abdomen	Head
12% O_2	0.95 ± 0.03] *	0.69 ± 0.01	0.12 ± 0.01] **	0.10 ± 0.01] ***
20% O_2	0.85 ± 0.01]	0.71 ± 0.01	0.10 ± 0.02] ***	0.07 ± 0.01]
80% O_2	0.84 ± 0.02	0.71 ± 0.01	0.15 ± 0.02]	0.07 ± 0.01

Clearance constants are presented as mean ± S.E. over 6 experiments. *p < 0.1, **p < 0.05, ***p < 0.001.

Whole body γ-irradiation also retarded the signal decay of carbamoyl-PROXYL in the abdomen of mice, indicating that the reducing capacity in the cardiovascular system is susceptible to γ-irradiation (Utsumi et al., 1992).

Effect of oxidative stress and antioxidants on radical reduction in whole mice

Table 5 demonstrates decay constants in the head and abdomen of mice that were exposed to different oxygen concentrations (Miura et al., 1992). The decay constants of both hydroxy-TEMPO and carbamoyl-PROXYL under 12% oxygen were significantly larger than those under 20% oxygen in both regions. We previously reported that nitroxyl radical loses its paramagnetism more rapidly by the treatment of microsomes under hypoxic conditions (Utsumi et al., 1989; Miura et al., 1990). A hypoxic condition may also favour reduction of nitroxyl radicals in living mice.

The decay constant of carbamoyl-PROXYL in the abdomen under 80% oxygen was significantly greater than that under 20% oxygen ($p < 0.001$). Active oxygen species such as O_2^-, ·OH, and H_2O_2 are reported to be generated in the liver under hyperoxia (Nishiki et al., 1976), and the nitroxide radical loses its paramagnetism by interaction with active oxygen species (Samuni et al., 1988). In fact, the pre-load of antioxidants such as Trolox, uric acid, and glutathione retarded the enhancement of signal decay under hyperoxia (Tab. 6, Miura et al., 1995), and their retardations corresponded to those estimated with TBA-reactive substances. Ascorbic acid

Table 6. Effect of various antioxidants on decay constants for carbamoyl-PROXYL in abdomen under normoxia and hyperoxia. (From Miura et al., 1995)

	Dose (mg/kg)	20% Oxygen	(/min) 80% Oxygen
Control 1[1]		0.102 ± 0.012	0.138 ± 0.014
Control 2[2]		0.105 ± 0.015	0.132 ± 0.012
Trolox	1	0.106 ± 0.010	0.114 ± 0.015[4]
Uric acid	10	0.109 ± 0.011	0.104 ± 0.021[5]
Glutathione	10	0.115 ± 0.009[3]	0.109 ± 0.014[6]
Ascorbic acid	10	0.105 ± 0.016	0.135 ± 0.008

Kinetic constants are presented as mean ± S.D. over 5 or 6 experiments.
[1]Saline (0.2 ml) administered.
[2]Saline containing ethanol (1.4% v/v, 0.2 ml) administered.
[3]$p < 0.1$, different from control 1.
[4]$p < 0.1$, different from control 2.
[5]$p < 0.01$, different from control 1.
[6]$p < 0.005$, different from control 1.

administered intraperitoneally at the dose of 10 mg/kg body weight did not show any effect *in vivo*, although it can reduce nitroxyl radical quickly *in vitro*.

The decay constants of hydroxy-TEMPO and carbamoyl-PROXYL in the head were significantly smaller than those in the abdomen under various oxygen concentrations, which suggests that the mechanism of reduction of nitroxyl radicals might differ between head and abdomen (Tab. 5, Miura et al., 1992). The following might be one explanation for this. Nitroxyl reduction systems in liver hepatocytes exist in mitochondria and microsomes (Iannone et al., 1989; Utsumi et al., 1989). Clark et al. (1976) investigated oxygen affinity of mitochondria and demonstrated that the affinity of rat brain mitochondria was five times that of liver mitochondria. The high oxygen affinity of brain might contribute to slow reduction of hydroxy-TEMPO in the head under hypoxia.

Ischemia-reperfusion also enhanced the *in vivo* signal decay (Masuda et al., 1992; Utsumi et al., 1993). Figure 7 shows the influence of femoral ischemia-reperfusion on the reduction of amino-TEMPO. Spin probe was first injected into the left thigh and ESR spectra were measured until any signal became undetectable. Then the same amount of the probe was injected into the right thigh with and without prior treatment of ischemia-reperfusion. Again, the clearance constant in the right thigh was measured, and the ratio of the clearance constant in the right thigh to those in the left one was used to estimate the effect of ischemia-reperfusion on radical reduction. The ratio in the group treated with ischemia-reperfusion was significantly larger than that without ischemia-reperfusion ($p < 0.05$), and the increase in the ratio was inhibited by treatment with SOD or allopurinol, suggesting that the generation of $O_2^{-\cdot}$ contributes signals to reduction of amino-TEMPO by ischemia-reperfusion injury of mouse thigh.

Conclusions

In this chapter, we demonstrate that *in vivo* ESR spectroscopy with nitroxyl radical as probe makes it possible to estimate free radical reactions in living animals non-invasively, since the nitroxyl radicals are sensitive to both biological redox state and active oxygens. It was found that the *in vivo* signal decay of nitroxyl radicals is influenced by physiological and pathological phenomena such as aging, γ-irradiation, oxygen concentration, ischemia-reperfusion injury, etc. The present paper strongly suggests that *in vivo* ESR measurement with nitroxyl radical as a probe should be a very useful technique to estimate the influence of antioxidants on biological radical reactions in the living body.

Recently, drug delivery system (DDS) has been actively investigated in the field of pharmaceutical sciences using radio-active compounds. *In vivo* ESR technique is non-invasive and real-time,

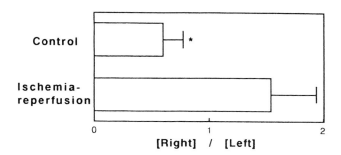

Figure 7. Ratio of the reduction constants of amino-TEMPO in right to left thigh of mice with and without treatment of ischemia-reperfusion. The right thigh in the group of ischemia-reperfusion was treated with 20 min occlusion of the base of thigh followed by reperfusion ($p < 0.05$, $n = 5$). (From Masuda et al., 1992).

and should be a powerful method for DDS investigations, if nitroxyl radicals are used as probes (Yamaguchi et al., 1995; Ohguchi et al., 1995). Imaging pictures of DDS can also be constructed with an ESR-CT system (Matsumoto et al., 1995), demonstrating the advantage of *in vivo* ESR-CT technique in pharmacodynamic investigations, especially the evaluation of DDS in the living body.

Acknowledgements
This work was supported by a Grant-in-Aid for Scientific Research from the Ministry of Education, Science and Culture of Japan, the Scientific Research Promotion Fund from Japan Private School Promotion Foundation, and Special Coordination Funds of the Science and Technology Agency of the Japanese Government.

References

Bacic, G., Nilges, M.J., Magin, R.L., Walczak, T. and Swartz, H.M. (1989) *In vivo* localized ESR spectroscopy reflecting metabolism. *Magn. Reson. Med.* 10, 266–272.
Clark, J.B., Nicklas, W.J. and Degn, H. (1976) The apparent Km for oxygen of rat brain mitochondrial respiration. *J. Neurochem.* 26: 409–411.
Eaton, G.R., Eaton, S.S. and Ohno, K. (1991) *EPR Imaging and* In vivo *EPR*. CRC Press, Boca Raton.
Ferrari, M., Colacicchi, S., Gualtieri, G., Santini, M.T. and Sotgiu, A. (1990) Whole mouse nitroxide free radical pharmacokinetics by low frequency electron paramagnetic resonance. *Biochem. Biophys. Res. Commun.* 166: 168–173.
Gomi, F., Utsumi, H., Hamada, A. and Matsuo, M. (1993) Aging retards spin clearance from mouse brain and food restriction prevents its age-dependent retardation, *Life Science* 52(25): 2027–2033.
Iannone, A., Hu, H., Tomasi, A., Vannini, V. and Swartz, H.M. (1989) Metabolism of aqueous soluble nitroxides in hepatocytes: effects of cell integrity, oxygen, and structure of nitroxides. *Biochim. Biophys. Acta* 991: 90–96.
Ishida, S., Kumashiro, H., Tsuchihashi, N., Ogata, T., Ono, M., Kamada, H. and Yoshida, E. (1989) *In vivo* analysis of nitroxide radicals injected into small animals by L-band ESR technique. *Phys. Med. Biol.* 34: 1317–1323.
Masumizu, T., Tatebe, T., Masuda, S., Muto, E., Utsumi, H. and Hamada, A. (1991) Trial-manufacture of ESR-CT apparatus. *Jap. Mag. Reson. Med.* (in Japanese) 2: 63–68.

Masuda, S., Utsumi, H. and Hamada, A. (1991) *In vivo* ESR study on diffusion of spin labeled compounds in femoral muscle. *Jap. Mag. Reson. Med.* (in Japanese) 2: 69–74.

Masuda, S., Utsumi, H. and Hamada, A. (1992) *In vivo* ESR studies on radical reduction in femoral ischemia-reperfusion of whole mice. *In:* K. Yagi, M. Kondo, E. Niki, and T. Yoshikawa (eds): *Oxygen Radicals*, Excerpta Medica, Amsterdam, pp 175–178.

Matsumoto, K., Hamada, A. and Utsumi, H. (1995) ESR-CT Imaging for coexisting plural radical species; *submitted.*

Miura, Y., Utsumi, H., Kashiwagi, M. and Hamada, A. (1990) Effect of oxygen on the membrane structure and the metabolism of lipophilic nitroxide in rat liver microsomes. *J. Biochem.* 108: 516–518.

Miura, Y., Utsumi, H. and Hamada, A. (1992) Effects of inspired oxygen concentration on *in vivo* redox reaction of nitroxide radicals in whole mice, *Biochem. Biophys. Res. Commn.* 182: 1108–1114.

Miura, Y., Utsumi, H. and Hamada, A. (1993) Antioxidant activity of nitroxide radicals in lipid peroxidation of rat liver microsomes. *Arch. Biochem. Biophys.* 300: 148–156.

Miura, Y., Hamada, A. and Utsumi, H. (1995) *In vivo* ESR studies of antioxidant activity on free radical reaction in living mice under oxidative stress. *Free Rad. Res.* 22: 209–214.

Nilsson, U.A., Olsson, L.-I., Carlin, G. and Bylund-Fellenius, A.-C. (1989) Inhibition lipid peroxidation by spin labels. *J. Biol. Chem.* 264: 11131–11135.

Nishiki, K., Jamieson, N., Oshino, N. and Chance, B. (1976) Oxygen toxicity in the perfused rat liver and lung under hyperbolic conditions. *Biochem. J.* 160: 343–355.

Nishikawa, H., Fujii, H. and Berliner, L.J. (1985) Helices and surface coils for low-field *in vivo* ESR and EPR imaging applications. *J. Mag. Res.* 62: 79–86.

Ohguchi, K., Takeshita, K., Hamada, A. and Utsumi, H. (1995) Non-invasive measurements of the pharmacokinetics and the imaging of liposomes in living mice; *submitted.*

Oyanagui, Y., Sato, S. and Okajima, T. (1988) Suppressions of ischemia paw oedema in mice, rats and guinea pigs by superoxide dismutases from different sources. *Free Rad. Res. Comms.* 4: 385–396.

Rauckman, E.J., Rosen, G.M. and Griffeth, L.K. (1984) Enzymatic reactions of spin labels. *In:* J.L. Holtzman (ed.): *Spin-Labeling in Pharmacology.* Academic Press, pp 175–190.

Samuni, A., Krishna, C.M., Riesz, P., Finkelstein, E. and Russo, A. (1988) A novel metal-free low molecular weight superoxide dismutase mimic. *J. Biol. Chem.* 263: 17921–17924.

Sano, H., Chignell, C.F. and Utsumi, H. (1995) Influence of radical structure and anesthesia on the *in vivo* reduction of nitroxyl-probes in living mice; *submitted.*

Subczynski, W.K., Lukiewica, S. and Hyde, J.S. (1986) Murine *in vivo* L-band ESR spin-label oximetry with a loop-gap resonator. *Magn. Reson. Med.* 3: 747–754.

Swartz, H.M. (1987) Measurement of pertinent oxygen concentrations in biological systems. *Acta Biochim. Biophys. Hung.* 22: 277.

Takeshita, K., Utsumi, H. and Hamada, A. (1991) ESR measurement of radical clearance in lung of whole mouse. *Biochem. Biophys. Res. Commun.* 177: 874–880.

Takeshita, K., Utsumi, H. and Hamada, A. (1992) ESR study on radical clearance system in lung of whole mouse. *In:* K. Yagi, M. Kondo, E. Niki, and T. Yoshikawa (eds): *Oxygen Radicals*, Excerpta Medica, Amsterdam, pp 171–174.

Takeshita, K., Utsumi, H. and Hamada, A. (1993) Whole mouse measurement of paramagnetism-loss of nitroxide free radical in lung with a L-band ESR spectrometer. *Biochem. Mol. Bio. Int.* 29: 17–24.

Tomasi, A., Santis, G.D., Iannone, A. and Vannini, V. (1990) Spin trapping of free radicals in rat muscle: A novel model system of ischemia-reperfusion. *Free Rad. Biol. Med.* 9: 36.

Utsumi, H., Shimakura, A., Kashiwagi, M. and Hamada, A. (1989) Localization of the active center of nitroxide radical reduction in rat liver microsomes: Its relation to cytochrome P-450 and membranes fluidity. *J. Biochem.* 105: 239–244.

Utsumi, H., Hamada, A. and Kohno, M. (1990a) Electron spin resonance spectrometer. *Pharm. Tech. Japan* (in Japanese) 6:1329–1337.

Utsumi, H., Muto, E., Masuda, S. and Hamada, A. (1990b) *In vivo* ESR measurement of free radicals in whole mice. *Biochem. Biophys. Res. Commun.* 172: 1342–1348.

Utsumi, H., Masuda, S., Muto, E. and Hamada, A. (1991) *In vivo* ESR studies on pharmacokinetics of nitroxide radicals in whole mice. *In:* K.J.A.Davies (ed.): *Oxidative Damage & Repairs*, Pergamon Press, New York, pp 165–170.

Utsumi, H., Kawabe, H., Masuda, S., Takeshita, K., Miura, Y., Ozawa, T., Hashimoto, T., Ikehira, H., Ando, K., Yukawa, O. and Hamada, A. (1992) *In vivo* ESR studies on radical reaction in whole mice. Effect of radiation exposure. *Free Rad. Res. Commun.* 16: 1.5.

Utsumi, H., Takeshita, K., Miura, Y., Masuda, S. and Hamada, A. (1993) *In vivo* EPR measurement of radical reaction in whole mice. Influence of inspired oxygen and ischemia-reperfusion injury on nitroxide reduction. *Free Rad. Res. Commun.* 19: s219–225.

Yamaguchi, T., Itai, S., Hayashi, H., Soda, S., Hamada, A. and Utsumi, H. (1995) Non-invasive analysis of parenteral lipid emulsion in mice by *in vivo* ESR spectroscopy; *submitted.*

Subject index

(The page number refers to the first page of the chapter in which the keyword occurs)

Analysis of Free Radicals in Biological Systems

Edited by
A.E. Favier, *CHU Albert Michallon, Grenoble, France*
J. Cadet, *CEA, Grenoble, France*
B. Kalyanaraman, *Medical College of Wisconsin, Milwaukee, WI, USA*
M. Fontecave, J.L. Pierre, *LEDSS II, St. Martin d'Hères, France*

1995. Approx. 300 pages. Hardcover. ISBN 3-7643-5137-3

The main aim of this book is to provide a comprehensive survey on recent methodological aspects of the measurement of damage within cellular targets, information which may be used as an indicator of oxidative stress.

In the introductory chapters, emphasis is placed on the chemical properties of reactive oxygen species and their role in the induction of cellular modifications together with their links to various diseases. The central part of the book is devoted to the description of selected methods aimed at monitoring the production of free radicals in cellular systems. In addition, several assays are provided to assess the chemical damage induced by reactive oxygen species in critical cellular-targets in vitro and in humans in vivo.

Thus both practical aspects and general considerations, including discussions on the applications and limitations of the assays, are critically reviewed. One of the major features of the book is the description of new experimental methods. These include the measurement of oxydized DNA bases and nucleosides, new techniques for the determination of LDL oxidation using spin-trap agents, and the use of salicylate as an indicator of oxidative stress. In addition, more classical though significantly improved techniques devoted to the measurement of hydroperoxides and aldehydes are described.

This book will serve a large scientific community including biologists, chemists, and clinicians working on the chemical and biological effects of oxidative stress. It may also be of interest to investigators in the fields of drug, cosmetic and new food research.

Birkhäuser Verlag • Basel • Boston • Berlin

Oxygen Free Radicals in Tissue Damage

Edited by
M. Tarr / F. Samson
University of Kansas Medical Center, Kansas City, KS, USA

1993. 296 pages. Hardcover • ISBN 3-7643-3609-9

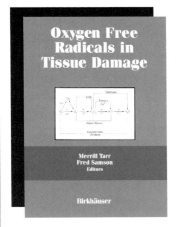

This important volume provides insights into current thinking and controversies regarding oxygen free radicals and reactive oxygen species in tissue dysfunction and pathology. Bringing together contributions from scientists actively involved in reactive oxygen species research, this volume presents these contributors' ideas and concepts regarding new advances in this rapidly expanding field of research. To readers new to this research field, this volume provides an introduction and a state-of-the-art overview of important topics relevant to understanding oxygen free radicals. Readers knowledgeable in oxygen free radicals should gain new insights.

Included in this authoritative and timely text are chapters detailing the chemistry of oxygen free radicals as well as discussing methods for detecting and generating these highly reactive and short-lived chemicals. Chapters also discuss the roles played by oxygen free radicals in damage to the heart, the brain, the microvascular system, and the immune system.

A discussion of the properties of various chemicals which can provide protection against these potentially destructive substances is also presented.

**This book will be valuable to all biomedical researchers ~
especially to those investigating the mechanisms of tissue damage
~ and to libraries that serve basic medical sciences.**

With contributions by:
K.L. Audus, J.S. Beckman, R. Bolli, D.C. Borg, J. Chen, K.A. Conger, R.A. Floyd, I. Fridovich, J.I. Goldhaber, D.N. Granger, E.D. Hall, J.H. Halsey Jr., N. R. Harris, C.J. Hartley, H. Ischiropoulos, M.O. Jeroudi, S. Ji, J.R. Kanofsky, M. E. Layton, X.Y. Li, T.L. Pazdernik, F. Samson, M. Tarr, J. P. Uetrecht, D.P. Valenzeno, J.N. Weiss, B.J. Zimmerman, L. Zu, M. Zughaib.

Birkhäuser Verlag • Basel • Boston • Berlin

Free Radicals:
From Basic Science to Medicine

Edited by
G. Poli / E. Albano / M.U. Dianzani
Univ. di Torino, Italy

1993. 528 pages. Hardcover
ISBN 3-7643-2763-4 (MCBU)

This book is a compilation of reviews covering the major areas of free radical science. Physiological and pathological aspects of free radical reactions are considered. While the contributions reflect the differing backgrounds and interests of investigators involved in basic or applied studies, all take a „vertical" approach to the main fields in which free radicals are assumed to play an important role, e.g. aging, cancer, metabolic disorders, inflammation, radiation, and mutagenesis. Special emphasis is placed on the medical aspects of free radicals, an area which until recently has been accorded only scant attention by the medical profession. Finally, the book reports comprehensively on prevention and therapy of free radical based pathology by means of various antioxidants.

Free Radical Research • Biochemistry • Pathology

Birkhäuser Verlag • Basel • Boston • Berlin